Functional Molecular Materials

Functional Molecular Materials

An Introductory Textbook

Matteo Atzori | Flavia Artizzu

Pan Stanford Publishing

Published by

Pan Stanford Publishing Pte. Ltd.
Penthouse Level, Suntec Tower 3
8 Temasek Boulevard
Singapore 038988

Email: editorial@panstanford.com
Web: www.panstanford.com

British Library Cataloguing-in-Publication Data
A catalogue record for this book is available from the British Library.

Functional Molecular Materials: An Introductory Textbook

ISBN 978-981-4774-76-5 (Hardcover)
ISBN 978-1-351-23366-8 (eBook)

Contents

Preface

The field of molecular materials represents an exciting playground for the design, tailoring, and combination of chemical building blocks as carriers of physical properties and aims at the understanding and development of novel functional molecular devices. Thanks to the countless possibilities to associate diverse chemical assemblies and physical properties, infinite different materials can be, in principle, foreseen.

Within this extraordinarily widespread framework, the realization of materials with the desired functionalities can only be achieved through a rational design strategy based on a solid understanding of the chemical/physical features of each constituting building block. This required background knowledge can be acquired thanks to an extensive existing literature providing exhaustive reviews on important topics and classes of molecular materials, owing to the contributions of authors who are recognized authorities in quite specific fields. While these volumes are, indeed, of great value to expert scientists, as the contents are addressed at a high and rigorous scientific level, they might be nonetheless challenging reading for early-stage researchers entering this field. The basic idea behind this book is therefore to provide a general overview of molecular materials, discussing their key features in a simple and organic way and focusing more on basic concepts rather than on specialized descriptions, in order to supply the nonexpert reader with the immediate fundamental tools and hints to understand and develop research in this field.

With this view, this textbook is intended as a step-by-step guide toward the preparation of functional molecular materials, where the knowledge and understanding so far attained by the scientific community through the investigation of significant archetypical examples are deconstructed down to the fundamentals and then presented from the bottom to the top.

After a brief introduction to the key features of molecular materials (Chapter 1), basic concepts of molecular magnetism, conductivity, and molecular photophysics are discussed (Chapter

2). Different classes of functional molecular materials are presented by taking advantage of selected relevant compounds reported in the literature (Chapter 3). The reader will be then guided to design multifunctional molecular materials with predesigned functionalities, taking into account the chemical/structural features and the physical properties of specific molecular building blocks (Chapter 4). Finally, selected applications of molecular materials for device preparation and challenging experiments in the field are discussed (Chapter 5).

The original purpose behind material design is analyzed and the physical properties explained on a fundamental basis throughout the book. In addition, the correlation between structure and properties is thoroughly discussed with the aim to provide evidence of the importance of considering the interplay between these aspects in obtaining a desired material. Relevant and recent examples have been selected so that experts in this field also may retrieve useful information.

Functional Molecular Materials: An Introductory Textbook is intended as a general, easy-to-read, and practical reference textbook the authors themselves would have liked to read in the early stages of their research career, and it has been written in the hope of providing a useful starting guide for future scientists in the field.

Matteo Atzori

Flavia Artizzu

2018

Chapter 1

Introduction to Molecular Materials

The aim of this chapter is to introduce the reader to the concept of molecular materials by defining what they are and what distinguishes them from the more conventional atom-based systems. The general ideas that drive the approach to the design and preparation of molecular materials will be presented, shedding light on the enormous potentialities that this field can offer in the development of new technologically advanced applications.

1.1 What Are Molecular Materials?

A *molecular material* can be defined as a material built by molecules responsible for specific functionalities that can be potentially used for a certain technological application. Given the large number of diverse physical properties that a molecule can feature and the wide range of research fields to which they belong, it appears clear that the number of classes and types of systems that can be categorized under the term "molecular materials" is enormous. One classification can be related to the nature of the physical properties shown by such systems, more specifically, associating them to the areas of solid-state physics, that is, *magnetism*, *optics*, and *electronics*. Therefore, we can talk about molecular materials that show magnetic, optical, or electrical properties, independently of the nature and complexity of the molecules that originate them (Fig. 1.1).

Functional Molecular Materials: An Introductory Textbook
Matteo Atzori and Flavia Artizzu

ISBN 978-981-4774-76-5 (Hardcover), 978-1-351-23366-8 (eBook)
www.panstanford.com

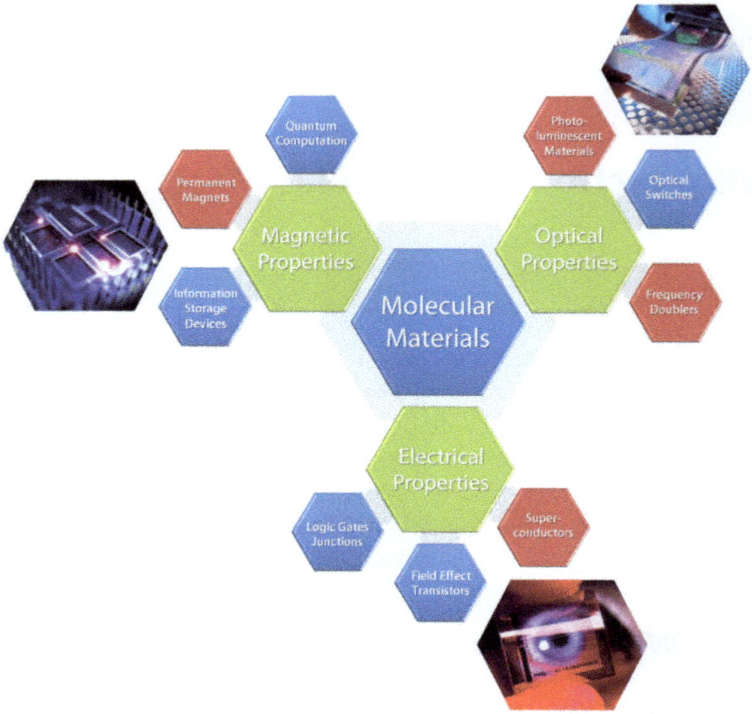

Figure 1.1 Some examples of molecular materials with relevant technological applications categorized as a function of their physical properties.

As far as the nature of these materials is concerned, we can define as molecular materials all those systems based on molecular units, both organic and inorganic, or a combination of them (*hybrid systems*), whose physical properties arise from the peculiar nature of the single molecular entity or from particular interactions occurring between molecules with specific structural and electronic features.

The first case is referred to as a single-molecule property, that is, one molecule able to show the same physical behavior as that observed for the bulk material at the molecular scale, whereas the second case is related to a cooperative property that arises from key interactions between a number of molecules that, if isolated, do not show any relevant physical property or, in some cases, a different one. From these considerations it may appear evident that molecular materials mostly belong to the fields of molecular inorganic/coordination chemistry, supramolecular chemistry, and organic chemistry. A variety of molecular architectures, such as transition metals or lanthanide-based complexes (mononuclear or multinuclear, homometallic or heterometallic, homoleptic or heteroleptic), coordination or supramolecular polymers with various dimensionalities (1D, 2D, or 3D), and redox-active and/or π-delocalized organic molecules, can be counted among this class.

One of the most intriguing potentialities of molecular materials lies in the extraordinary versatility of their chemistry, which allows, in principle, the tailoring of their physical behavior and the opportunity to afford systems showing multiple functionalities at the molecular or supramolecular level. Systems exhibiting a single physical property are commonly called *monofunctional molecular materials*, while those exhibiting any combination of the aforementioned functionalities are *multifunctional molecular materials*. Although the coexistence of multiple physical properties in a single molecular system is not uncommon, generally, the most pursued strategy to achieve multifunctional molecular materials consists of the proper combination of selected monofunctional molecular materials exploiting the principles of molecular self-assembly.[1] Monofunctional molecular materials showing the additional feature of being suitable to be combined with each other can be called *building blocks*, after their usage as molecular bricks for the construction of more sophisticated architectures, whether molecular, supramolecular, or nanostructured, which multifunctional molecular materials are based on (Fig. 1.2).

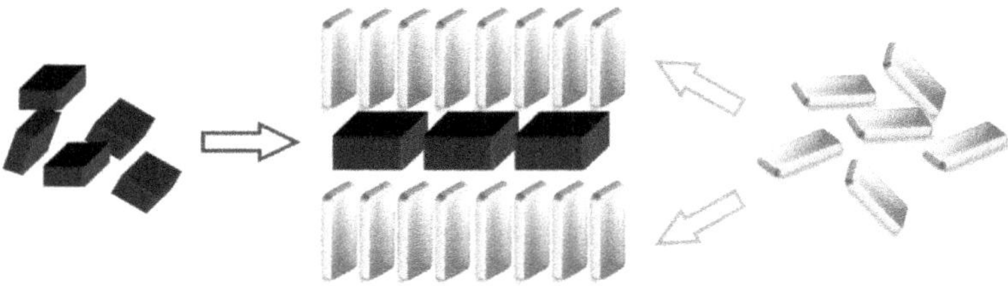

Figure 1.2 Schematic representation of two types of generic molecular building blocks (left and right) that self-assemble in a more sophisticated architecture (middle).

[1]Molecular self-assembly defines a process in which preformed molecular building blocks initially disposed in a random configuration interact with each other through specific interactions, not dictated by external factors, leading to an organized molecular or supramolecular structure.

The possibility of combining two or more different molecular building units, each one the carrier of a specific physical property, allows the achievement of advanced materials where these properties coexist or interact. This opportunity distinguishes molecular materials from more conventional atom-based inorganic materials (metals and alloys, metal oxides, chalcogenides, metal oxoanion–based compounds, etc.) that usually exhibit a single physical property and rigid structural features that make them rather unflexible and hamper further manipulations. With this view, one would say that molecular materials can promote the research in material science to an upgraded level.

1.2 The Versatility of Molecular Chemistry

As already briefly pointed out, thanks to the extraordinary versatility of coordination, organic, and supramolecular chemistry, the design of molecular materials can spread over a large variety of molecular architectures with different degrees of complexity that atom-based materials cannot intrinsically achieve because of the limited number of inequivalent atoms composing their structures. This structural flexibility results in a unique opportunity of designing and preparing molecules with the desired size, shape, charge, polarity, and electronic properties through composition control and rational synthetic strategies. Since the physical properties of molecule-based materials can be significantly affected by subtle changes in the molecular structure of their molecular components, the functionality of molecular materials can be easily modulated according to tailored synthetic approaches based on "soft routes." In fact, molecular materials are usually obtained through solution chemistry, under mild conditions, in contrast to the high pressures and temperatures that are instead usually needed to obtain some conventional inorganic solids.

1.3 Top-Down and Bottom-Up Approaches

The concept of molecular materials, intended as materials built by functional molecules that retain their properties down to the molecular scale, also changed the way to conceive the construction of technological devices. Back in the 1980s, the unceasing miniaturization of electronic circuits led to the idea that the down-scaling achievable in bulk materials was inadequate to satisfy the upcoming technological demand and a novel approach for designing and preparing new materials, suitable to provide smaller devices, was required. In contrast to the top-down approach, which consists in decreasing the dimension of bulk materials to prepare smaller objects without losing the material's performance, the so-called bottom-up approach represents the alternative strategy deriving from the concept of molecular materials. This approach consists in the assembly of chemically and physically controlled molecular building blocks, starting from the smallest available building units (atoms), to realize devices at the molecular scale. This will provide, in principle, smaller and more performing devices than those achievable through bulk miniaturization (Fig. 1.3).

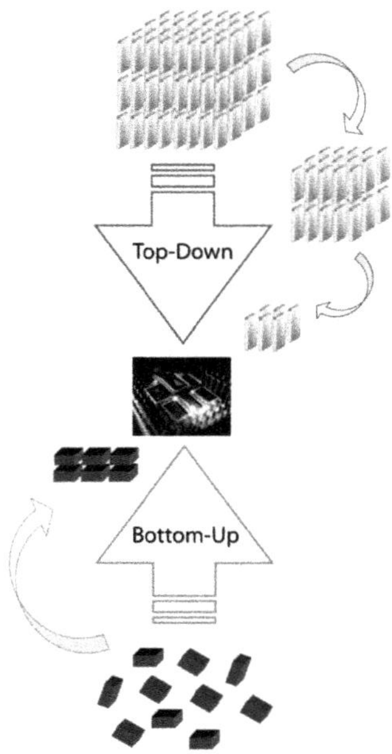

Figure 1.3 Schematic representation of the top-down and bottom-up approaches toward devices' miniaturization. (Top) A generic bulk material whose dimensions are subsequently reduced to afford a smaller object with unchanged composition, structure, and properties is shown. (Bottom) An ensemble of generic molecular building blocks that assemble together to afford a molecule-based material is, instead, depicted.

1.4 Dimensions and Dimensionalities

The physical properties of conventional atom-based inorganic materials, such as metals, alloys, and oxides, are strongly dependent on the dimensions of each discrete object constituting the material, be it a crystal, a particle, etc. Materials in the bulk form, that is, materials having macroscopic dimensions and a number of atoms constituting the volume that are several orders of magnitude higher than those composing the surface, are characterized by certain physical properties that can undergo deep modifications when the size of the objects is reduced at the nanoscale or, more interestingly, additional properties with respect to those of the bulk can emerge. This is related to the fact that in nanomaterials, that is, materials in which at least one of the three space dimensions belongs to the nanometer scale (e.g., nanoparticles, nanowires, nanorods, nanotubes, nanostructured surfaces, and nanostructured 3D materials), the nanosize of each discrete object induces a specific and unique behavior of a given physical property. Consider, for example, the case of magnetic domains, whose sizes are comparable to those of nanoparticles. The down-scaling to nanodimensionality of these materials determines a drastic change in the magnetic structure, which goes from a multidomain to a single domain, with important implications for the resulting magnetic properties.

In general, the properties displayed by atom-based materials (electrical conductivity, long-range magnetic ordering, superparamagnetism, etc.) arise from wide-range interactions and are strongly dependent on the number of atoms composing each finite object constituting the material. In the case of the smallest objects (nanomaterials) this number is of the order of 10^2–10^4 atoms. As a consequence, the dimensions of such entities can be reduced (with concomitant change in the material properties) until a critical number of atoms is reached, which determines the threshold below which the extent of interatomic interactions is too low to give rise to relevant physical properties. It is then evident that this represents a severe limitation to further material miniaturization. However, this boundary can be indeed overcome by molecular materials.

The dimensionality scale that can be achieved with molecular materials is noticeably smaller than that with atom-based inorganic materials. Since the physical properties of molecular materials usually originate from the peculiar nature of a single and discrete molecular entity (single-molecule property), and these molecular units are usually composed of no more than 10^1–10^2 atoms, it is evident that they virtually set no limits on material miniaturization. This is the case, for example, with single-molecule magnets (SMMs) or light-emitting lanthanide-based complexes (see further chapters), where the characteristics of a single molecule are retained up to the macroscopic scale. This is due to the fact that, in such systems, the displayed physical properties are independent from any interactions between equivalent molecules. Nonetheless, molecular materials can also show physical properties related to cooperative phenomena, in analogy to those properties observed in 3D bulk inorganic materials, such as long-range magnetic ordering and electrical conductivity, but with a wider and, indeed, more tunable range of dimensionalities. In fact, due to the intrinsically higher anisotropic shapes of molecules with respect to those of atoms, the specific interactions occurring between molecules, be they covalent or noncovalent, allow for the preparation of materials with alternative dimensionality than the usual 3D form, such as monodimensional (1D) and bidimensional (2D) architectures, thus introducing a higher structural flexibility that can ultimately affect the physical properties.

1.5 A Guide to Design Functional Molecular Materials

The ultimate aim that motivates scientists to strive for novel molecular materials is the exploitation of their properties for a specific functionality of technological interest. However, before these materials become suitable to be used for real-world applications, several conceptual and practical steps need to be accomplished to allow their technological use.

In this last section of this introductory chapter, we aim to point out which are the steps that, from the authors' viewpoint, should be followed by a material scientist who starts to deal with the design and preparation of novel molecular materials with potential technological interest. As a matter of fact, these steps are the common thread behind the organization of this book.

First of all, one should be aware that a deep knowledge and understanding of the fundamental concepts that are at the basis of a desired physical property is crucial to be able to rationally design a new material with improved functionality.[2] For this reason, a recall of the basic concepts of physics and molecular physics needed to fully understand the design approach to molecular materials cannot be neglected, and this matter is thoroughly addressed in the second chapter of this book.

With the fundamental concepts in mind, the successive step is to find out which are the chemical, structural, and electronic features that a molecular system should have to allow for a certain property to be displayed. With this view, the aim of the third chapter is to describe the general guidelines that should be followed to build a molecular material showing specific features associated with some of the most investigated properties. In this regard, since science advancements are always heirs to foregoing research experience, the most useful approach to the clear presentation of the "how to design" criteria is to illustrate them on the basis of selected archetypal examples.

One of the subsequent logical stages in the field of molecular materials is represented by the possibility to design multifunctional molecular materials through self-assembling of selected molecular building blocks with diverse functionalities. As evidenced previously, this is a peculiarity of molecule-based materials, and some of the approaches and design strategies that can be adopted for the achievement of molecule-based systems showing two or more superimposed or cooperative properties are addressed in the fourth chapter of this book. This topic will be once again developed through the description of the most relevant examples of multifunctional materials and related physical properties, with the aim to provide the reader useful hints for the further expansion of this field.

Once these conceptual steps have been addressed, the aim of the final chapter is to provide a brief overview on how molecular materials can be fruitfully and practically employed for the preparation of

---

[2]It should be mentioned, however, that in the course of research activity unknown properties can be discovered and unexpected materials can be obtained. Although this possibility can be even more interesting to an experimental scientist, it goes beyond a rational design approach and cannot logically be predicted. Nonetheless, it is of utmost importance for the experimental scientist to achieve a solid fundamental background in order to be able to recognize novel phenomena and properties still unexplored.

technological devices with advanced functional properties. This is indeed the ultimate goal to be pursued in this field. Yet, at this stage of research advancement, some classes of molecular materials are not suitable to be employed for practical technological applications. The relevant challenges that scientists are undertaking to fill the gap between fundamental science and technological applications represent the final message that is delivered to the reader of this book, bearing in mind that the field of molecular materials is still largely unexplored and it extends well beyond the contents briefly presented herein.

Chapter 2

Background

This chapter provides a summary of those basic notions of magnetism, electrical conductivity, and photophysics that are required for nonexpert readers to get a full understanding of the physical properties of molecular materials addressed in the chapters that follow. For a neat presentation, the chapter is organized into three main parts. Firstly, an overview of the basic principles of magnetism, magnetochemistry, and molecular magnetism needed to understand and discuss the magnetic properties of molecular materials is provided. In the second part the fundamental physical principles of electrical conductivity in molecular materials are discussed. Finally, in the third part, an introduction to molecular photophysics is addressed.

It must be remarked, however, that a complete and thorough description of these subjects is beyond the aim of this chapter, for which the reader is invited to consult more specialized texts.

2.1 Basic Concepts of Magnetism, Magnetochemistry, and Molecular Magnetism

Before entering into a description of some basic concepts of magnetism, a clarification of the topics concerning three strongly correlated disciplines—magnetism, magnetochemistry, and

Functional Molecular Materials: An Introductory Textbook
Matteo Atzori and Flavia Artizzu

ISBN 978-981-4774-76-5 (Hardcover), 978-1-351-23366-8 (eBook)
www.panstanford.com

molecular magnetism—may be helpful for the reader. Magnetism is the field of physics dealing with the study of the magnetic properties of the matter by analyzing the behavior of materials when they are exposed to the action of a magnetic field. This is one of the most ancient scientific subjects and one of the most difficult to understand. Closely connected with magnetism is magnetochemistry. This is a field of chemical physics that deals with magnetic measurements performed on chemical compounds, both molecular and atom based, as a tool to get information on their chemical structure. Finally, molecular magnetism can be defined as a relatively new field of physics born to describe and understand the magnetic properties shown by magnetic molecular materials.[1] A deep knowledge of these three disciplines allows one to design new materials with expected magnetic properties and to understand their magnetic behaviors. The following sections describe their fundamental principles, with particular attention to those classes of materials that will be the theme of Chapter 3.

2.1.1 Magnetic Field

One of the fundamental principles in magnetism is the concept of "magnetic field." A magnetic field is produced whenever there is a charge in motion. This can be associated with an electrical current flowing in a conductor, as was first discovered in 1819 by Oersted, or it can be generated by a permanent magnet. In the latter case there is no conventional electric current, but the orbital and spin motions of the electrons forming the permanent magnet lead to a magnetization within the material and to magnetic field generation.

When a magnetic field is generated in a certain volume of space, an energy gradient is established and a force is produced. This force can be evidenced by:

- The acceleration of a charge moving in the field
- The electromotive force acting on a current-carrying conductor
- The torque moment on a magnetic dipole
- The electronic spin reorientation in certain types of atoms or ions

[1]These are not universally agreed definitions, especially those of magnetochemistry and molecular magnetism, and reflect the personal points of view of the authors.

The force of a magnetic field **H** generated by an electrical current can be calculated through the Biot–Savart law.

2.1.2 Magnetic Induction

When a magnetic field **H** is generated in a medium by a current *i*, the response of the medium to this perturbation is the magnetic induction **B**, also called flux density. Accordingly, whenever a magnetic field is present, there will be a magnetic flux Φ. The magnetic induction **B** represents the amount of magnetic flux Φ per unit area A ($\mathbf{B} = \Phi/A$) and is measured in webers per meter square (Wb m^{-2}; weber = V s). A flux density of 1 Wb m^{-2} corresponds to a magnetic induction of 1 tesla (T). The extent of the flux generated by a magnetic field depends on the properties of the medium and varies from one medium to another.

The physical parameter that correlates the magnetic field **H** and the magnetic induction **B** is called magnetic permeability of the medium, μ. In several media the magnetic induction is a linear function of the magnetic field. In particular, in free space:

$$\mathbf{B} = \mu_0\mathbf{H} \tag{2.1}$$

where μ_0 is the magnetic permeability of the free space (vacuum permeability), which is a universal constant. The magnetic permeability represents the tendency of a medium to be magnetized. To evaluate the response of a medium to a magnetic field, it is convenient to refer to the relative magnetic permeability, μ_r, as the ratio μ/μ_0 (μ_r takes a unitary value for free space). In other words, if μ_r is close to 1, **B** will be only slightly different from **H** and the magnetic field will have a strong influence on the material, whereas for large μ_r values, **B** will significantly differ from **H** and the magnetic field will not have a strong influence on the material.

2.1.3 Magnetic Moment

The most elementary objects that can generate a magnetic field are (Fig. 2.1):

- A closed circuit of a conductor carrying an electric current
- A magnetic dipole, that is, two magnetic monopoles of different signs separated by a distance *l*

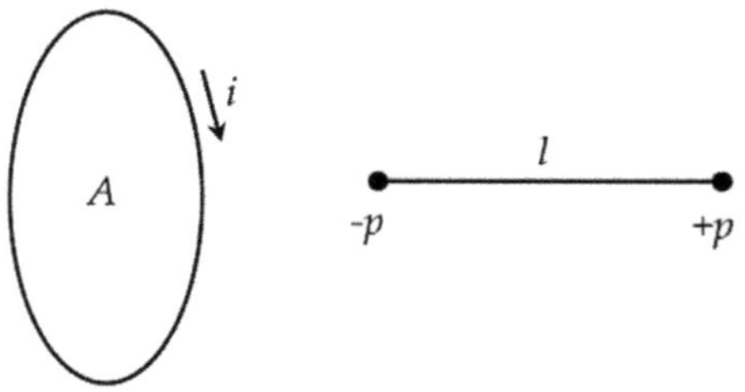

Figure 2.1 Schematic representation of the most elemental objects in magnetism: a closed circuit carrying an electric current (left) and a linear magnetic dipole (right).

In each case, a magnetic moment **m** associated with the elementary unit of magnetism is present. In the closed circuit case the magnetic moment is equal to the product of the current i and the area of the loop, A:

$$\mathbf{m} = i \times A \tag{2.2}$$

In the dipole moment case the magnetic moment is equal to the product of the poles' strength p and the distance between them, l:

$$\mathbf{m} = p \times l \tag{2.3}$$

When a magnetic induction **B** is generated by an external magnetic field **H**, the magnetic dipole tries to align itself so that the magnetic moment **m** lies parallel to the magnetic induction. This generates a torque $\boldsymbol{\tau}$ on the magnetic dipole moment **m** (Fig. 2.2), given by the cross product:

$$\boldsymbol{\tau} = \mathbf{m} \times \mathbf{B} \tag{2.4}$$

which in free space is:

$$\boldsymbol{\tau} = \mu_0 \mathbf{m} \times \mathbf{H} \tag{2.5}$$

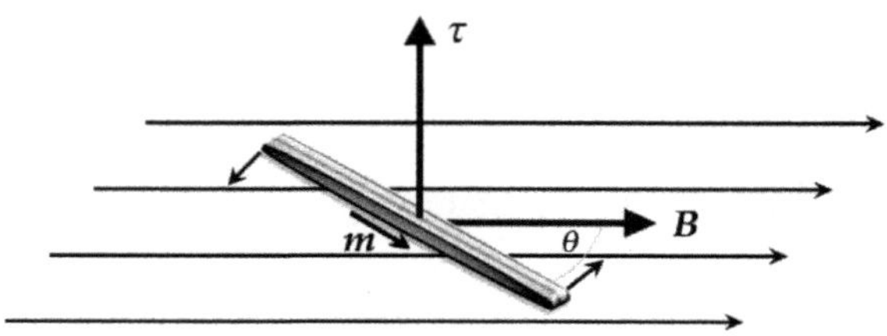

Figure 2.2 The torque **τ** on a bar magnet in an external magnetic field **H**. The magnetic moment vector tends to align itself along the **B** direction.

The magnetic moment can thus be expressed as the maximum torque on a magnetic dipole moment, $\boldsymbol{\tau}_{max}$, divided by **B**:

$$\mathbf{m} = \frac{\tau_{\text{max}}}{\mathbf{B}} \tag{2.6}$$

Hence in free space:

$$\mathbf{m} = \frac{\tau_{\text{max}}}{\mu_0 \mathbf{H}} \tag{2.7}$$

The mechanical work of the magnetic force (the torque) on the alignment of the dipole moment under a magnetic induction defines the energy associated with a magnetic moment **m** in the presence of magnetic induction **B** as:

$$E = -\mathbf{m} \times \mathbf{B} = -\mathbf{mB} \cos \theta \tag{2.8}$$

It is clear that for θ values of 0° and 180° the energy assumes its minimum values.

2.1.4 Effect of Magnetic Field on Matter

While considering magnetic materials, it is first necessary to define the physical quantities that represent the response of these materials to the magnetic field. These are magnetization and susceptibility.

2.1.4.1 Magnetization

The concept of magnetization describes the correlation between the magnetic properties of a certain material and the magnetic induction **B**. The magnetization **M** can be defined as the magnetic moment per volume unit of a solid:

$$\mathbf{M} = \frac{\mathbf{m}}{V} \tag{2.9}$$

From the relation between magnetic moment **m** and magnetic flux $\boldsymbol{\Phi}$, $\mathbf{m} = (\boldsymbol{\Phi} l)/\mu_0$. A simple relation between **M** and **B** can be found:

$$\mathbf{M} = \frac{\mathbf{B}}{\mu_0} \tag{2.10}$$

It is noteworthy that **B** and **M** are correlated as **B** and **H** (Eq. 2.1). This means that both magnetization **M** and magnetic field **H** contribute to the magnetic induction **B** in a similar way. The magnetic induction in free space is $\mathbf{B} = \mu_0\mathbf{H}$, while the contribution from the magnetization of the material is $\mathbf{B} = \mu_0\mathbf{M}$. Thus magnetic induction **B** is simply the vector sum of the two contributions:

$$\mathbf{B} = \mu_0(\mathbf{H} + \mathbf{M}) \tag{2.11}$$

where **B** is given in T and **M** and **H** in A m^{-1}. If both magnetization and magnetic field are present, their contributions can be added up to give the global value of magnetic induction.

2.1.4.2 Magnetic permeability and susceptibility

It is now possible to define the magnetic permeability μ as:

$$\mu = \frac{\mathbf{B}}{\mathbf{H}} \tag{2.12}$$

and the magnetic susceptibility χ as:

$$\chi = \frac{\mathbf{M}}{\mathbf{H}} \tag{2.13}$$

Since **B** and **M** may or may not be linear functions of **H**, depending on the type of material or medium, it should be noted that the permeability and susceptibility may or may not be constant.

This quantity is particularly useful as it provides an evaluation of the magnetization of a certain material that is independent of the applied magnetic field used to perform the measurement.

Although the magnetic susceptibility χ is a dimensionless unit, it is actually common in experimental practice to refer to magnetic susceptibility per mass unit χ_m (cm^3 g^{-1}) or to molar magnetic susceptibility χ_{mol} (cm^3 mol^{-1}).

The values assumed by the magnetic susceptibility can be used to classify the conventional classes of magnetic materials. Materials having a small and negative χ, typically $\chi \approx -10^{-5}$, are called diamagnets, and their magnetic response opposes the magnetic field. Superconductors are a special group of diamagnets for which $\chi \approx -1$. Materials having a small and positive χ, typically $\chi \approx 10^{-3}$–10^{-5}, are paramagnets. The magnetization of paramagnets is weak but aligned parallel with the direction of the applied magnetic field. A third type of magnetic materials is of that ferromagnets, for which the susceptibility is positive and much greater than 1; typical χ values are found in the 10^1–10^5 range.

It should be mentioned that experimental susceptibility measurements for all magnetic systems, pure diamagnets excluded, provide a value of χ that results from the sum of two contributions:

- The paramagnetic susceptibility that arises from the unpaired electrons χ_p
- The diamagnetic susceptibility that arises from the diamagnetic response of the paired electrons χ_d

$$\chi = \chi_\mathrm{p} + \chi_\mathrm{d} \tag{2.14}$$

The diamagnetic susceptibility is approximately constant for any given atom, independently of the nature of the system involved, and is temperature independent. Since the diamagnetic susceptibility is negative, the effect on the global susceptibility value resulting from an experimental measurement is to reduce χ. To obtain the value of the paramagnetic susceptibility χ_p corrected from the diamagnetic contribution χ_d, the sum of the different diamagnetic contributions deriving from each atom, χ_dA, and each bond, χ_dB, must be added to the experimental value χ (Eq. 2.15).

$$\chi_\mathrm{d} = \sum_i \chi_\mathrm{dA} + \sum_i \chi_\mathrm{dB} \tag{2.15}$$

The diamagnetic susceptibility values for atoms, ions, and bonds are commonly referred to as Pascal's constants (Bain and Berry, 2008). Empirical practice tells us that a rough estimation of χ_d for molecular compounds can be obtained by taking the numerical value of half the molecular weight of the investigated compound by expressing it in 10^{-6} $\mathrm{cm^3\ mol^{-1}}$ units (Kahn, 1993). This approximation works well for paramagnetic compounds of relatively low molecular weight (ca. <2000 g $\mathrm{mol^{-1}}$) where the diamagnetic contribution is small and becomes almost negligible at low temperatures. If high-molecular-weight compounds with only a few paramagnetic centers per molecule are concerned, the above approximation is not valid since paramagnetic and diamagnetic contributions can assume comparable values (Kahn, 1993).

For the rest of this book the symbol χ will refer to the paramagnetic component of susceptibility already corrected for the diamagnetic contribution.

2.1.5 Electronic Magnetic Moments

The magnetic properties of electrons are of fundamental importance in magnetism because they determine the magnetic properties of an atom. The combination of electronic magnetic moments, originating from their angular momenta, gives the so-called atomic magnetic moment.

The magnetic moments associated with atoms or ions on which magnetic materials are based can have three origins: the electronic orbital motion, which causes an orbital magnetic moment; the electronic spin motion, which causes a spin magnetic moment; and the change in the orbital motion of electrons caused by an applied magnetic field.

The first two contributions dominate in the case of atoms or ions having unpaired electrons, whereas the third one achieves significance only in those species that have paired electrons and is at the origin of the diamagnetic susceptibilities of diamagnets. Indeed, the magnetic moments originating from the orbital and spin motions of paired electrons are cancelled by themselves as they have the same magnitude but opposite signs due to their antiparallel orientation within the same orbital.

We will now focus on the first two contributions, which will be described at first on the basis of a classical model and then by means of a quantum mechanical approach.

2.1.5.1 Classical model of magnetic moments

In the classical model the angular momentum of electrons can be used to describe the electronic magnetic moment by invoking the concept of electrical charge in motion. The electronic magnetic moment is given as a vector sum of two contributions:

- The orbital magnetic moment due to the orbital angular momentum
- The spin magnetic moment due to the electronic spin angular momentum

2.1.5.1.1 *Electronic orbital magnetic moment*

Assume that an electron of charge e moves in a circular orbit around an atomic nucleus having area A and period τ. This motion would produce a current i given by:

$$i = \frac{e}{\tau} \tag{2.16}$$

On the basis of the previously described definitions, the orbital magnetic moment $\mathbf{m}_0$ associated with this system is:

$$\mathbf{m}_0 = iA = -\frac{eA}{\tau} \tag{2.17}$$

The moment of inertia I of such an electron in orbital motion is that of a particle of mass m_e orbiting the nucleus, $I = m_e r^2$. Since the angular momentum of the particle is $\mathbf{p}_0 = I\boldsymbol{\omega}$, where $\boldsymbol{\omega}$ is the angular velocity:

$$\mathrm{p}_0 = m_e r^2 \boldsymbol{\omega} \tag{2.18}$$

where r is the orbit radius. Recalling that $A = \pi r^2$ and $\boldsymbol{\omega} = 2\pi/\tau$, the area of the orbit will be:

$$A = \left(\frac{1}{2}\right)\mathbf{p}_0 \frac{\tau}{m_e} \tag{2.19}$$

At this point an equation for describing the orbital magnetic moment $\mathbf{m}_0$ of the electron in terms of orbital angular momentum $\mathbf{p}_0$ can be written as follows:

$$\mathbf{m}_0 = -\left(\frac{e}{2m_e}\right)\mathbf{p}_0 \tag{2.20}$$

recalling that the magnetic moment vector points in the opposite direction to the angular moment vector since the electron has a negative charge (Jiles, 1998).

2.1.5.1.2 *Electronic spin magnetic moment*

The electronic spin angular momentum $\mathbf{p}_S$ generates an electronic spin magnetic moment $\mathbf{m}_S$. In this case the relation between $\mathbf{m}_S$ and $\mathbf{p}_S$ is:

$$\mathbf{m}_S = -\left(\frac{e}{m_e}\right)\mathbf{p}_S \tag{2.21}$$

It should be noted that for a given angular momentum the spin is formally twice the magnetic orbital moment. These results can be classically justified by relating this system to a rotating disk of mass m_e whose charge is localized on the circumference (Jiles, 1998).

In this case the moment of inertia is $I = ½\, m_e r^2$, while the spin angular momentum is again $\mathbf{p}_S = I\boldsymbol{\omega}$. This gives rise to a spin angular momentum of $\mathbf{p}_S = ½\, m_e r^2 \boldsymbol{\omega}$ and a magnetic moment $\mathbf{m}_S$:

$$\mathbf{m}_S = iA = -\frac{e\,\boldsymbol{\omega}\, r^2}{2} \tag{2.22}$$

2.1.5.1.3 *Total electronic magnetic moment*

The total magnetic moment per electron can be described as the vector sum of the spin and orbital magnetic moments:

$$\mathbf{m}_{tot} = \mathbf{m}_O + \mathbf{m}_S \tag{2.23}$$

or

$$\mathbf{m}_{tot} = -\left(\frac{e}{2m_e}\right)\mathbf{p}_O - \left(\frac{e}{2m_e}\right)2\mathbf{p}_S \tag{2.24}$$

The terms on the right side can be combined to give the following expression:

$$\mathbf{m}_{tot} = -\mathbf{g}\left(\frac{e}{2m_e}\right)\mathbf{p}_{tot} \tag{2.25}$$

where $\mathbf{p}_{tot}$ is the total angular momentum of the electron. The term $\mathbf{g}$ is called the Landè splitting factor. It assumes a value of 2 for the spin component and 1 for the orbital component. The value of $\mathbf{g}$ thus changes as a function of the relative contribution of the spin and the orbit component to the total angular momentum.

Although this equation relating the angular momentum to the magnetic moment of an electron works quite well, this classical interpretation has to be abandoned in favor of a quantum mechanical approach.

2.1.5.2 Quantum mechanical model of magnetic moments

In the classical model the angular momenta $\mathbf{p}_O$ and $\mathbf{p}_S$ can assume any value. Actually, this is not realistic since the possible values of electronic angular momenta are restricted by quantum mechanics. Consequently, the magnetic moments are also quantized.

2.1.5.2.1 *Electronic orbital magnetic moment*

The orbital angular momentum can be expressed in $h/2\pi$ units (where h is the Planck constant). Then its expression becomes:

$$\mathbf{m}_\mathrm{O} = -\left(\frac{eh}{4\pi m_\mathrm{e}}\right)\left(\frac{2\pi \mathbf{p}_\mathrm{O}}{h}\right) \tag{2.26}$$

and since $\mathbf{p}_\mathrm{O} = \mathbf{l}(h/2\pi)$:

$$\mathbf{m}_\mathrm{O} = -\left(\frac{eh}{4\pi m_\mathrm{e}}\right)\mathbf{l} \tag{2.27}$$

where **l** is the orbital angular momentum quantum number.

From quantum theory, the electronic orbital angular momentum $\mathbf{p}_\mathrm{O}$ is a multiple integer of $h/2\pi$, so the contribution to the magnetic moment of the orbital angular momentum is a multiple integer of $eh/4\pi m_\mathrm{e}$. This quantity is known as the Bohr magneton μ_B, which has the value of 9.27×10^{-24} A m^2 in SI units. Consequently, the previous equation can be rewritten as:

$$\mathbf{m}_\mathrm{O} = -\mu_\mathrm{B}\mathbf{l} \tag{2.28}$$

2.1.5.2.2 *Electronic spin magnetic moment*

The electronic spin always takes the value ½. Then the spin angular momentum $\mathbf{p}_\mathrm{S}$ is a multiple integer of $h/2\pi$, $\mathbf{p}_\mathrm{S} = \mathbf{s}(h/2\pi)$, where **s** is the spin quantum number. The magnetic moment $\mathbf{m}_\mathrm{S}$ associated with the spin is:

$$\mathbf{m}_\mathrm{S} = -\left(\frac{eh}{2\pi m_\mathrm{e}}\right)\left(\frac{2\pi \mathbf{p}_\mathrm{S}}{h}\right) \tag{2.29}$$

Replacing $(2\pi\,\mathbf{p}_\mathrm{s}/h)$ with the spin quantum number **s**:

$$\mathbf{m}_\mathrm{S} = -2\left(\frac{eh}{4\pi m_\mathrm{e}}\right)\mathbf{s} \tag{2.30}$$

Here it is found that for a given angular momentum, the spin gives twice the magnetic moment with respect to the orbital contribution to the total electronic moment, as already found with the classical model. The difference in the relation between the spin and the orbital magnetic moments and their respective angular momenta is attributed to relativistic effects.

Replacing the quantity $eh/4\pi m_\mathrm{e}$ with the Bohr magneton μ_B:

$$\mathbf{m}_\mathrm{S} = -2\mu_\mathrm{B}\mathbf{s} \tag{2.31}$$

Since **s** must be +½ or −½, the spin magnetic moment of one electron corresponds to one Bohr magneton.

2.1.5.2.3 *Total electronic magnetic moment*

The total magnetic moment of an electron can be expressed in terms of multiples of the total angular momentum $(h/2\pi)\mathbf{j}$ as:

$$\mathbf{m}_{\text{tot}} = -\mathbf{g}\left(\frac{eh}{4\pi m_{\text{e}}}\right)\left(\frac{2\pi\mathbf{p}_{\text{tot}}}{h}\right) = -\mathbf{g}\mu_{\text{B}}\left(\frac{2\pi\mathbf{p}_{\text{tot}}}{h}\right) = -\mathbf{g}\mu_{\text{B}}\mathbf{j} \quad (2.32)$$

where $\mathbf{j}$ is the total angular momentum quantum number.

For an electron the total magnetic moment can be written as:

$$\mathbf{m}_{\text{tot}} = \mathbf{m}_{\text{O}} + \mathbf{m}_{\text{S}} \quad (2.33)$$

or

$$\mathbf{m}_{\text{tot}} = -(\mu_{\text{B}}\mathbf{l} + 2\mu_{\text{B}}\mathbf{s}) \quad (2.34)$$

For completeness, it should be taken into account that the assumption that the angular momenta are integer multiples of the quantum numbers $\mathbf{l}$ and $\mathbf{s}$ is not rigorous. Precisely, according to wave mechanics, the values are $\mathbf{p}_{\text{O}}^2 = (h/2\pi)^2\mathbf{l}(\mathbf{l} + 1)$ and $\mathbf{p}_{\text{S}}^2 = (h/2\pi)^2\mathbf{s}(\mathbf{s} + 1)$, and the corresponding magnetic moments become $\mathbf{m}_{\text{O}} = \mu_{\text{B}}\sqrt{\mathbf{l}(\mathbf{l}+1)}$ and $\mathbf{m}_{\text{S}} = \mu_{\text{B}}\sqrt{\mathbf{s}(\mathbf{s}+1)}$. The total magnetic moment $\mathbf{m}_{\text{tot}}$ needs an analogous correction since $\mathbf{p}_{\text{tot}} = (h/2\pi)\sqrt{\mathbf{j}(\mathbf{j}+1)}$ (Jiles, 1998).

2.1.6 Magnetic Properties of Free Atoms

As already pointed out, the magnetic properties of an atom are principally determined by the magnetic moments of its electrons, which can be calculated as the sum of spin and angular contributions. Since the net magnetic moment of a filled electron shell is zero, only the unfilled shell needs to be considered.

The atomic orbital (AO) angular momentum, denoted by the quantum number $\mathbf{L}$, is the vector sum of the orbital angular momenta of the electrons within an atom:

$$\mathbf{L} = \sum \mathbf{l}_i \quad (2.35)$$

The magnitude of the orbital angular momentum $\mathbf{L}$, $|\mathbf{L}|$, is:

$$|\mathbf{L}| = \sqrt{\mathbf{L}(\mathbf{L}+1)} \quad (2.36)$$

which is identical in form to the relation of |**l**| for a single electron. The atomic spin angular momentum **S** and its magnitude |**S**| can be obtained in a similar way as described for **L**:

$$\mathbf{S} = \sum \mathbf{s}_i \tag{2.37}$$

$$|\mathbf{S}| = \sqrt{\mathbf{S}(\mathbf{S}+1)} \tag{2.38}$$

The total atomic angular momentum **J** of an atom can be found by vectorially summing the spin and orbital angular momenta of its electrons. This can be done in two ways:

- By independently summing up the orbital and spin angular momenta of the unpaired electrons:

$$\mathbf{J} = \sum \mathbf{l}_i + \sum \mathbf{s}_i \tag{2.39}$$

  This method is known as Russell–Saunders coupling. Within this approach the **l** vectors of the electrons are coupled together (**l**–**l**), as are the **s** vectors (**s**–**s**), under the assumption that these couplings are stronger than the coupling between the resultant **L** and **S** vectors.
- By summing up the total angular momenta for each electron, which are, in turn, obtained by summing its **l** and **s** values:

$$\mathbf{J} = \sum \mathbf{j}_i = \sum (\mathbf{l}_i + \mathbf{s}_i) \tag{2.40}$$

  This method is known as **j**–**j** coupling, and it is preferably used whenever there is a strong spin–orbit coupling (**l**–**s**) for each electron. Under this circumstance, the coupling between **l** and **s** for an individual electron is taken as stronger than the coupling between the various **l** and **s** values for different electrons.

In general, the Russell–Saunders method is the most appropriate for describing the magnetic properties of first transition elements whereas the **j**–**j** coupling applies better to describing heavier atoms, especially third row transition metals and lanthanides.

It should be mentioned that a small contribution to the total angular momentum of an atom is due to the magnetic moment of the nuclear spin. This contribution is about 3 orders of magnitude smaller than the electronic spin. The nuclear magnetic moment is measured in nuclear magnetons μ_N units. μ_N has the value of 5.05×10^{-27} A m^2, compared to 9.27×10^{-24} A m^2 for the Bohr magneton.

2.1.6.1 Hund's rules: Occupancy of available electronic states (Russell–Saunders scheme)

A set of empirical rules known as *Hund's rules* allows for the determination of the occupancy of the available electronic states within an atom. These rules identify how the possible electronic states are filled and can be therefore used to determine the energy state of an atomic species by calculating the **L**, **S**, and **J** values for the electronic configuration of its unfilled shell. For the fundamental, that is, the most stable, energy state of the atom three rules apply:

- The maximum value of the total atomic spin angular momentum $\mathbf{S} = \Sigma \mathbf{m}_s$ is obtained without violating the Pauli exclusion principle.
- The maximum value of the total AO angular momentum $\mathbf{L} = \Sigma \mathbf{m}_l$ is determined as a function of the state occupancy given by **S**.
- The value of the total atomic angular momentum **J** is equal to $|\mathbf{L} - \mathbf{S}|$ when the shell is less than half full and is equal to $|\mathbf{L} + \mathbf{S}|$ when the shell is more than half full. When the shell is exactly half full $\mathbf{L} = 0$ so that $\mathbf{J} = \mathbf{S}$.

The electronic configuration obtained following Hund's rules is associated with the lowest-energy state of the system that is identified by a term symbol:

$$^{2\mathbf{S}+1}\Gamma_{\mathbf{J}}$$

where $2\mathbf{S} + 1$ is the spin multiplicity and the symbol Γ = S, P, D, F, . . . indicates the value of $\mathbf{L} = 0, 1, 2, 3, \ldots$, respectively.

The electronic energy state corresponding to a specific term symbol is further constituted by degenerate terms. In particular, the degeneracy of the states associated with **L** and **S** quantum numbers is $(2\mathbf{S} + 1)(2\mathbf{L} + 1)$. Whenever the spin–orbit coupling becomes relevant, further degeneracy of energy levels of $2\mathbf{J} + 1$ terms must be also considered.

A Practical Example

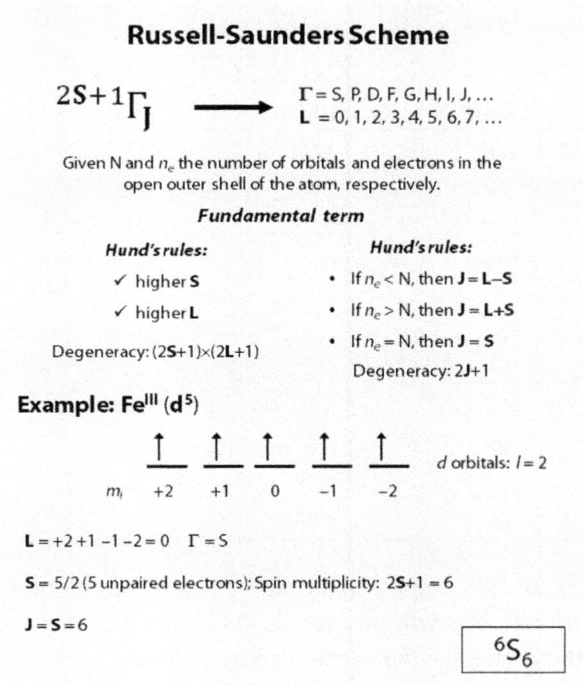

It must be reminded now that any given energy level occupied by an electron in the presence of a magnetic field is further split into energy states associated with different values of m_S, the projection of the spin quantum number on the axis of quantization.

2.1.6.2 Quenching of the orbital angular momentum L

For 3d elements it has been experimentally found that the values of the atomic magnetic moments disagree with those calculated taking into account both spin (**S**) and orbital (**L**) contributions. Instead, when considering only the spin component, that is, **J** = **S**, a good agreement between experimental and theoretical values is found, as shown in Table 2.1. This implies that **L** = 0, and thus the orbital angular momentum is "quenched."

Table 2.1 Calculated magnetic moments for $\mathbf{L} \neq 0$ and $\mathbf{L} = 0$ and experimental magnetic moments **m** for 3d ions in μ_B units

Ion	Configuration	$\mathbf{m} = g\sqrt{J(J+1)}$ (μ_B)	$\mathbf{m} = g\sqrt{S(S+1)}$ (μ_B)	m (μ_B)
Ti(III), V(IV)	$3d^1$	1.55	1.73	1.8
V(III)	$3d^2$	1.63	2.83	2.8
Cr(III)	$3d^3$	0.77	3.87	3.8
Mn(III), Cr(II)	$3d^4$	0	4.90	4.9
Fe(III), Mn(II)	$3d^5$	5.92	5.92	5.9
Fe(II)	$3d^6$	6.70	4.90	5.4
Co(II)	$3d^7$	6.63	3.87	4.8
Ni(II)	$3d^8$	5.59	2.83	3.2
Cu(II)	$3d^9$	3.55	1.73	1.9

We can, in fact, picture that under certain conditions, the plane associated with the orbit motion of the electron moves frequently, so the orbital angular momentum contribution is averaged to zero. In these elements **J** = **S** and **L** = 0. Thus the total atomic magnetic moment is given by the equation:

$$\mathbf{m} = 2\mu_B\sqrt{\mathbf{S}(\mathbf{S}+1)} \quad (2.41)$$

the "spin-only magnetic moment" formula.

For example, the magnetic moment for an Fe^{II} ion (d^6) calculated with the spin-only formula is 4.90 μ_B. This reasonably agrees with the observed value of 5.40 μ_B, whereas the calculated value without quenching of the orbital contribution is 6.70 μ_B, which is clearly an overestimation.

It must be remarked that this phenomenon does not occur for elements having unpaired electrons on f orbitals or for heavy transition metals, where the spin–orbit coupling is indeed relevant, and both spin and orbital contributions are involved to give the observed magnetic moment (Table 2.2).

Table 2.2 Calculated magnetic moments for **L** ≠ 0 and **L** = 0 and experimental magnetic moments **m** for rare earth ions in μ_B units

Ion	Configuration	S	L	J	g_J	$m = g_J\sqrt{\lvert J(J+1)\rvert}$ (μ_B)	$m = g\sqrt{\lvert S(S+1)\rvert}$ (μ_B)	m (μ_B)
Ce(III)	$4f^1$	1/2	3	5/2	6/7	2.54	1.73	2.3–2.5
Pr(III)	$4f^2$	1	5	4	4/5	3.58	2.83	3.4–3.6
Nd(III)	$4f^3$	3/2	6	9/2	8/11	3.62	3.87	3.5–3.6
Pm(III)	$4f^4$	2	6	4	3/5	2.68	4.90	–
Sm(III)	$4f^5$	5/2	5	5/2	2/7	0.85	5.92	1.4–1.7
Eu(III)	$4f^6$	3	3	0	–	0	6.93	3.3–3.5
Gd(III)	$4f^7$	7/2	0	7/2	2	7.94	7.94	7.9–8.0
Tb(III)	$4f^8$	3	3	6	3/2	9.72	6.93	9.5–9.8
Dy(III)	$4f^9$	5/2	5	15/2	4/3	10.63	5.92	10.4–10.6
Ho(III)	$4f^{10}$	2	6	8	5/4	10.60	4.90	10.4–10.7
Er(III)	$4f^{11}$	3/2	6	15/2	6/5	9.57	3.87	9.4–9.6
Tm(III)	$4f^{12}$	1	5	6	7/6	7.63	2.83	7.1–7.6
Yb(III)	$4f^{13}$	1/2	3	7/2	8/7	4.54	1.73	4.3–4.9
Lu(III)	$4f^{14}$	0	0	0	–	0	0	0

Moreover, it should be highlighted that these calculations, both in the spin-only and, in particular, in the total magnetic moment case, apply only to isolated paramagnetic (noninteracting) atoms or ions (vide infra).

2.1.7 The Curie Law

The susceptibilities of a number of paramagnetic solids were measured over a wide temperature range by Pierre Curie. He found that the susceptibility varied inversely with temperature, as shown in Fig. 2.3. This dependence can be described using the *Curie law*:

$$\chi = \frac{C}{T} \tag{2.42}$$

where C is the Curie constant.

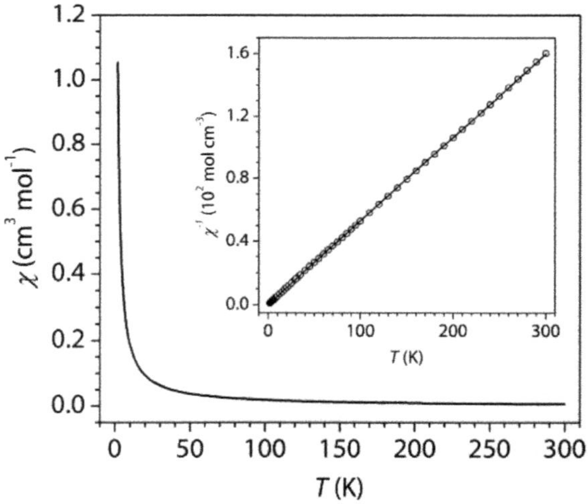

Figure 2.3 Temperature dependence of paramagnetic susceptibility for a Curie paramagnet.

The χ variation proportional to the inverse of T comes from the competitive effects of alignment of magnetic moments, due to the action of the external magnetic field and the randomization of their orientation as a consequence of thermal agitation. Materials that obey this law are those in which the magnetic moments are localized at the atomic or ionic sites. These can be considered as "diluted" magnetic materials, in which the magnetic atoms are surrounded by a number of nonmagnetic atoms, like in mononuclear transition metal complexes where the magnetic ion is usually surrounded by diamagnetic ligands.

On the basis of Langevin theory of paramagnetism, the magnetization of a diluted paramagnet is:

$$\mathbf{M} = \frac{N_A \mu_0 \mathbf{m}^2 \mathbf{H}}{3k_B T} \tag{2.43}$$

where N_A is Avogadro's number, μ_0 is the magnetic permeability of free space, $\mathbf{m}$ is the atomic magnetic moment, and k_B the Boltzmann constant. Replacing $\mathbf{M}$ in the general expression of χ (Eq. 2.13), an expression that defines the Curie constant can be immediately obtained:

$$\chi = \frac{\mathbf{M}}{\mathbf{H}} = \frac{N_A \mu_0 \mathbf{m}^2}{3k_B T} = \frac{C}{T} \tag{2.44}$$

For systems that follow the spin-only approximation, the spin-only magnetic moment expression can be combined with the Curie law to obtain an expression that describes the thermal variation of their paramagnetic susceptibility:

$$\chi = \frac{N_A \mu_B{}^2 \mathbf{g}^2}{3k_B T} \mathbf{S}(\mathbf{S}+1) \tag{2.45}$$

where the expression of the Curie constant for the spin-only approximation is:

$$C = \frac{N_A \mu_B{}^2 \mathbf{g}^2}{3k_B} \mathbf{S}(\mathbf{S}+1) \tag{2.46}$$

Constants and Units

The $N_A\mu_B{}^2/3k_B$ constant assumes the value 0.125048612 $cm^3\ mol^{-1}$ in cgs units. The cgs system is the most employed for discussing magnetic data, rather than the SI, and will be used throughout this book. Since the Boltzmann constant k_B and the Bohr magneton μ_B are two fundamental constants in magnetism, it is useful to state here that they assume the values 0.69503877 $cm^{-1}\ K^{-1}$ and 4.66864374 × $10^{-5}\ cm^{-1}\ emu^{-1}$, respectively, if the energy is expressed in cm^{-1}.

2.1.8 The Curie–Weiss Law

The thermal variation of the susceptibilities of a number of paramagnetic materials obeys a modified or generalized version of the Curie law, the *Curie–Weiss law*. The susceptibility of these materials was found to vary with temperature, as shown in Fig. 2.4, and it can be represented in the paramagnetic region by the equation:

$$\chi = \frac{C}{T - T_C} \tag{2.47}$$

where C is, again, the Curie constant and T_C is another constant with the dimension of the temperature.

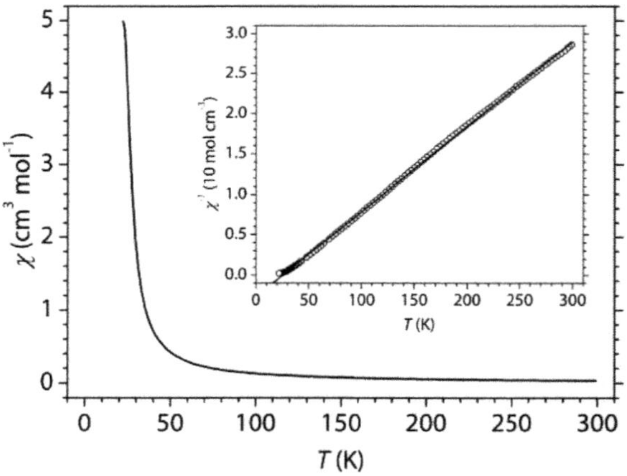

Figure 2.4 Thermal variation of the susceptibility for a paramagnetic material that undergoes a transition to ferromagnetism at the Curie temperature T_C.

T_C can be positive, negative, or zero. $T_C = 0$ corresponds, obviously, to the Curie law. Materials having $T_C > 0$ undergo a paramagnetic to ferromagnetic transition and are called ferromagnets, whereas materials having $T_C < 0$ undergo a paramagnetic to antiferromagnetic transition and are called antiferromagnets. The T_C constant is also indicated with the symbol θ and assumes the name of Weiss constant. The sign and magnitude of θ are more easily retrievable by plotting the inverse susceptibility, χ^{-1}, as a function of T (Fig. 2.5).

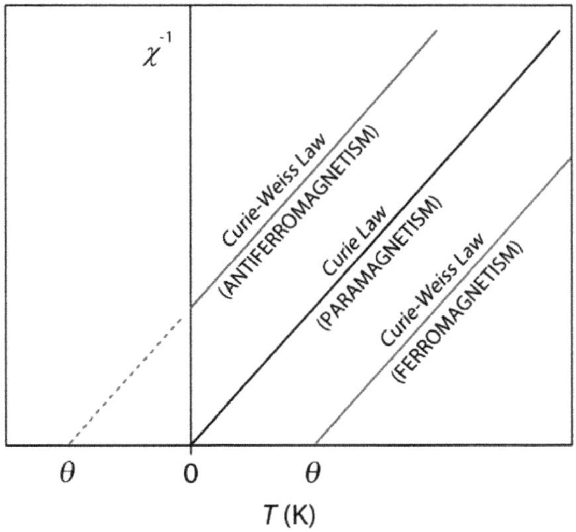

Figure 2.5 Thermal variation of the reciprocal susceptibility for a paramagnet that follows the Curie law ($\theta = 0$) (black line), a paramagnet that undergoes a paramagnetic to ferromagnetic transition and follows the Curie–Weiss law ($\theta > 0$), and a paramagnet that undergoes a paramagnetic to antiferromagnetic transition and follows the Curie–Weiss law ($\theta < 0$); the slope of each line is C^{-1}.

The real transition temperatures from a paramagnetic to a ferromagnetic or an antiferromagnetic state are both positive and are called Curie temperature and Néel temperature, respectively. It should be remembered that the susceptibility of ferromagnets and antiferromagnets only follows the Curie–Weiss law in the paramagnetic region. For lower temperatures, once the material becomes ordered, the susceptibility behaves in a more complicated way and no longer has a unique value for a given field strength.

In conclusion, the Curie–Weiss law emerges as an extension of the Curie law to account for deviations from the paramagnetic behavior that come out when magnetic ordering phenomena (see Chapter 3) take place in a certain material as a result of the magnetic coupling between magnetic centers. The critical temperature T_c defines the boundary between the paramagnetic (disordered) and ferromagnetic or antiferromagnetic (ordered) states of magnetic materials.

The origin and strength of the magnetic coupling will be discussed further in this chapter (cf. Sections 2.1.12 and 2.1.13), but at this point it is important to point out that coupling is not a synonym for ordering. Magnetic ordering can be observed in magnetic materials featuring 2D or 3D structures, as a result of magnetic coupling phenomena extended through microscopic portions of the structure called magnetic domains. Magnetic coupling is, instead, simply the interaction between magnetic moments and then it is observed whenever two or more magnetic moments, separated by a certain distance, interact with each other. This interaction does not necessarily lead to magnetic ordering.

2.1.9 Deviations from the Curie Law

Although most paramagnetic molecular materials are based on discrete molecules well separated from each other in the crystal structure, they are rarely perfectly isolated from the magnetic point of view. When intermolecular magnetic exchange coupling interactions occur between paramagnetic molecules, deviations from the Curie law behavior of an isolated ion can be observed, especially at low temperatures. To account for these interactions, the Curie law is modified in the "molecular" or "main field approximation," where a constant *zj* allows one to quantify the average interaction *j* with the *z* nearest-neighbor magnetic species, similarly to the constant θ in the Curie–Weiss law (Kahn, 1993). In the main field approximation the Curie law assumes the form:

$$\chi = \frac{N_A {\mu_B}^2 \mathbf{g}^2 \mathbf{S}(\mathbf{S}+1)}{3k_B T - zj\mathbf{S}(\mathbf{S}+1)} \tag{2.48}$$

where:

$$\frac{zj\mathbf{S}(\mathbf{S}+1)}{3k_B} = \theta \tag{2.49}$$

The sign of the coupling constant *j* is correlated with the type of coupling interaction (ferromagnetic or antiferromagnetic) as the sign of θ is associated with ferromagnetic or antiferromagnetic ordering.

Deviations from the Curie law can also emerge as a result of an intrinsic property of isolated magnetic ions. Consider again the simplest case of a metal ion or an ensemble of ions that obeys the spin-only approximation. Its spin state **S** is associated with different values of m_S. They assume the values **S**, **S**–1, **S**–2, ... –**S**, and are the (2**S** + 1)-fold projections of the spin quantum number on the axis of quantization. In the presence of a magnetic field, these states are split into energy states of energy $E(m_S) = m_S \mathbf{g} \mu_B \mathbf{H}$ according to what is commonly known as Zeeman splitting. The population of these states will be governed by a Boltzmann distribution.

Actually, for those systems with **S** > ½, the 2**S** + 1 m_S states are most often not degenerate in a zero field as a result of magnetic dipolar (MD) interactions between unpaired electrons and by interaction with higher-energy states through spin–orbit coupling. This phenomenon is called zero-field splitting (ZFS) and can be responsible for deviation from the Curie law at low temperatures, where perturbation of thermal population of the different energy states can result in deviation of the expected susceptibility value (Robertson and Yee, 2010). The form of the ZFS depends on the symmetry of the coordination sphere around the magnetic center, and in a first approximation, its magnitude is defined by the axial ZFS parameter indicated by the symbol D (Fig. 2.6).

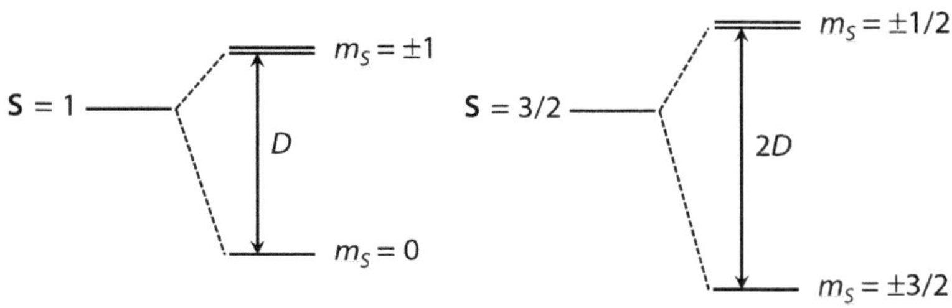

Figure 2.6 Zero-field splitting of the m_S states for two generic **S** = 1 and **S** = 3/2 ions. Note that the sign of D is positive and negative for **S** = 1 and **S** = 3/2, respectively.

D can be positive or negative. If the components with the highest $|m_S|$ lie lowest in energy, D is negative, whereas, on the contrary, D is positive whenever the components with the lowest $|m_S|$ are the most stabilized. Systems characterized by large negative D parameters show easy-axis anisotropy, whereas systems characterized by large positive D parameters show easy-plane anisotropy. This has important implications in the description of the magnetic behavior of peculiar magnetic systems as single-molecule magnets and single-ion magnets (see Chapter 3).

2.1.10 Effective Magnetic Moment and χT Value

The expression that describes the thermal variation of the paramagnetic susceptibility on the basis of the Curie law and the spin-only approximation can be rewritten as the product of susceptibility and temperature, χT, as follows:

$$\chi T = \frac{N_A \mu_B{}^2 \mathbf{g}^2}{3k_B} \mathbf{S}(\mathbf{S}+1) \tag{2.50}$$

The χT product has great importance in experimental susceptibility measurements because its value at high temperature is associated with the magnetic moment per mole of the system and then furnishes quantitative information about the chemical structure of the investigated compound. In particular, a plot of χT as a function of T (Fig. 2.7) is a key for the interpretation of the structure/property characteristics of the studied material in view of the following points:

- For a Curie paramagnet the variation of χT as a function of temperature is constant and appears in the whole temperature range as a straight line, except at very low temperature where ZFS effects (vide supra) or coupling phenomena (vide infra) can be observed.
- For those systems that undergo a magnetic phase transition or that show coupling phenomena, the variation of χT as a function of temperature is not constant and deviations from the linearity can be easily observed. An increase in χT as the temperature decreases is associated with ferromagnetic ordering and/or coupling, whereas a decrease of χT on cooling is related to antiferromagnetic interactions.

The χT product is associated with another quantity used in experimental practice, the effective magnetic moment μ_{eff}, defined as:

$$\mu_{\text{eff}} = \mathbf{g}\sqrt{\mathbf{S}(\mathbf{S}+1)} = \sqrt{8\chi T} \tag{2.51}$$

in the case of the spin-only approximation and expressed in Bohr magneton units. The factor 8 arises from the combination of the various constants mentioned earlier (Robertson and Yee, 2010).

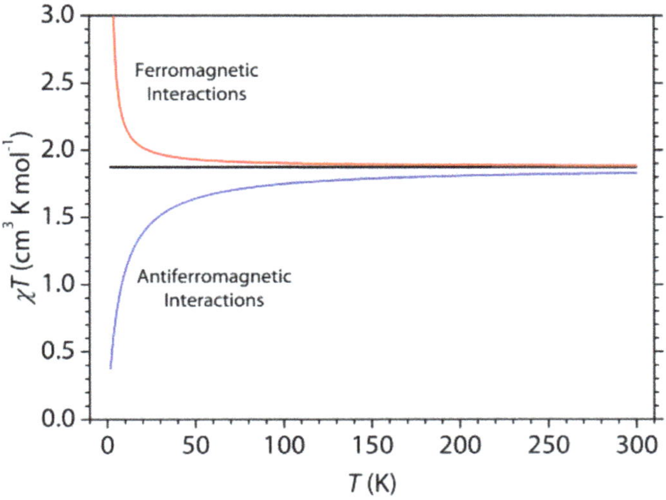

Figure 2.7 Thermal variation of χT for a generic Cr^{III} complex. The room temperature value of χT (1.88 cm^3 K mol^{-1}) is in agreement with the calculated value for a d^3 metal ion in the spin-only approximation (**S** = 3/2) 1.875 cm^3 K mol^{-1}. Red and blue curves are simulations in the case of ferromagnetic (red curve) or antiferromagnetic (blue curve) interactions occurring between magnetic centers at a low temperature.

Table 2.3 shows the calculated μ_{eff} and χT values for different numbers of unpaired electrons in the case of spin-only approximation.

Table 2.3 Calculated values of χT and μ_{eff} for different numbers of unpaired electrons in the case of spin-only approximation

e	μ_{eff} (μ_B)	χT (cm^3 K mol^{-1})
1	1.73	0.375
2	2.83	1.000
3	3.87	1.875
4	4.90	3.000
5	5.92	4.375

It should be highlighted that these values are associated with high-temperatures states, that is, around room temperature, in which thermal energy is higher than the coupling interactions, and the system behaves as a pure paramagnet.

The χT product has the additional advantage that it can be summed up, whereas μ_{eff}, being proportional to $\sqrt{\chi}$, is not additive. Thus, if a molecule has two spin centers, A and B, the calculated value of χT is simply the sum of the contributions given by the two metal ions:

$$\chi T = \frac{N_A \mu_B^{\ 2}}{3k_B}\left[\left(\mathbf{g}_A^{\ 2}\mathbf{S}_A\left(\mathbf{S}_A+1\right)\right)+\left(\mathbf{g}_B^{\ 2}\mathbf{S}_B\left(\mathbf{S}_B+1\right)\right)\right] \tag{2.52}$$

where $\mathbf{g}_A$ and $\mathbf{g}_B$ are the Landé splitting factors for the two different metal ions.

Whereas the expression of μ_{eff} is common in the older literature, the χT product is currently more often employed to describe the magnetic behavior of molecular materials and will be used for the rest of the book.

2.1.11 Field Dependence of the Magnetization

To fully characterize the static magnetic properties of a magnetic material, useful information about the ground spin state, that is, the thermally populated state at low temperatures, can be obtained by studying the field dependence of the magnetization **M** at a given temperature.

For a Curie paramagnet, the expression of the magnetization is given by the equation:

$$\mathbf{M} = N_A \mathbf{g}\mathbf{S}\mu_B B_f(\mathbf{H}) \tag{2.53}$$

where N_A is Avogadro's number, **S** is the ground spin state, **g** is the Landé factor, μ_B is the Bohr magneton, and $B_f(\mathbf{H})$ is the Brillouin function (Kahn, 1993). The Brillouin function, which defines the **H** dependence of the magnetization, is written as:

$$B_f(\mathbf{H}) = \frac{2\mathbf{S}+1}{2\mathbf{S}}\coth\left(\frac{2\mathbf{S}+1}{2\mathbf{S}}\mathbf{H}\right) - \frac{1}{2\mathbf{S}}\coth\left(\frac{1}{2\mathbf{S}}\mathbf{H}\right) \tag{2.54}$$

This measurement allows the determination of the ground spin state of a system when subjected to the action of a magnetic field strong enough to orientate all magnetic moments along the same direction (Fig. 2.8). In fact, if the magnetization is expressed as μ_B mol^{-1} units, its value at high fields, the saturation magnetization

$\mathbf{M}_S$, is simply given by the product **gS**, where **g** can be considered equal to 2.0 in a first approximation. This is due to the fact that the Brillouin function for high values of **H** is equal to 1 for each value of **S**, so only the values of **g** and **S** contribute (cf. Eq. 2.53).

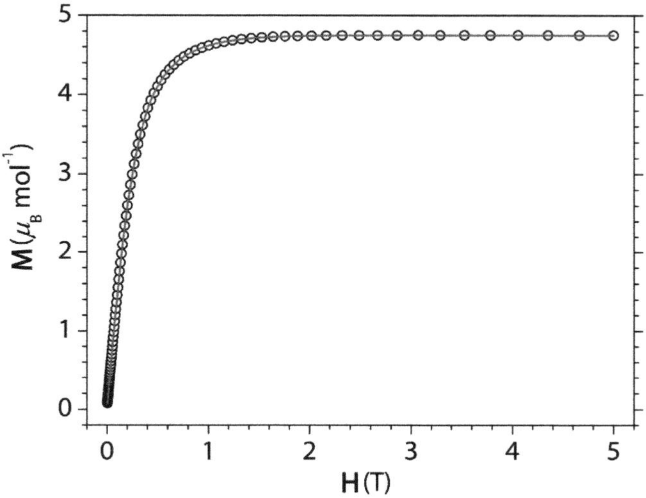

Figure 2.8 Variation of **M** as a function of the field **H** for an Fe^{III} (**S** = 5/2) mononuclear complex (black circles). The magnetization values at high fields tend toward the calculated value for $\mathbf{M}_S$ of 5 μ_B mol^{-1}. The red curve is the best fit using the **S** = 5/2 Brillouin function.

Figure 2.9 shows the calculated field dependence of **M** (μ_B mol^{-1} units) for different values of **S**, assuming **g** = 2.0.

Deviations from this trend indicate that the system does not behave as a pure Curie paramagnet.

To compare the magnetization curves obtained at different temperatures for the same system, it is useful to plot **M** as a function of **H**/T instead of **H** (Fig. 2.10).

Units

As magnetic measurements usually provide magnetization values in emu (cm^3) units, a useful way to obtain the value of **M** in μ_B mol^{-1} units from cm^3 mol^{-1} data is to multiply by the conversion factor 1.791 × 10^{-4}.

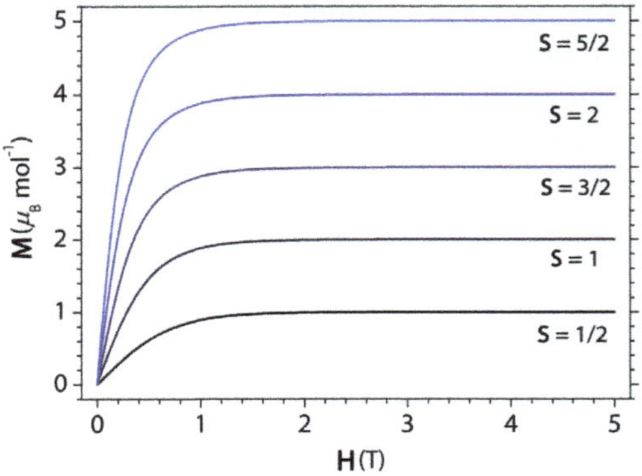

Figure 2.9 Calculated field dependence of **M** for **S** = 1/2, . . . , 5/2 ground spin states assuming **g** = 2.0.

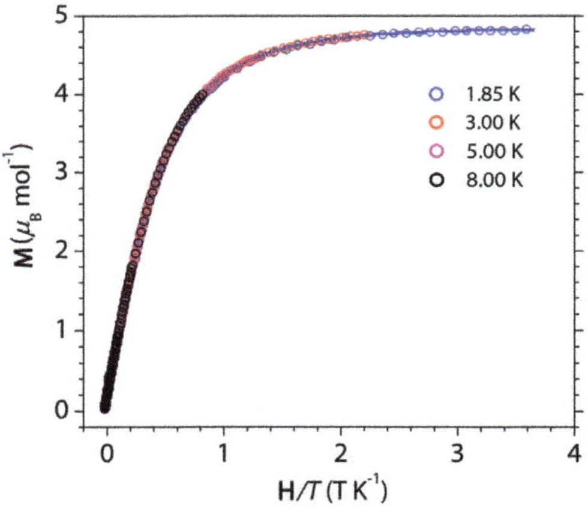

Figure 2.10 Variation of **M** as a function of **H**/*T* for an Fe^{III} (**S** = 5/2) mononuclear complex (circles) at different temperatures (see legend). The magnetization values at high fields tend toward the calculated value for $\mathbf{M}_S$ of $5\mu_B$ mol^{-1} for all temperatures. The blue curve is the best fit using the **S** = 5/2 Brillouin function.

2.1.12 Magnetic Exchange Coupling

The cooperative magnetic phenomena, which lead to a magnetically ordered state, result from the presence of exchange interactions between atomic magnetic moments. The mechanisms responsible for these interactions will be now described through a phenomenological approach. The knowledge of these mechanisms allows one to understand and predict the magnetic properties of new magnetic compounds on the basis of their electronic configuration and their molecular structure.

Magnetic moments can interact through space via dipolar interactions, but these interactions are too weak to justify the relatively strong interactions observed in magnetic materials where the magnetic centers are quite afar from each other. On the other hand, magnetic moments can interact through exchange interactions, based on orbital overlapping. The exchange constant J_{ij} quantifies the strength of the exchange interaction between two atomic centers i and j (vide infra). These interactions can occur through a direct overlap between the magnetic orbitals (AOs or molecular orbitals [MOs] containing an unpaired electron) of two close-lying magnetic ions in the so-called direct exchange coupling. However, most magnetic materials have magnetic centers separated by nonmagnetic atoms, as, for example, transition metal oxides or complexes. In these cases a mechanism based on indirect exchange coupling is required to describe magnetic interactions. This mechanism implies the presence of a mono- or polyatomic species bound to the metal centers, that is, a bridging ligand that is responsible for the communication between the magnetic centers in a so-called superexchange interaction.

Let's now consider two metal ions M^{n+} connected in a linear way through a covalent bond with an O^{2-} oxide ion acting as a bridging ligand. The metal d orbital containing the unpaired electrons responsible for the magnetic properties overlaps with the ligand p orbitals containing paired electrons. This leads to a partial donation of the electron density from the ligand to the metal sites. Suppose that the unpaired electron of the M_A metal ion has a spin +½ (↑). Then, on the basis of the Pauli exclusion principle, the electron density transferred from the ligand p orbital must have the opposite spin, −½ (↓). To preserve the spin neutrality of the ligand p orbital, 0 (↑↓), the electron density donated simultaneously to the other metal center M_B must have the opposite spin sign, +½ (↑). This induces the unpaired electron on the M_B metal site to have a spin −½ (↓), which is opposite to the one associated with the M_A metal center, as depicted in Fig. 2.11 (Orchard, 2003).

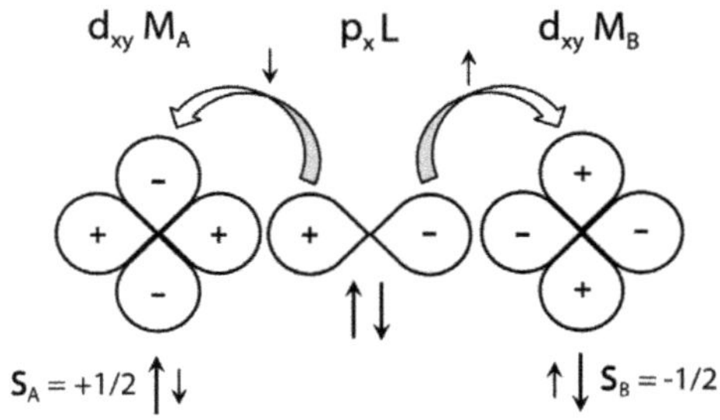

Figure 2.11 Superexchange interaction between two metal ions connected in a linear way through a ligand.

The final result of such a superexchange interaction is an antiferromagnetic coupling between the metal centers. It should be noted that if metal ions have different magnetic moments the interaction will be ferrimagnetic (see Chapter 3).

If the magnetic ions are connected through the same type of ligand but in a 90°-angled spatial configuration, the magnetic interaction will be different (Fig. 2.12).

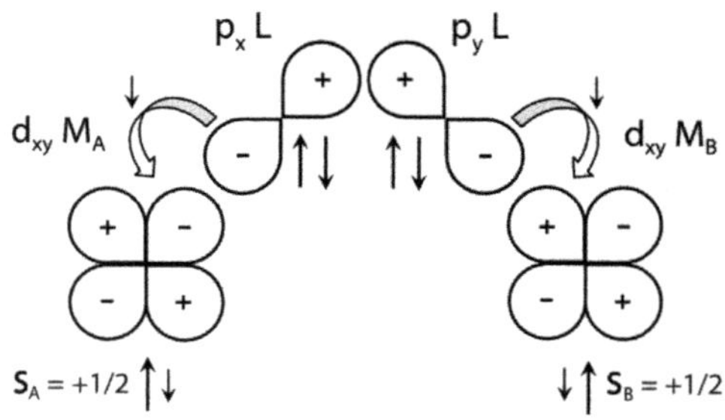

Figure 2.12 Superexchange interaction between two metal ions connected through a bridging ligand in an angular way.

In this case the d orbitals overlap with different p orbitals of the same ligand. The spin correlation within the ligand, which always favors the parallel alignment of spins in different orbitals, allows the ligand to donate electron density of the same sign to the metal ions. The Pauli exclusion principle, as in the previous case, forces the spin orientation of the unpaired electrons on the metal centers to be antiparallel with respect to the electron density given by the ligand. This gives rise to a parallel alignment of the spins on the metal centers, causing ferromagnetic coupling (Orchard, 2003).

It should be noted that the superexchange interactions can lead to different types of magnetic couplings as a function of the M-L-M geometry and, especially, the symmetry of the interacting ligand and metal orbitals. The type of superexchange interaction, ferromagnetic or antiferromagnetic, can be predicted on the basis of the symmetry of the interacting magnetic orbitals and their occupancy following the *Goodenough–Kanamori–Anderson rules* (Anderson, 1963; Goodenough, 1966). Briefly, antiferromagnetic coupling is expected if the interacting metal ions have partially occupied orbitals having the same symmetry (e.g., Cr^{III} t_{2g}^3, V^{II} t_{2g}^3), whereas ferromagnetic coupling is expected for interaction between metal ions having partially occupied orbitals with different symmetries (e.g., Cr^{III} t_{2g}^3, Ni^{II} $t_{2g}^6e_g^2$). If a certain metal ion has more than one type of partially occupied orbitals having different symmetries (e.g., Mn^{II} $t_{2g}^3e_g^2$), the interaction with another metal ion whose orbitals are partially occupied for a certain symmetry and empty for the other type of symmetry (e.g., Cr^{III} $t_{2g}^3e_g^0$) can be ferromagnetic or antiferromagnetic. This is a consequence of the competition between antiferromagnetic interactions between partially occupied orbitals having the same symmetry (t_{2g}^3) and ferromagnetic interactions between partially occupied orbitals having different symmetries (e.g., Mn^{II} e_g^2, Cr^{III} t_{2g}^3).

It must be highlighted that these are semiempirical rules that are not always evidenced since other factors, such as the bond geometry and the ligand orbital symmetry, contribute to the overall sign of the interaction.

2.1.13 Magnetic Exchange Coupling in a Dinuclear Compound

Magnetic exchange interactions between several metal centers are at the origin of some of the most interesting behaviors exhibited by magnetic materials. Now we will treat the simplest case of isotropic magnetic interactions in a generic dinuclear compound, with the aim to qualitatively and quantitatively evaluate their effect on the resulting magnetic properties.

If we consider two isolated metal ions indicated as M_A and M_B having a single unpaired electron in their d orbitals, their ground spin states will be $\mathbf{S}_A = \mathbf{S}_B = ½$. If these metal centers are part of the same molecular unit and separated by a ligand that provides a way for a superexchange interaction (as described in the previous section) the individual spin quantum numbers $\mathbf{S}_A = \mathbf{S}_B = ½$ do not describe well the magnetic properties of the system as they do not take into account their interaction. The spin quantum numbers that best describe this system are $\mathbf{S} = \mathbf{S}_A - \mathbf{S}_B = 0$ and $\mathbf{S} = \mathbf{S}_A + \mathbf{S}_B = 1$, a singlet state and a triplet state, respectively. In the singlet state the magnetic moments are antiferromagnetically coupled (antiparallel aligned) and the spin state will be diamagnetic. Instead, in the triplet state, the magnetic moments are ferromagnetically coupled (parallel aligned) and the spin state will be magnetic (Kahn, 1993).

Each spin state is associated with an energy, $E(\mathbf{S} = 0)$ and $E(\mathbf{S} = 1)$, that is expected to differ for the two states as a consequence of the different electrostatic forces that are involved when two electrons are parallel or antiparallel aligned. The energy gap between these two states is represented by the J_{AB} coupling constant:

$$J_{AB} = E(\mathbf{S}_{AB} = 0) - E(\mathbf{S}_{AB} = 1) \tag{2.55}$$

which is usually expressed in cm^{-1} units, a practical energy unit to evaluate the difference in energy between the magnetic states.

If $\mathbf{S}_{AB} = 0$ is the ground state, that is, $E(\mathbf{S}_{AB} = 0) < E(\mathbf{S}_{AB} = 1)$, the interaction is antiferromagnetic and J will be characterized by a negative value. If $\mathbf{S} = 1$ is the ground state, that is, $E(\mathbf{S}_{AB} = 0) > E(\mathbf{S}_{AB} = 1)$, the interaction is ferromagnetic and J will be characterized by a positive value.

In this simple case the thermal variation of the susceptibility is represented by the analytical expression:

$$\chi = \frac{2N_A \mathbf{g}^2 \mu_B^2}{k_B T\left[3 + e^{(-J/k_B T)}\right]} \tag{2.56}$$

which was derived for the first time by Bleaney and Bowers (1952) and can be used to obtain the value of the coupling constant J from experimental susceptibility data.

Figure 2.13 shows the thermal variation of χ for the $J = 0$, $J = \pm100$, and $J = \pm50$ cm^{-1} cases.

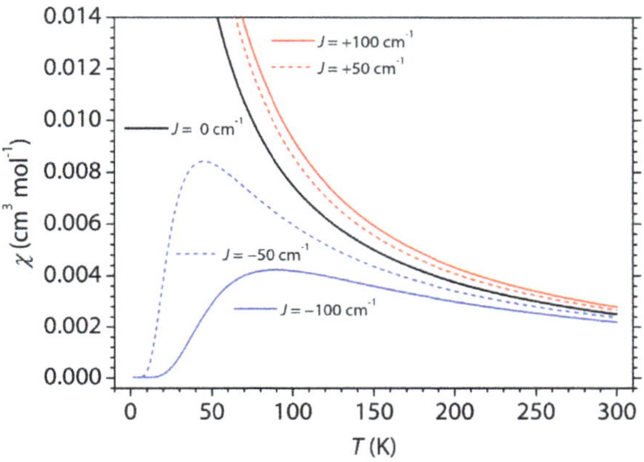

Figure 2.13 Calculated thermal variation of χ for a generic dinuclear compound ($\mathbf{S}_A = \mathbf{S}_B = ½$, $\mathbf{g}_A = g_B = 2.0$) for positive, negative, or zero values of the coupling constant J.

The room temperature values of χ are similar for all cases as the thermal energy favors the paramagnetic (noninteracting) state. For $J = 0$, the magnetic centers do not interact at all and the χ versus T plot is the one expected for a Curie paramagnet. For $J < 0$, χ presents a maximum and then tends to zero as the temperature approaches zero, indicating that at low temperatures only the diamagnetic $\mathbf{S} = 0$ state is thermally populated. When comparing the curves associated with different J constants it becomes clear that this maximum is associated with the strength of the antiferromagnetic coupling. The higher is the temperature of the maximum, the higher is the antiferromagnetic interaction. In this case T_{max} and J are correlated through the equation $|J|/k_B T_{max} = 1.599$, with k_B expressed in cm^{-1} K^{-1} units (Kahn, 1993). For $J > 0$, χ increases faster than for $J = 0$ as a result of the augmented response of a ferromagnetically coupled system to the action of the static magnetic field. Since the $J = 0$ and $J > 0$ cases are not clearly distinguishable, the best way to reveal a ferromagnetic interaction is to plot, as described before, the thermal variation of the χT product (Fig. 2.14).

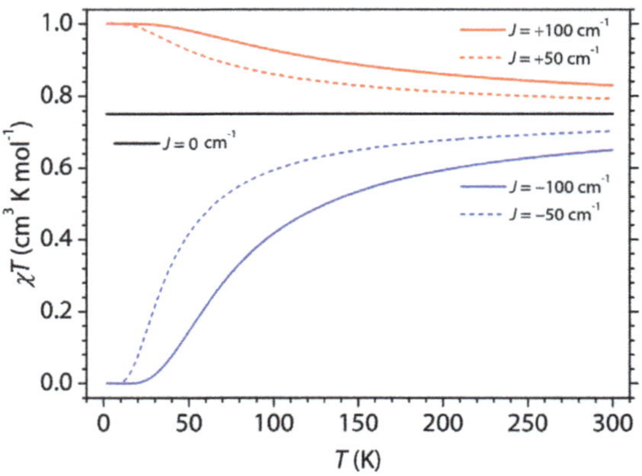

Figure 2.14 Calculated χT versus T curves for a generic dinuclear compound ($\mathbf{S}_A = \mathbf{S}_B = ½$, $\mathbf{g}_A = \mathbf{g}_B = 2.0$) for positive, negative, or zero values of the coupling constant J.

Again, for $J = 0$ the thermal variation of χT is a straight line in the whole temperature range with a χT value equal to the sum of two uncorrelated paramagnetic contributions of $\mathbf{S} = ½$ spin states. For $J > 0$, χT assumes higher values than those obtained in the noninteracting case as a result of the depopulation of the diamagnetic $\mathbf{S} = 0$ highest state in favor of the $\mathbf{S} = 1$ ground state, tending, at low temperature, to the value of χT that results from a fully populated triplet state and a fully depopulated singlet state. For $J < 0$ χT shows a continuous decrease with temperature, reaching the zero value at the lowest temperature.

2.1.14 Multicenter Magnetic Systems and Spin Hamiltonian Approach

The case of two interacting metal centers in a dinuclear compound described in Section 2.1.12 is the simplest case of isotropic interaction between two paramagnetic metal centers. Although in this case the derived analytical expression for the variation of χ as a function of T well describes the experimental magnetic properties and allows one to obtain a reliable value of the coupling constant J, the coupling between the two magnetic ions should be more rigorously described through the spin Hamiltonian formalism:

$$\mathcal{H} = -J\hat{\mathbf{S}}_A \cdot \hat{\mathbf{S}}_B \tag{2.57}$$

where $\mathcal{H}$ is a phenomenological Hamiltonian generally called the Heisenberg–Dirac–Van Vleck Hamiltonian, J is the coupling constant that quantifies the energy gap between the singlet and triplet energy states, and $\hat{\mathbf{S}}_A$ and $\hat{\mathbf{S}}_B$ are the local spin operators (Kahn, 1993). Defining the spin Hamiltonian is important in order to describe the number and type of coupling interactions shown by a system. In addition, it is of fundamental importance to enable the comparison of the different results reported in the scientific literature in view of the fact that there is no universal agreement on the definition of J, which can be defined either as $\pm J$ or as $\pm 2J$. Here we used the convention that the energy gap is J and its negative value implies an antiferromagnetic interaction (vide supra).

Consider now a slightly more complex system than that reported previously, composed of three generic magnetic centers M_A, M_B, and M_c interacting through two coupling constants J_{AB} and J_{BC}, with the third coupling constant J_{AC} assumed to be negligible for structural reasons (Fig. 2.15).

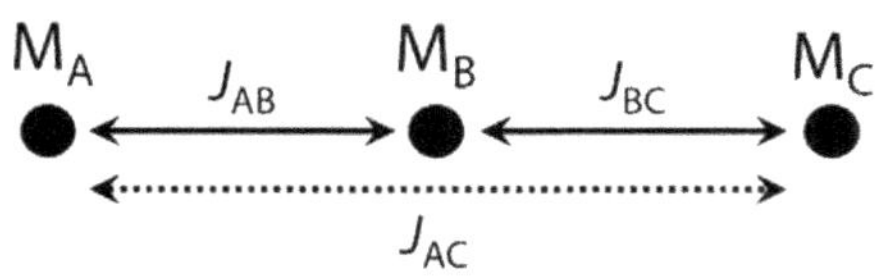

Figure 2.15 Schematic depiction of three generic magnetic centers M_A, M_B, and M_C and related coupling constants describing the magnetic superexchange interactions. While J_{AB} and J_{BC} between adjacent atoms are significant, the coupling constant J_{AC} can be considered negligible due to the spatial separation between M_A and M_C. Note that the same consideration holds whether the structural arrangement is linear or bent and bond lengths are alike or not and whatever the nature of the coordinated metal centers and bridging ligands.

The spin Hamiltonian that describes these isotropic interactions in the absence of an external magnetic field is defined as follows:

$$\mathcal{H} = -J_{AB}\hat{\mathbf{S}}_A \cdot \hat{\mathbf{S}}_B - J_{BC}\hat{\mathbf{S}}_B \cdot \hat{\mathbf{S}}_C \tag{2.58}$$

These considerations bring to attention the fact that for each system under investigation the appropriate spin Hamiltonian must be defined and the energy of each spin state must be calculated accordingly by means of a proper matrix calculation. The matrix expression can be written and the values of the energy states calculated quite easily if both the number of magnetic centers and the number of different coupling constants are restricted to a small value. However, as the complexity of the system increases it is common practice to employ software that, on the basis of the experimental χ versus T or χT versus T values, calculates the energy values of each coupled spin state and the involved coupling constants.[2]

2.2 Transport Properties of Molecular Materials

2.2.1 Electrical Conductivity

Electrical conductivity is the physical property that quantifies the ability of a given material to transport charge carriers when subjected to the action of an electric potential.

When a conducting material is subjected to an electric potential, or bias, denoted as ΔV, an electrical current of intensity i flows through it. The current intensity value depends on the applied electric potential and the material resistance R, as stated by the first Ohm's law:

$$i = \frac{\Delta V}{R} \tag{2.59}$$

The resistance value is correlated to the intrinsic nature of the material, that is, its atomic and electronic structure, and to geometrical factors not correlated to its composition, such as its length l and its section s, as stated by the second Ohm's law:

$$R = \rho \frac{l}{s} \tag{2.60}$$

where ρ is the resistivity or specific resistance and quantifies the resistance of a material independently from its geometrical features. It can be expressed in ohm cm, that is, Ω cm (1 ohm = 1/siemens).

[2]A further discussion of this topic is beyond the scope of this book, and we refer to the book *Molecular Magnetism* (Kahn, 1993), in which more examples and an exhaustive discussion of this can be found.

2.2.2 Classes of Conducting Materials

Conductivity or specific conductance, σ, expressed in S cm^{-1}, is the reciprocal of resistivity, as defined by the following expression:

$$\sigma = \frac{1}{\rho} = \frac{l}{R\,s} = \frac{i\,l}{\Delta V\,s} \tag{2.61}$$

The value of conductivity allows the classification of materials as insulators ($\sigma < 10^{-6}$ S cm^{-1}), semiconductors ($10^{-6} < \sigma < 10$ S cm^{-1}), metals ($10 < \sigma < 10^{7}$ S cm^{-1}), and superconductors ($\sigma > 10^{7}$ S cm^{-1}) (Fig. 2.16).

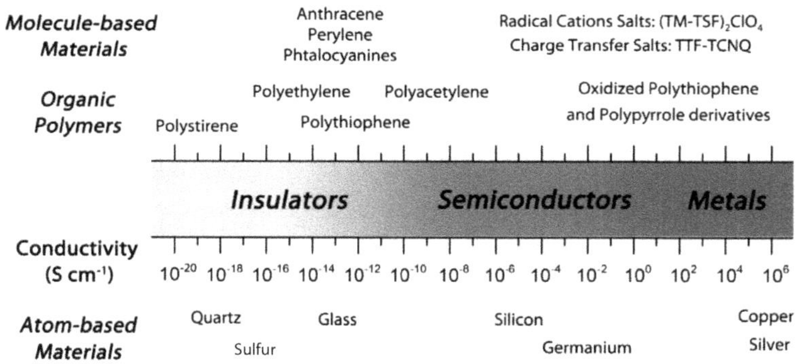

Figure 2.16 Conductivity values (in S cm^{-1}) for some atom-based and molecule-based materials and corresponding classification as insulators, semiconductors, and metals.

2.2.3 Energy Bands and Origin of Transport Properties

The origin of electrical conductivity in molecular conductors can be explained by means of the band theory of solids, as for the case of traditional conductors, considering the interactions between MOs instead of AOs. From a general point of view, the combination of two different orbitals leads to the formation of a bonding orbital and an antibonding orbital having lower and higher energy values, respectively, with respect to the energies of the noninteracting orbitals. This means that the combination of N orbitals leads to the formation of $N/2$ bonding orbitals and $N/2$ antibonding orbitals. The combination of a large number of orbitals does not allow for well-separated energy levels but gives rise to continuous energy levels, described as an energy band (Fig. 2.17).

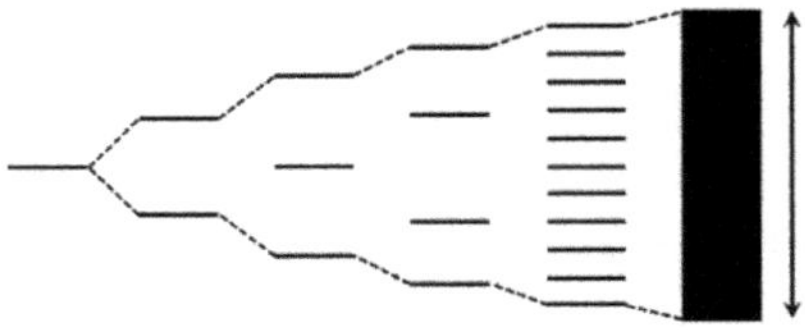

Figure 2.17 Origin of energy bands within a solid represented by the combination of an increasing number of orbitals.

The bandwidth that is, the energy difference between the highest- and lowest-energy levels, depends on the interactions between the involved orbitals.

An energy gap E_g can be found between the band formed by the bonding orbitals and the band formed by the antibonding ones and represents forbidden energy values for the electrons in the electronic structure of the material. The conducting properties of a material are strongly dependent on the level of the filled energy bands and the energy gap between two bands. The Fermi energy represents the highest occupied level at 0 K.

When the material has a high E_g value (>5 eV), no electrons have enough energy to be promoted to the empty higher-energy band, and the material behaves as an insulator (Fig. 2.18a). If the E_g value has an energy value comparable with the thermal energy (k_BT), a certain number of electrons have enough energy to be promoted from the lower-energy band, the valence band, to the higher-energy band, the conduction band, which becomes partially occupied. In this case, the electrical conductivity is thermally activated and the material behaves as a semiconductor (Fig. 2.18b). If no E_g separates the valence and conduction bands, it is said that they are bordering, and a small amount of thermal energy will be sufficient to promote the electrons from the valence band to the conducting one. In the typical case of semioccupied bordering bands, a metal-like conducting behavior is expected (Fig. 2.18c), whereas for occupancies around 90% of the valence band and 10% of the conduction band, the material behaves as a semimetal (Fig. 2.18d) (West, 1999).

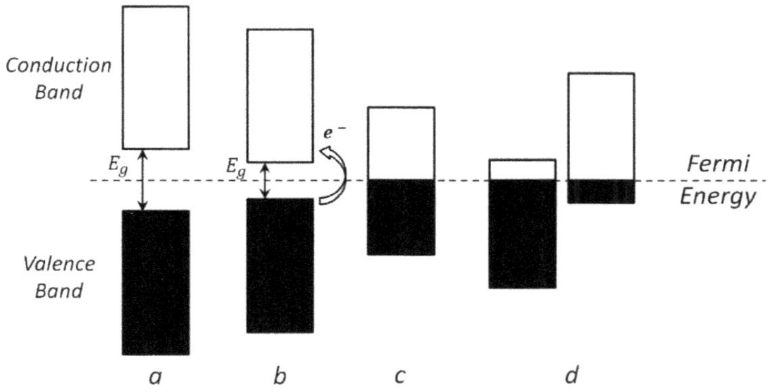

Figure 2.18 Energy bands for (a) an insulator, (b) a semiconductor, (c) a metal, and (d) a semimetal.

2.2.4 Thermal Dependence of Conductivity

The electrical conductivity of a material also depends on the number of charge carriers n (electron or holes) and their mobility μ. The relation between these quantities is:

$$\sigma = n\mu e \quad (2.62)$$

where e is the charge of the carrier.

For metal-like conductors n does not change as a function of the temperature, whereas μ decreases as the temperature increases because of diffusing phenomena acting on the carriers resulting from reticular vibrations. For this reason the conductivity of a metal increases as the temperature decreases. For semiconductors μ varies as for a metal but the number of carriers n increases as the temperature increases (vide supra); therefore, the conductivity is thermally activated. Semimetals show intermediate features between metals and semiconductors. The conducting behavior is similar to that shown by semiconductors, with the difference that n remains almost constant as the temperature changes, as for the metals. Superconductors represent a special class of metal-like conductors showing infinite conductivity below a certain critical temperature, T_c, at which their resistivity drops to zero. This behavior is temperature independent. Figure 2.19 shows the thermal variation of the resistivity for the most typical conducting behaviors shown by molecular materials:

- **Semiconductors**: The resistivity increases as the temperature decreases, until a maximum value at low temperatures corresponding to the resistivity of an insulator.
- **Metal-like conductors**: The resistivity decreases on cooling in a monotonous way, reaching its minimum value (not zero) at the lowest temperature.
- **Superconductors**: The resistivity decreases as the temperature decreases, until it reaches an almost zero value at the critical temperature T_c. Below that temperature the resistivity does not change anymore and the behavior is that of a superconductor, with almost null resistivity and infinite conductivity.

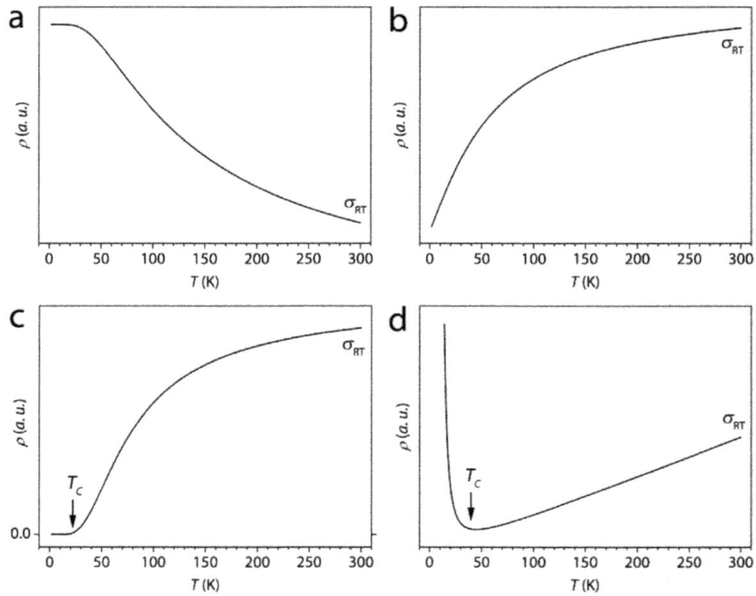

Figure 2.19 Temperature dependence of the resistivity ρ for (a) a semiconductor, (b) a metal, (c) a metal that behaves as a superconductor below a critical temperature T_c, and (d) a metal that undergoes a metal–insulator transition below a critical temperature T_c.

- **Metal-like conductors subjected to metal–insulator phase transition**: The resistivity decreases on lowering the temperature until a critical temperature T_c. Below that temperature, the resistivity suddenly increases, assuming higher values than observed at room temperature, and the system behaves as an insulator.

It has to be remembered that in semiconductors the conductivity is thermally activated as the thermal energy counterbalances the energy gap between the valence and conduction bands. In this case the thermal dependence of ρ can be described through an Arrhenius-type law:

$$\rho = \rho_0 \, e^{\left(\frac{E_A}{k_B T}\right)} \tag{2.63}$$

This means that a linear fit of a plot of lnρ as a function of $1/T$ provides the value of the activation energy E_A expressed in kelvin. This quantity can be easily converted into other energy units (through the Boltzmann constant), as, for example, electron volts, which are more frequently used for reporting energy gap values.[3]

2.2.5 Types of Semiconducting Materials: Chemical and Electrochemical Doping

The class of semiconducting materials deserves some more discussion on the origin of their conducting properties. The semiconductors described so far are a subgroup of all the semiconducting materials. More precisely, they are classified as intrinsic semiconductors. In intrinsic semiconductors the electrons are promoted to the conduction band and corresponding holes are created in the valence band, simply as a result of the thermal energy. Pure silicon is the most representative inorganic material that shows the conducting properties of an intrinsic semiconductor.

Electron promotion to the conduction band and hole creation in the valence band can also be achieved by doping a certain material with a small amount of atoms having one or two more or less valence electrons than those of the undoped material. The extra valence electrons, added with the inclusion of an electron-rich doping element into the semiconductor material, will inevitably occupy

[3] 1 K = 0.0862 meV.

a level close in energy to the conduction band. This will boost the conducting properties of the material since these electrons are free to move in the almost empty conduction band. Because negative charges are added, these materials are called n-type extrinsic semiconductors (Fig. 2.20a). Conversely, doping a pure system with electron-poor elements means that some vacancies or holes are created in the valence band, so the resulting positive charges can move into the valence band if an electric potential is applied. These materials are called p-type extrinsic semiconductors (Fig. 2.20b).

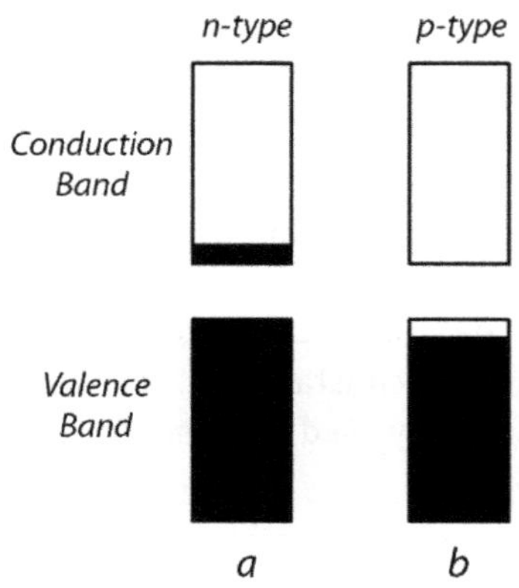

Figure 2.20 Schematic representation of the energy band filling for (a) n-type and (b) p-type extrinsic semiconductors.

These types of chemical doping processes are particularly suited for atom-based extended systems where defects can be easily created on the crystal structure. As far as molecule-based materials are concerned, it is evident that the constituting molecules, due to their complex chemical nature, cannot be easily replaced with others able to satisfy the conditions mentioned above. However, an alternative doping process can be realized through electrochemical methods.

The most effective way to create holes in a fully populated valence band of a molecular system is its partial oxidation. As for the chemical p-type doping, some electrons are removed from the valence band and the conductivity is activated by the ability of holes to move under the action of an electric potential. On the contrary, chemical reduction has the same effect of adding electrons to the conduction band of an n-type doped material. These effects are at the origin of the conducting properties of some classes of molecular materials and will be described in more detail in the next chapter.

The electrons of an intrinsic semiconductor can be promoted from the valence band to the conduction band also as a result of light irradiation (the energy of the beam must correspond to the band gap). This phenomenon is called photoconductivity.

2.2.6 Band Theory of Solids: A Closer Look at the Band Structure

The representation of the energy bands reported in the previous paragraphs is an oversimplification of the shape and number of bands shown by real materials, either atom based or molecule based. A further explanation of the energy distribution of the energy bands within a solid as a function of the particular directions of the crystal structure is then needed. This will be helpful to familiarize with the real band structures of molecular materials and to get information from them.

Different approaches have been developed to describe the band structure of solids. Two of them are the nearly free-electron model and the tight-binding approximation model (Hoffmann, 1989; Canadell et al., 2012). The first approach is based on considering a lattice of atoms that interacts with valence electrons almost completely free to move in a solid. It is based, in the first instance, on the assumption that the electrons behave as free electrons in a box (free-electron model) and then a periodic potential originated from the atoms lattice is introduced to account for the interactions between valence electrons and atoms (nearly free-electron model). On the contrary, the second approach considers discrete atoms or molecules as building blocks of a solid having isolated and noninteracting electrons and then takes into account orbital overlap interactions between discrete building blocks to explain the origin of the bands, as anticipated in Section 2.2.3. While the first approach satisfactorily describes the properties of atom-based solids, especially metals of groups I and II, the second approach works better for describing the properties of less electropositive elements and for molecular solids. Since we are focused on this last class of materials, we will consider the tight-binding approximation model with some mathematical considerations derived from the nearly free-electron model.

When considering the simplest case of a 1D system (Fig. 2.21), the energy of electrons subjected to a potential energy $V = 0$ as a function of the potential well of length L is given by:

$$E = \frac{n^2\hbar^2\pi^2}{2m_e L^2} \quad (2.64)$$

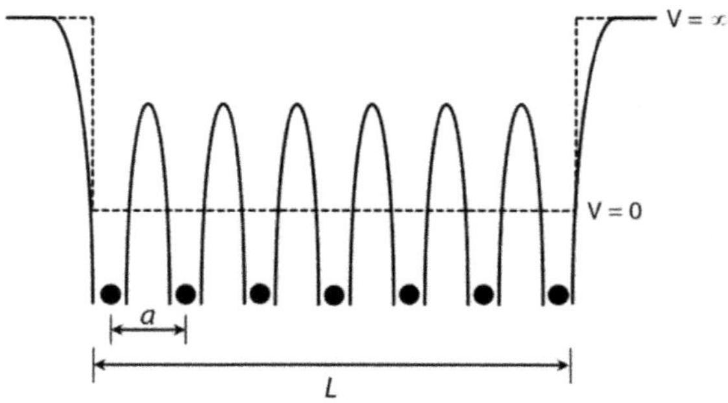

Figure 2.21 Schematic depiction of a 1D potential well of length *L*. Dashed lines indicate the potential profile in the case of the free-electron model (electron in a box). When a periodic lattice of atoms (black spheres) separated by a distance *a* is introduced (nearly free-electron model), the atomic nucleus contributes to the overall potential experimented by the electrons. The overall system can be then described as a series of potential wells (continuous lines).

Since the energy expression is inversely proportional to the distance, it is convenient to introduce the reciprocal distance k by replacing $n^2\pi^2/L^2$ with k^2. The reciprocal distance has the dimension of 1/distance and is proportional to the frequency of the wavefunction. This allows one to describe the energy dependence with a quantity that is proportional to the distance rather than inversely proportional. As a consequence, all the discussion holds in the reciprocal space instead of the direct space.

Equation 2.64 becomes:

$$E = \frac{k^2\hbar^2}{2m_e} \quad (2.65)$$

In three dimensions, the expression of the energy in terms of k is:

$$E = \frac{\hbar^2}{2m_e}\left(k_x{}^2 + k_y{}^2 + k_z{}^2\right) = \frac{\hbar^2}{2m_e}\mathbf{k}^2 \quad (2.66)$$

where **k** is the wave vector of the reciprocal space. This is the case of the free-electron model, where no restrictions due to the lattice are present. Since the energy is proportional to $\mathbf{k}^2$, E versus **k** gives a parabolic function. When introducing the potential due to an infinite array of atoms or molecular building blocks separated by a distance a, **k** becomes equal to π/a (Tanner, 1995).

In both models, the nearly free-electron model and the tight-binding approximation model, the solution of the wavefunction that describes the system is a periodic function in the form of cos($\mathbf{k}a$). In the tight-binding approximation this can be seen as the energy expression for the linear combination of atomic orbitals (LCAO) in the framework of the *Hückel theory*:

$$E_n = \alpha + 2\beta\cos(\mathbf{k}a) \tag{2.67}$$

where α is the Coulomb integral and describes the binding energy of isolated AOs and β is the resonance integral and is related to the strength of the interactions between neighbors separated by a distance a in the reciprocal space defined by the wave vector **k**. This means that α defines the energy value of the center of the band, whereas β represents the energy distribution of the band. Given the cosinusoidal periodicity of the energy with **k**, the meaningful values of **k** are 0 and $\pm\pi/a$, with minimum or maximum energy values for cos($\mathbf{k}a$) equal to ± 1. This means that the energy bandwidth can be expressed in terms of 4β and that the energy band distribution is fully described between $\pm\pi/a$. This region of the reciprocal space is known as the first Brillouin zone.

These conclusions become evident when considering the simplest case of orbital overlap between an infinite number of atomic s, or molecular σ, orbitals in a 1D array. For these types of orbitals the energy minimum will correspond to the completely constructive LCAO (k = 0, no nodes between orbitals) whereas the maximum energy will correspond to the completely destructive LCAO where a node is present between each atom or molecule ($k = \pm\pi/a$) (Fig. 2.22) (Hoffmann, 1989; Canadell et al., 2012).

An energy band will be created for each combination of a specific type of orbitals between adjacent atoms. Therefore, a higher-energy band with respect to that generated by atomic s or molecular σ orbitals will be formed as a result of the interaction between higher-energy atomic p_z or molecular σ_p orbitals along the σ-bonding direction. In both these cases the band profile as a function of k will be characterized by an energy minimum for the completely destructive LCAO ($k = \pm\pi/a$, maximum number of nodes) and an energy maximum for the completely constructive LCAO (k = 0, no nodes) (Fig. 2.22). The interaction between p_x and p_y AOs, or π MOs, will have an energy distribution of the opposite phase with respect to that of the lower-energy band (Fig. 2.22). In general, an alternation of energy distributions with the opposite phase (concave and convex energy distributions with respect k = 0) is obtained due to the symmetry of the orbital wavefunctions as the energy increases.

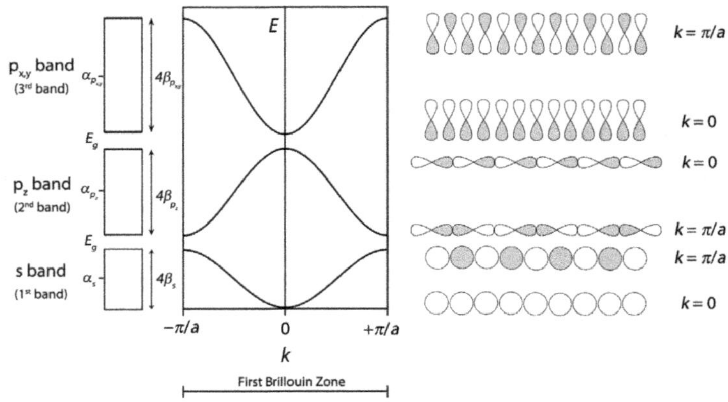

Figure 2.22 Schematic representation of the band structure for the first Brillouin zone for a 1D system in which the alternate energy profiles for different bands are highlighted (center). Orbital overlap for atomic s and p orbitals that define the minimum and maximum energy levels for each band (right). The resulting energy bands and the corresponding bandwidths are also represented (left).

The graphical representation of this energy distribution for the electrons within a solid without considering the dependence of the wave vector **k** is the energy band. The bandwidth being a measure of the interaction between the orbitals (4β), the higher its value, the larger will be the resonance integral and orbital overlap. The separation between the bands defines forbidden energy levels for the electrons (Fig. 2.22).

The principles of energy band formation for 2D or 3D systems are exactly the same as those illustrated for the 1D case. The only difference in their description is that a wider region of space must be considered to have a full account of the band structure. This means that, in the simplest cases, the first Brillouin zone will be a square for a 2D system and a cube for a 3D system. Fortunately, the symmetry of the crystal structure helps us in their representation. Indeed, it is sufficient to analyze the band profile along high symmetry directions to have a full description of the band structure of a given system. The reciprocal space defined by the high-symmetry directions is called the irreducible Brillouin zone (IBZ), and its complexity depends on the symmetry of a given system. The extremes of the band are identified with capital letters: Γ indicates the center of the first Brillouin zone, X the center of a face, M the center of a corner, etc. (Fig. 2.23).

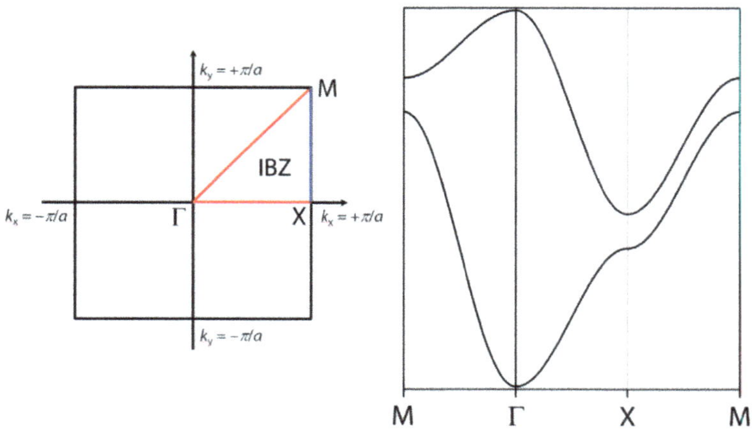

Figure 2.23 Schematic representation of the band structure (two bands) for the first Brillouin zone of a generic 2D system. The irreducible Brillouin zone is indicated with colored lines that define the high-symmetry directions.

For a 3D system the band structure appears more complicated because of the higher number of high-symmetry directions and the multitude of bands due to the orbital overlap of several atomic or MOs of different energy values. In this case the IBZ is a polyhedron.

Figure 2.24 shows the real band structures near the Fermi level for two types of molecule-based conducting materials.

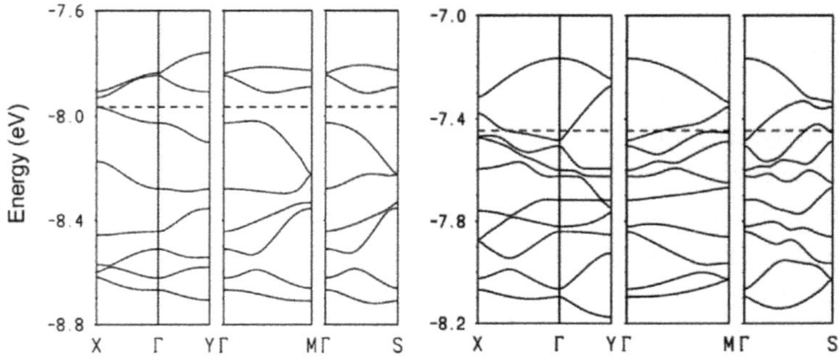

Figure 2.24 Band structures for a molecule-based semiconductor (left) and a molecule-based metal (right). The dashed lines indicate the Fermi energy.

From the energy distribution and the electron filling of the band structure of a solid, it can be understood whether a material behaves as a semiconductor, a metal, or an insulator at a given temperature. Taking into account the band structures reported in Fig. 2.24 we can easily get information about the transport properties of these two materials. The band structure reported on the left shows the Fermi energy (dashed line) located between two bands, that is, in an energy gap level. This means that valence electrons will occupy the energy bands below this energy value. The valence band is completely full, and the conduction band is completely empty. This material will then be a semiconductor or an insulator, depending on the magnitude of the energy gap between the valence band and the conduction band. Since this particular example refers to a calculated energy gap of ca. 35 meV (E_g << 5 eV) it will behave as a semiconductor. Instead, the band structure reported on the right shows a Fermi energy intersecting an energy band along all high-symmetry directions. This means that the valence and conduction bands are contiguous and partially occupied. In this case the electrons are almost free to move within the band in the presence of an electric potential and the material will behave as a metal.

2.2.7 Structural Anisotropy and Peierls Distortion

In general, in a molecular conductor the intermolecular interactions that are at the origin of the band structure are weaker than the interactions between atoms found in a classical conductor. This is due to the fact that the overlap between MOs is usually more complicated and less effective than that involving AOs, resulting, in most cases, in lower conductivity for molecular systems when compared to atom-based materials. Anyway, whenever electronic and structural parameters are well designed, some molecular materials can have conductivity values comparable with those of inorganic materials, as well as peculiar behavior not retrievable in atom-based materials.

It must be additionally considered that the conducting properties of molecular conductors are also strongly dependent on their structural anisotropy. These materials are often characterized by 1D or 2D stacks, along which conductivity has the maximum value, due to a favorable overlap between the MOs of the building units. The interactions between different stacks are instead usually weaker, resulting in lower conductivity in the direction perpendicular to them.

Conducting molecular materials are also subjected to the Peierls distortion, that is, dimerization phenomena between building blocks at low temperature (Peierls, 1955). This results in an increased energy gap between the valence and conduction bands and consequent lower conductivity (Fig. 2.25).

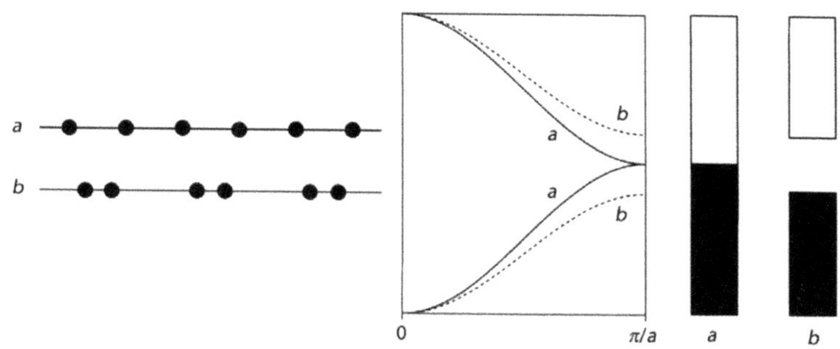

Figure 2.25 Comparison between the expected band structure for a monodimensional regular array of atoms or molecules (black spheres) separated by a distance *a* (a) and the same array in the case of Peierls distortion (b).

The conducting properties of materials subjected to Peierls distortion are in fact often characterized by metal–insulator or metal–semiconductor transitions due to these phenomena (cf. the fourth example in Section 2.2.4). To reduce or avoid the Peierls distortion, experimental evidence and theoretical calculations suggest that the material dimensionality should be increased by favoring interactions between different stacks. This would lead to less deviation from the metal-like behavior, favoring conductivity and superconductivity. For this reason, the most performing building blocks that show these properties have a planar structure and highly polarizable atoms, such as S and Se. More details on the nature of these building blocks and the type of molecular materials showing conducting properties will be presented in Chapter 3.

2.3 Principles of Molecular Photophysics

This section will provide a short description and brief recall of the fundamental notions of photophysics necessary to introduce the various classes of molecular materials with optical properties that will be presented in the next chapters.

Light–matter interaction is the concept at the heart of molecular photophysics. When light interacts with matter, it perturbs the irradiated material, inducing a change in its state that can be responsible for a number of optical phenomena. Unlike magnetic and conducting properties, which are mostly related to the ground or fundamental state of matter, photophysical properties involve excited states at higher energies. When dealing with the photophysical and optical properties of molecular materials, it is then crucial to keep in mind that these properties are related to species that are actually different to those in the ground state. For this reason, to predict and understand the photophysical/optical behavior of molecular materials it is often necessary to consider the evolution of their nature upon interaction with light. An optically excited molecule can also transition to a state suitable for undergoing chemical reactions otherwise not thermodynamically accessible. Light-induced chemical reactions are the subject of photochemistry, which is indeed an extensive field of study of utmost importance both in synthetic chemistry as well as for applicative purposes, whose review is however beyond the scope of this book.

2.3.1 Electromagnetic Radiation and Light–Matter Interaction

Light is, in general terms, electromagnetic radiation, consisting of synchronized sinusoidal oscillations of electric and magnetic fields (electromagnetic waves), which are perpendicular to each other and perpendicular to the direction of light propagation, as shown in Fig. 2.26.

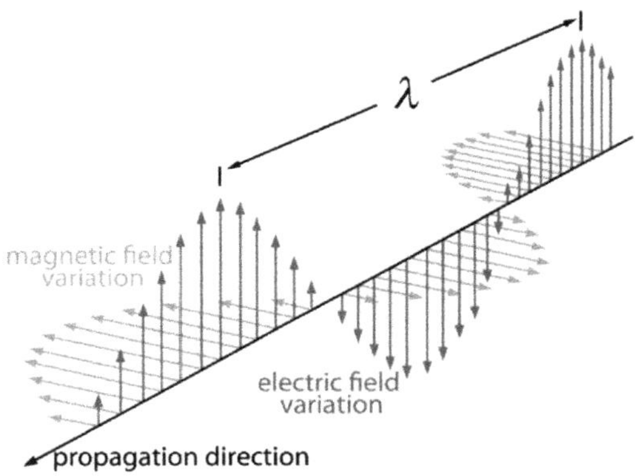

Figure 2.26 Schematic depiction of an electromagnetic wave.

The oscillating electric field of the electromagnetic radiation can interact with atoms and molecules by causing a displacement of their charge distribution and consequently inducing an oscillating dipole moment (polarization) in the irradiated chemical species. Whenever the incident radiation reaches resonance with this induced oscillating dipole at a certain frequency, a quantum of electromagnetic radiation, or photon, is absorbed and the energy of the species is raised to an excited state. Light can interact not only with electrons but also with all the charges constituting the matter. This kind of interaction is determined by how fast the charges can move to keep up with the frequency of the oscillating electric field of the radiation. Whereas the lightweight electrons can be easily displaced under the effect of a highly energetic electromagnetic field, the heavier atomic nuclei can only be polarized by low-frequency light, so molecular vibrations and revolving motions (rotations) are excited by radiation in the low-energy side of the electromagnetic spectrum (cf. Fig. 2.27). Hence, in the field of linear optics (LO),[4] light–matter interaction is a quantized process that can affect the material in different ways depending on the energy of the electromagnetic radiation. In general, for molecular materials, high-energy radiation, that is, γ- and X-rays and vacuum ultraviolet (UV) radiation, can cause a permanent change not only in the state but also in the nature of the material itself,[5] whereas lower-energy light (from the near-ultraviolet [NUV] to the microwave region; cf. Fig. 2.27) induces transient perturbations that allow the material to relax in its original state. As far as optical molecular materials are concerned, the most interesting phenomena arise from transient electronic transitions involving electrons in the outer energy levels of the molecule. These phenomena are generally induced by electromagnetic radiation in the UV–visible–near infrared (NIR) spectral region, so only these types of interactions will be considered in the following discussion.

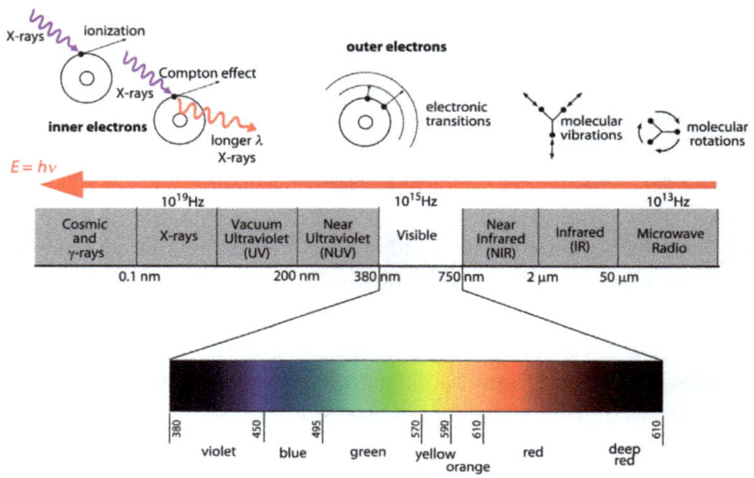

Figure 2.27 Electromagnetic spectrum and related light–matter interactions.

Following the previous considerations, we can summarize that a chemical species in its fundamental ground state can reach a

[4]We will refer to the field linear optics (LO) throughout the general discussion that follows until nonlinear optics (NLO) is introduced in a dedicated paragraph at the end of this section, where the distinction between LO and NLO phenomena will be also clarified.

[5]γ-rays and X-rays have enough energy to break molecular bonds, form highly reactive radicals, induce nuclear reactions, etc.

quantized excited state as a consequence of the interaction with the electric field of the incident radiation and subsequent absorption of a finite amount of energy, that is, a photon, $h\nu$ (Grotthus's law)[6]. For optical phenomena involving electrons, this process consists in the transfer of one electron from an energy state to a higher one, that is, an electronic transition. In technical terms, it is said that the upper excited state becomes populated. The excited species, usually labeled with the symbol *, then dissipates the acquired extra energy by relaxing to its ground state in a radiative way, that is, with photon emission, or in a nonradiative way, for example, through thermal exchange with the environment. In a simplified two-energy-level diagram, the main radiative photophysical processes related to electronic transitions can be then pictured as shown in Fig. 2.28.

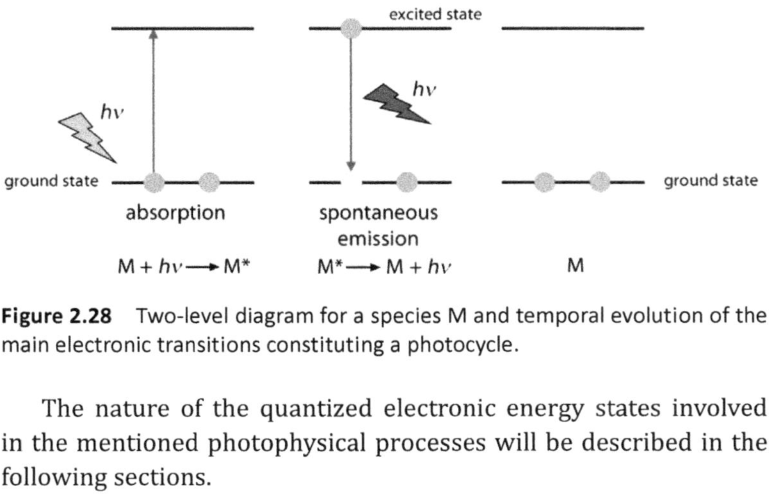

Figure 2.28 Two-level diagram for a species M and temporal evolution of the main electronic transitions constituting a photocycle.

The nature of the quantized electronic energy states involved in the mentioned photophysical processes will be described in the following sections.

2.3.2 Electronic Energy States

2.3.2.1 Electronic, rotational, and vibrational states

A free molecule in space possesses different types of energies. Besides the energy related to the electronic configuration, it also holds rotational energy, associated with the revolving motion of

[6]More complicated mechanisms involving more than one photon, such as two-photon absorption and stimulated emission, are nonetheless possible under specific conditions.

the molecule itself around its barycenter, and vibrational energy, originating from the periodic oscillation of its atoms around the equilibrium position (cf. Fig. 2.27). These two forms of energy are quantized into vibrorotational levels. For a simple diatomic molecule, the electronic energy states can be effectively exemplified by the Morse curve, shown in Fig. 2.29, which represents the potential energy curve for the molecule as a function of the interatomic distance *r*. The deviation from the theoretical parabolic shape reflects the anharmonicity of the oscillation around the internuclear equilibrium distance[7] and the existence of a bond-breaking point corresponding to the dissociation energy D_e.

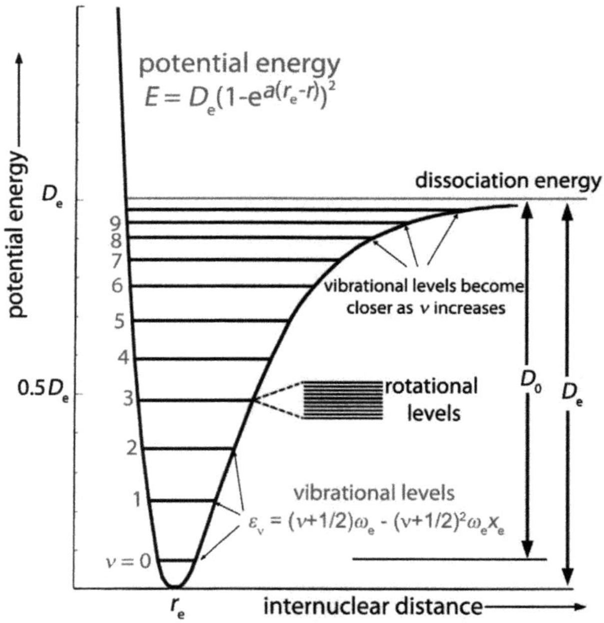

Figure 2.29 Typical Morse curve for a diatomic molecule as a function of the interatomic distance r, where r_e is the equilibrium distance, v is the vibrational quantum number, ε_v is the energy of the vibrational level, ω_e the vibrational wavenumber, and x_e the anharmonicity constant. D_0 is the real dissociation energy, which is lower than the theoretical value D_e because of the nonzero energy of the lowest vibrational level.

---

[7]In simple terms, the anharmonicity is related to the fact that a bond can be much more easily elongated than compressed.

All the electronic states of a given molecule, either ground or excited state, which are stable with respect to molecular dissociation, can be represented by similar Morse curves, whose characteristics (vibrational frequency, equilibrium bond distance, dissociation energy, etc.) will differ depending on the diverse electronic distribution and configuration of the ground or excited species.

The energy gaps ($\Delta\varepsilon$) between vibrational levels are much lower than those separating the electronic levels, whereas rotational levels are, in comparison, extremely close in energy, so it is possible to approximate:

$$\Delta\varepsilon_{\text{elec}} \approx \Delta\varepsilon_{\text{vib}} \times 10^3 \approx \Delta\varepsilon_{\text{rot}} \times 10^6 \ (\text{cm}^{-1}) \qquad (2.68)$$

As a result of the significant differences in the $\Delta\varepsilon$ values, it is reasonable to assume that the electron distribution and nuclear motions of a molecule can be treated separately and the electronic, vibrational, and rotational energies (E_{elec}, E_{vib}, E_{rot}) are independent from each other (Born–Oppenheimer approximation). The total energy E_{T} of a molecule can then be considered as the linear sum of the three main energy components:

$$E_{\text{T}} = E_{\text{rot}} + E_{\text{vib}} + E_{\text{elec}} \qquad (2.69)$$

2.3.2.2 Absorption of radiation

Whenever a molecule undergoes an electronic transition induced by the absorption of radiation in the UV–visible–NIR spectral region, it is inevitable that vibrational and rotational levels also become involved. In light of the above-mentioned Born–Oppenheimer approximation, a change in the energy state of a molecule as a consequence of an electronic transition can be simply written as:

$$\Delta E_{\text{T}} = \Delta E_{\text{rot}} + \Delta E_{\text{vib}} + \Delta E_{\text{elec}} \qquad (2.70)$$

However, since the rotational energy is many orders of magnitude lower than the electronic and vibrational energies (cf. Eq. 2.68) ΔE_{rot} can be often neglected when dealing with electronic and vibrational transitions. Therefore, molecular electronic transitions are often referred to as vibronic transitions.

The two fundamental requirements that must be met for the occurrence of an energy transition between an initial and a final state, described by the wavefunctions Ψ_{i} and Ψ_{f} (see box below), with energies E_{i} and E_{f}, respectively, are the following:

- The Bohr frequency condition, simply expressed by the equation $E_f - E_i = h\nu$, that is, the radiation energy, has to be equal to the difference in energy of the two states.
- There must be a variation of the dipole moment $\boldsymbol{\mu}$ between the initial and final states, and the transition dipole moment $\boldsymbol{\mu}_{if}$ associated with the induced charge oscillation must be different from zero:

$$\mu_{if}^{r} = \int \Psi_f \mu \Psi_i d\tau = -e \int \Psi_f r \Psi_i d\tau \neq 0 \qquad (2.71)$$

where the dipolar moment operator $\boldsymbol{\mu} = -e\mathbf{r}$ and $d\tau$ indicates integration all over the coordinates in space.

Whereas the first point is rather obvious and most likely does not need further explanation, the latter implies different circumstances, depending on the type of transition involved. Rotational transitions are in fact only allowed for molecules possessing a permanent dipolar moment, whereas for vibrational ones, a change of the dipolar moment associated with the involved oscillation is required. In the case of electronic transitions, the change of $\boldsymbol{\mu}$ is usually ensured by the electronic redistribution of the excited state with respect to the ground state (Fig. 2.30).

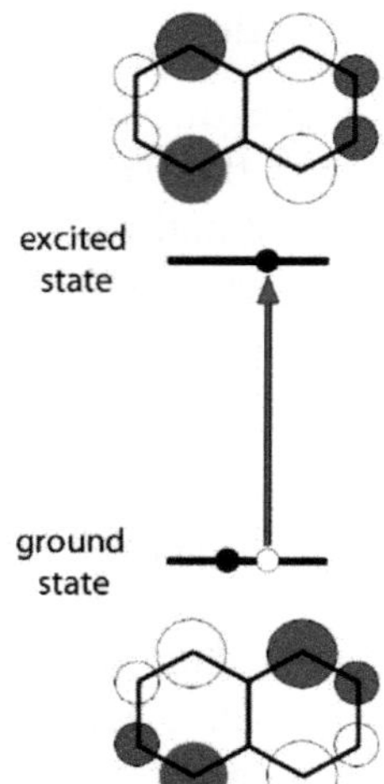

Figure 2.30 Electronic distribution of the ground and excited states in the naphthalene molecule.

Molecular Orbitals and Electronic States

For hydrogen-like atoms, Schrödinger's wavefunction Ψ describes the electron distribution, depending on the four quantum numbers n, l, m_l, and s, and the related AO. By definition, an electron occupies a certain AO whenever it is associated to the quantum numbers of a specific Ψ. The monoelectronic wavefunction is constituted by a spatial function, which, in turn, can be subdivided into the radial $R(r)$ and angular $Y(\tau, \phi)$ terms, plus a spin function S:

$$\Psi_{n,l,m_l,s} = R_{n,l} Y_{l,m_l} S_s$$

For multielectron species, the total wavefunction is, in principle, a function of the spatial coordinates of all the electrons of the atom. However, to simplify the resulting equation, the orbital approximation can be reasonably applied, so the total Ψ is considered as the product of the single monoelectronic wavefunctions of each electron in the atom:

$$\Psi = \psi_1 \psi_2 \psi_3 \ldots$$

Similar considerations can be also applied to more sophisticated systems, such as molecules. However, to describe the MOs through mathematical equations that can be solved, some approximations must be applied, so MOs are expressed through linear combinations of AOs (LCAO method, cf. Section 2.2.6):

$$\Psi = \sum_i c_i \phi_i$$

In this equation, ϕ_i represents the normalized wavefunctions of AOs of suitable symmetry and the c_i coefficients are the weights of the associated AO in the resulting MO.

The wavefunction Ψ represents the entire configuration of a certain molecular system. Therefore, if the ground state of a molecule is described by Ψ_i, the excited state formed after the absorption of radiation will be described by a different wavefunction $\Psi_f{}^*$, which is actually related to a species chemically different to its corresponding ground state.

This concept is of utmost importance, especially in chemistry (photochemistry). In fact, the newly generated excited species can possess suitable features, such as electronic distribution and configuration, to undergo chemical reactions otherwise not accessible through thermally activated states. This is the case of many light-induced reactions common in the fields of organic and coordination chemistry, such as photocyclization, photo-oxidation/photoreduction, and ligand photosubstitution.

2.3.2.3 Transition moment integral and selection rules

As already pointed out, energy states represent a specific electron configuration described by the quantum numbers associated to Schrödinger's wavefunction. Although there are multiple possible configurations associated with energetically different excited states, not all are available to the electron, and only certain transitions are allowed.

One of the fundamental selection rules for monoelectronic transitions in hydrogen-like species, either absorption or emission transitions, can be simply deducted on the basis of the law of conservation of the angular momentum. In fact, a photon is characterized by an intrinsic spin angular momentum $s = 1$.[8] For this reason, *any electronic transition induced by the absorption or emission of one photon must be accompanied by an angular momentum variation* $\Delta\mathbf{l} = \pm 1$. This means that in atoms, s ↔ p, p ↔ d, and d ↔ f transitions are allowed, whereas transitions between like orbitals, such as p ↔ p, d ↔ d, and f ↔ f, are forbidden.[9]

In a more general view, the transition moment integral, which can be separated into the electronic (Ψ_e), nuclear (Ψ_n)[10], and spin (Ψ_s) parts, defines the selection rules for electronic transitions in atoms and molecules and their amplitude (intensity):

$$\mu_{\mathrm{if}}^{\mathrm{r}} = \int \Psi_{\mathrm{e,f}}^{*} \mu \Psi_{\mathrm{e,i}} d\tau \int \Psi_{\mathrm{n,f}}^{*} \Psi_{\mathrm{n,i}} d\tau \int \Psi_{\mathrm{s,f}}^{*} \Psi_{\mathrm{s,i}} d\tau \tag{2.72}$$

The total integral will be nonzero if the three components of the overlap integral are also nonzero. The conditions by which the above statement is verified can be retrieved by simple mathematical considerations by taking into account the *parity* (even or odd) of the integrand functions. In mathematical terms, an even function is a function for which $f(x) = f(-x)$, whereas for an odd function $g(x) = -g(-x)$ or $g(-x) = -g(x)$. For the integrals of Eq. 2.72 to be nonzero, the integrand must correspond to an even function (if an integrand

---

[8] In quantum mechanics, particles with integer spin are *bosons* (e.g., photons and gravitons), while particles with half-integer spin (e.g., electrons and protons) are *fermions*.

[9] For polyelectronic species, the total angular momentum rule translates into $\Delta\mathbf{L} = 0, \pm 1$, where $\mathbf{L}$ results from the contribution of the orbital angular momenta of all the electrons in the atom or molecule.

[10] The nuclear integral part governs vibrational transitions (motions of nuclei) in molecules.

is odd, then the integral is zero because the contribution from $x > 0$ is cancelled by the contribution from $x < 0$). By remembering that an even function times an odd function gives an odd function, while an even function times an even function and an odd function times an odd function give an even function, the general rules for which each integral of Eq. 2.72 takes a nonzero value can be easily drawn.

Regarding the spin integral, as the spin is said to be of odd parity, the two spin functions, corresponding to the initial $\Psi_{s,i}$ and the final $\Psi_{s,f}^*$ states, which result from the product of the spins of all the electrons in the molecule, must be of the same parity. In practical terms, only those transitions between states with the same total spin quantum number **S** are spin-allowed. The physical explanation for this is quite obvious: since the electronic spin is related to magnetic effects, electronic dipole transitions will not alter it.

The remaining terms, that is, the electronic and nuclear integrals, are mainly governed by symmetry considerations. These are fully described by the group theory,[11] but they cannot be easily applied for complex molecular architectures typically encountered in the field of molecular materials. However, some general guidelines can be pointed out, which should allow the reader to understand the concepts behind the basic mechanisms of electronic transitions in molecules. In general, whereas vibrational transitions accompanying electronic transitions are always allowed for all Δv, with the exception of unsymmetrical normal modes in polyatomic molecules that are only permitted for certain values of Δv, in the case of electronic transitions, the first term of Eq. 2.72 is nonzero only if the initial- and final-state wavefunctions are of different parity. This rule mathematically originates from the fact that the dipolar moment operator $\boldsymbol{\mu}$ is an odd function, so if even functions are denoted with the letter "g" and the odd ones are labeled "u" (vide infra): g × u × g = u and u × u × u = u (not allowed); g × u × u = g and u × u × g = g (allowed). This means that only transitions from an even to an odd state and vice versa are allowed, whereas those involving both even or both odd ground- and excited-state wavefunctions are prohibited. In atoms and centrosymmetric molecules, that is, those molecules having an inversion center, this rule is often translated into Laporte's rule (or parity rule), which enunciates that *only transitions between states with different symmetries with respect to the inversion center*

[11]See, for example, (Cotton, 1990).

are allowed. According to the group theory, wavefunctions that are symmetric (even) with respect to the inversion operation are denoted with the letter "g," while the antisymmetric (odd) ones are labeled "u" from the German words *gerade* and *ungerade*, respectively. In atoms, s and d orbitals are g in symmetry, while p and f are u. Therefore this also explains why transitions within the same subshell are forbidden (e.g., d → d and f → f). In molecules, the irreducible representations of the ground and excited states of the corresponding character table of the molecule should be determined and only the transitions involving states of g and u subscripts are allowed (e.g., $A_g \rightarrow B_u$ in C_{2h} group). More specifically, the integrand product of the electronic integral must be or contain the total-symmetric representation (A_g or A_{1g}). This rule will be better explained through a simple example in the next paragraph.

In summary, for allowed electronic transitions:[12]

- The angular momentum variation for a single electron must be $\Delta \mathbf{l} = \pm 1$ (angular momentum conservation rule)[9].
- The spin multiplicity ($2\mathbf{S} + 1$) must not change, $\Delta \mathbf{S} = 0$ (spin rule).
- The ground and excited states must be of different parity with respect to the inversion center (parity or Laporte's rule).

However, in the field of molecular materials, the occurrence of molecules having an inversion center is quite a rare case, and the parity rule is more often taken into consideration when dealing with electronic transitions centered on the metal, that is, involving metal orbitals, such as d–d or f–f transitions (vide infra). On the other hand, the spin rule takes on an important role since in many cases it can crucially determine the nature of the transition, as will be discussed in detail further in this book.

2.3.2.3.1 *Symmetry allowed transitions in molecules*

Symmetric representations of electronic energy states in a molecule can be simply retrieved by considering the symmetry of MOs involved in the transition, often calculated through density functional theory (DFT) calculations, taking into account the signs of the orbitals with

[12]These rules are only valid for electric dipole transitions. Although these types of transitions are, by far, the most relevant in molecular photophysics, other types of transitions, such as electric multipole and magnetic dipole transitions, exist and can become significant in particular cases, for example, for lanthanide compounds.

respect to the axes coordinates. Let's consider the simple case of butadiene, which belongs to the C_{2h} symmetry point group, whose character table is reported in Table 2.4.

Table 2.4 Character table for the symmetry point group C_{2h}

C_{2h}	I	$C_2(z)$	$\sigma_h(xy)$	i	
A_g	+1	+1	+1	+1	*Rz*
A_u	+1	+1	−1	−1	*z*
B_g	+1	−1	−1	+1	*Rx, Ry*
B_u	+1	−1	+1	−1	*x, y*

This symmetry group has three symmetry operations leading to equivalent molecular configurations besides the identity (I), that is, a 180° rotation around the *z* axis (C_2), a reflection through the *xy* plane (σ_h), and the inversion (i) with respect to the origin of the axis. Given that the −1 character is assigned to a change of symmetry as a result of one operation and the +1 character indicates that the symmetry remains unchanged, it is then possible to easily assign the appropriate representations of the character table to the frontier MOs (HOMOs = highest occupied molecular orbitals and LUMOs = lowest unoccupied molecular orbitals) of butadiene, as shown in Fig. 2.31.

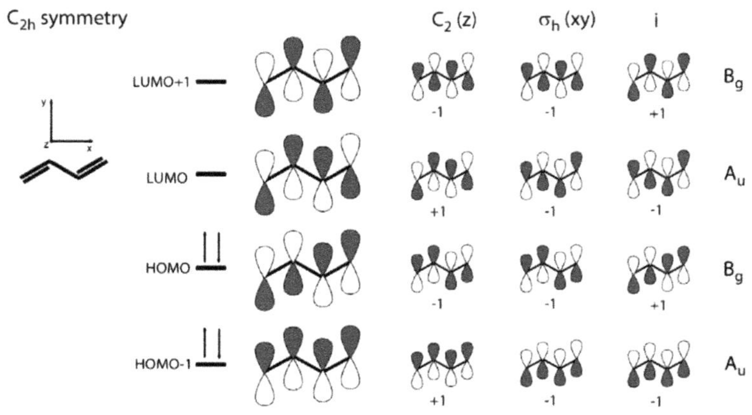

Figure 2.31 Frontier molecular orbitals and symmetry operations for butadiene.

Taking into account that the electric dipole moment operator transforms as *x*, *y*, and *z* (A_u and B_g in the C_{2h} point group), the integrand of the transition moment can be calculated as the direct product (character by character) of the appropriate irreducible representations. Thus, for the HOMO (B_g) → LUMO (A_u) transition:

$$\int \Psi^{*}_{e,f} \mu_z \Psi_{e,i} d\tau = A_u \times A_u \times B_g = B_g \text{ (forbidden)} \qquad (2.73)$$

and

$$\int \Psi^{*}_{e,f} \mu_{x,y} \Psi_{e,i} d\tau = A_u \times B_u \times B_g = A_g \text{ (allowed)} \qquad (2.74)$$

Since only the second integral gives the total symmetric representation A_g, only transitions polarized along the *x* and *y* axes are allowed.

MOs and Energy States

Although practical, the above-described procedure is far from rigorous as it roughly considers the MOs of the ground-state species as representing electronic energy states. As already pointed out, energy states are related to the entire electronic configuration of the molecule and certainly not to MOs only. In simple terms, the LUMO does not provide a true representation of the molecular excited state since it lacks information on the electronic configuration. However, in the case of typical diamagnetic molecules with all paired electrons, this assumption is often reliable since the ground state for species with completely filled orbitals is always a totally symmetric state A_{1g} and widely used for the sake of convenience. On the other hand, this simplification cannot be applied for paramagnetic species, for which the total representation is given by the product of the representations of each electron in the species.[13]

2.3.2.3.2 *Vibronic coupling and selection rules' relaxation*

The selection rules for electric dipole transitions can be somewhat relaxed (that is, become less strict) under certain conditions.

An electronically forbidden transition, such as a transition between states of the same parity (e.g., d–d transitions) can become partially allowed as the vibrational modes of the molecule couple with the electronic states, partially perturbing the molecular

---

[13]See Chapter 7 of *Chemical Applications of Group Theory* (Cotton, 1990).

symmetry. Under such a condition, the transition moment integral can be written as follows (neglecting the spin part):

$$\mu_{if}^{r} = \int \Psi_{e,f}^{*} \Psi_{n,f}^{*} \mu \Psi_{e,i} \Psi_{n,i} d\tau \tag{2.75}$$

If there are some vibrational normal modes that would allow the integrand product to contain the total-symmetric representation, then the transition moment integral can take nonzero values. This kind of interaction is called vibronic coupling, and a transition that is, in principle, electrically forbidden, can become vibronically allowed. A typical example in molecular materials is represented by metal-centered d–d and f–f transitions in coordination compounds. In this case, symmetry rules are also somewhat disrupted by the symmetry of the metal environment whenever the whole molecule does not belong to a purely centrosymmetric group (e.g., in the case of heteroleptic complexes, a ligand of high steric hindrance can cause a distortion of the molecular structure). The *mixing of states of opposite parity wavefunctions* is also often accounted for by the lifting of parity rules (vide infra).

The spin rule can be also overcome due to some quantum mechanics effects, such as the *heavy atom effect*. In simple words, the non-negligible spin–orbit coupling of heavy atoms can favor a "change of direction" of the spins toward a more energetically stable state due to the parallel alignment of the electrons (Hund's rule), resulting in a total spin multiplicity change $\Delta\mathbf{S} \neq 0$. This phenomenon is often encountered in the field of molecular materials based on metal complexes, as third-row transition metals and lanthanides are characterized by large spin–orbit coupling (vide supra). However, also organic molecules bearing heavy substituents, such as iodine, can undergo photophysical processes involving spin multiplicity change. Since most of the molecules are diamagnetic in their ground state, that is, are in a singlet state ($2\mathbf{S} + 1 = 1$) where all the electrons are paired, this phenomenon usually leads to the formation of more energetically stable triplet states ($2\mathbf{S} + 1 = 3$) (Fig. 2.32).

Furthermore, spin multiplicity change is also somewhat less prohibited in species with unpaired electrons (paramagnetic effect).

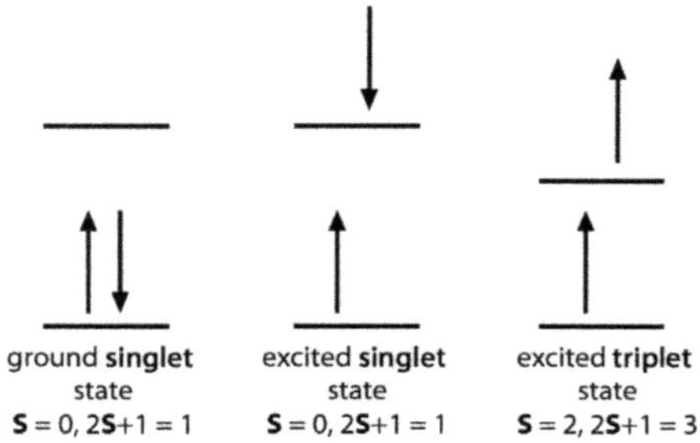

Figure 2.32 Singlet and triplet configurations for a molecule in the ground or excited state.

2.3.2.3.3 *Intensity of transitions*

Selection rules tell us whether a transition is allowed or not. However, they do not tell us the whole story on how probable or intense a transition is. Although absorption (and emission) intensity can certainly be related to the associated transition probability, according to the extent it is allowed or forbidden, it nonetheless depends on the absolute value of the transition dipole integral. For electronic transitions the intensity is associated with the degree of the variation of the electric dipole $\Delta\boldsymbol{\mu}$ between the ground and the excited state. Generally, the most intense transitions are related to a relevant spatial redistribution of the electron cloud between the ground and excited state and are said to have a charge transfer (CT) character, as, for example, in the case of 5,7-dichloro-8-hydroxyquinoline, whose frontier MOs are depicted in Fig. 2.33. The electron displacement will, in most cases, determine a precise direction of the transition dipole moment along a molecular axis. Since light absorption is determined by the interaction of the molecular dipole moment with the electric vector of the electromagnetic radiation and its extent is given by the square scalar product of these two vectors, it is straightforward that whenever the molecule is excited with linearly polarized light, the maximum absorption will coincide with their parallel alignment, whereas zero absorption can be even predicted for orthogonal vectors.

Another important factor is the spatial overlap between the orbitals in the ground and excited states. Since the electronic transition integral is actually an overlap integral all over the coordinates in space, *the orbitals of the initial and final states must overlap spatially if the corresponding transition moment is to be relevant.*

The "natural"[14] intensity of an electronic transition, given as a result of the transition moment integral, is called oscillator strength.[15]

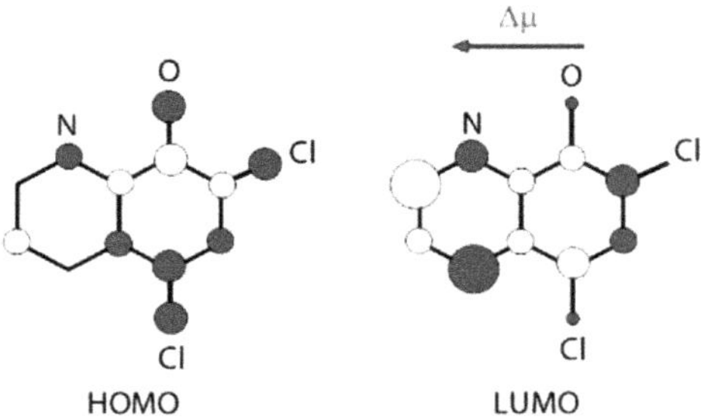

Figure 2.33 Depiction of the HOMO and LUMO orbitals of 5,7-dichloro-8-hydroxy-quinoline.

2.3.2.4 The Franck–Condon principle

It has been already pointed out that electronic transitions are inevitably accompanied by vibrational (and rotational) transitions, which are also normally promoted by thermal energy of molecules in the ground state. However, electrons' motion is so fast in comparison to the motion of the nuclei that it can be reliably inferred that *during an electronic transition the atoms constituting the molecule do not change their coordinates*, that is, they stand still (Franck–Condon principle). The consequence of the Franck–Condon principle is that any electronic transition can be visualized with a vertical arrow between the vibrational sublevels of the ground and excited states, as shown in Fig. 2.34. This assumption has important consequences for the nature of the excited state formed after photon absorption.

[14]In photophysics, the word "natural" is associated with ideal phenomena related to a single molecule or atom in the absence of any external interaction, for example, solvent or solid-state interactions, which can affect the intensity and nature of the considered transition. See Section 2.3.4.3 for a more specific definition.

[15]It was pointed out that light–matter interaction originates from the interaction of the electric field of the molecule with that of the electromagnetic radiation oscillating with a certain frequency. Therefore, in photophysics, any species able to absorb or emit electromagnetic radiation is called an oscillator.

First, it has to be recalled that since electronic excited states are often associated to MOs of antibonding character, the position of the corresponding Morse curve is usually shifted with respect to the ground state.[16] Absorption transitions take place with the highest probability from the $v' = 0$ sublevel of the ground state, according to the probability distribution functions of vibrational levels (Fig. 2.34). As a result, vertical absorption electronic transitions will most often produce an excited state at a superior vibrational level, which, in turn, can generate different deactivation paths, as will be explained in detail next.

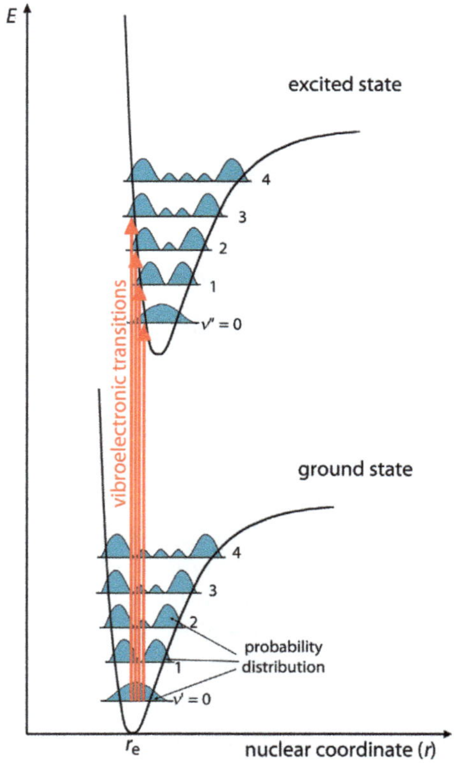

Figure 2.34 Schematic representation of electronic (absorption) transitions represented as vertical arrows between vibrational sublevels of the ground and excited states of a molecule. The blue areas indicate the probability density (population) of vibrational levels.

---

[16]The promotion of an electron in an antibonding orbital results in bond elongation; therefore, for excited states' potential energy curves, it is usually $r > r_e$.

The occurrence of multiple vibroelectronic transitions between the ground- and excited-state sublevels also explains the broadening of the absorption bands observed in electronic spectroscopy, which are actually a convolution of the multiple peaks related to each $v' \rightarrow v''$ transition (Frank–Condon progression). In some cases, especially when the molecule is embedded in a rigid environment somehow lowering vibrational modes (i.e., in solid samples or at low temperatures), a fine (vibronic) structure of the absorption band can be observed, as sketched in Fig. 2.35. It is also worth underlining that the shape (asymmetry) of absorption bands reflects the spatial dislocation of the equilibrium coordinates of the ground and excited states, as the intensity of each line depends on the density of probability of the vibrational levels of the starting and ending states (i.e., absorption transitions starting from the center of the level $v' = 0$ are the most intense, cf. Fig. 2.34).

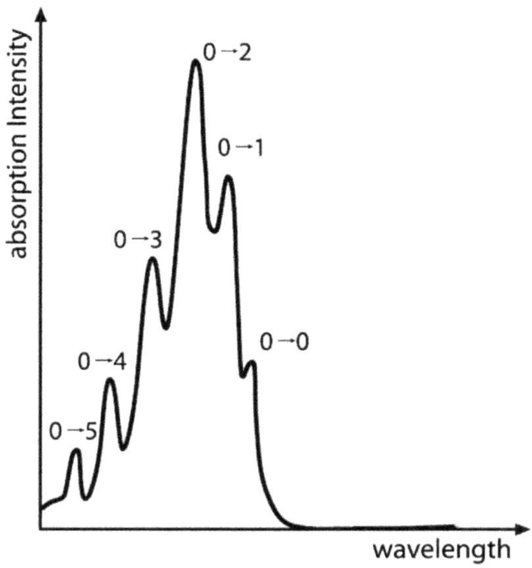

Figure 2.35 Typical absorption band resolved into a fine vibronic structure. The peaks relatable to $v' \rightarrow v''$ transitions are recognizable. In this example, the most intense, $0 \rightarrow 2$ transition takes place from the center of the vibrational level $v' = 0$ and ends on the border of the $v'' = 2$ level, where the vibrational population density is the highest.

It must be nevertheless pointed out that not all the transitions originate from the lowest-lying $v' = 0$ level. At temperatures higher than room temperature, the thermal energy may promote some electrons to vibrationally excited levels of the ground state. This would result in the appearance, in the absorption spectrum, of the so-called hot bands, which can, however, be easily recognized on the basis of the temperature dependence of their intensity.

2.3.3 Evolution of Excited States

2.3.3.1 Luminescence

Luminescence is, in general terms, the emission of light (radiative emission) from a chemical species in its excited state. It is natural to understand that this process, being a quantized phenomenon occurring with photon emission and originating from an electronic transition, follows the same rules as those described for absorption transitions.

In the following discussion we will exclusively refer to spontaneous emission, that is, that process by which a species in its excited state spontaneously undergoes a transition to a lower state (deactivation or decay) with the emission of one photon.

In Section 2.3.2.4 the formation of an electronic excited state subsequent to photon absorption was described. Obviously, thermodynamics tells us that such a high-energy condition cannot survive for long, so the fate of this excited species can be foretold to be doomed to extinction. The simplest evolution of a species that undergoes absorption and emission of one photon of light can be sequenced into different steps constituting the photocycle,[17] as depicted in Fig. 2.36. First, a vertical vibroelectronic transition occurs from the $v' = 0$ level of the ground state to a superior vibrational sublevel of the excited state (see Section 2.3.2.4). The "extra" vibrational energy is immediately dissipated nonradiatively (i.e., through thermal dissipation) by decaying along the potential energy (Morse) curve to the $v'' = 0$ level of the electronic excited state

[17]The photoemission process, that is, photoexcitation and subsequent luminescence, must be cyclic. If not, the excited state gets extinguished for some other reasons, for example, by undergoing photochemical reactions or by photogeneration of reactive species, so the molecule becomes chemically altered and the original ground state is not recovered. This phenomenon is called photobleaching.

(vibrational relaxation). From there, in the absence of any external interactions providing alternative deactivation channels, the excited species relaxes through radiative decay (luminescence) to one of the vibrational levels of the ground state. Should the resulting populated level be vibrationally excited, subsequent vibrational relaxation to the lowest-energy level restores the initial conditions and ends the photocycle.

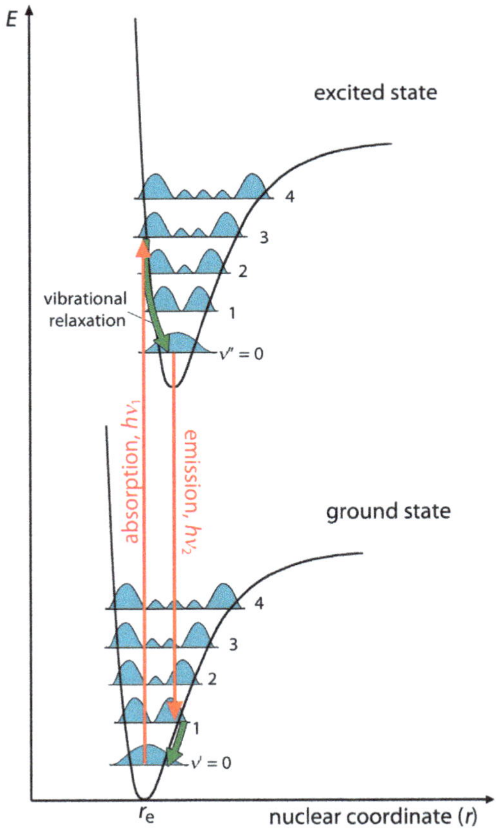

Figure 2.36 Simplified scheme of a typical photocycle in a molecular species, showing radiative transitions, that is, absorption and emission, as vertical arrows between the involved vibroelectronic states. Radiative emission is often preceded and followed by nonradiative vibrational relaxation (green curved arrows) to the lowest vibrational sublevels.

The reason radiative emission occurs from the lowest vibrational level of the excited state is that this process is much slower (10^{-9}–10^{-5} s) than vibrational relaxation (10^{-14}–10^{-11} s). In other words, the vibrationally excited state does not survive long enough to allow radiative deactivation. As a general rule, all the photophysical processes taking place in molecules should be considered as kinetically competitive, but this concept will become clearer throughout the text.

One of the important consequences of the existence of vibrational sublevels of electronic states is that the wavelength of the emitted light is longer than that of the absorbed radiation. In practice, absorption and emission spectra are usually shifted with respect to each other, and the emission band appears at higher wavelengths (Fig. 2.37). This effect is called the Stokes shift and is of great importance in the field of molecular materials, since it can allow light of a certain wavelength to be converted into radiation of a different wavelength. In certain species, such as emissive lanthanide complexes, the entity of the Stokes shift is so remarkable that these molecules are often called molecular light converters (see Section 3.3.4).

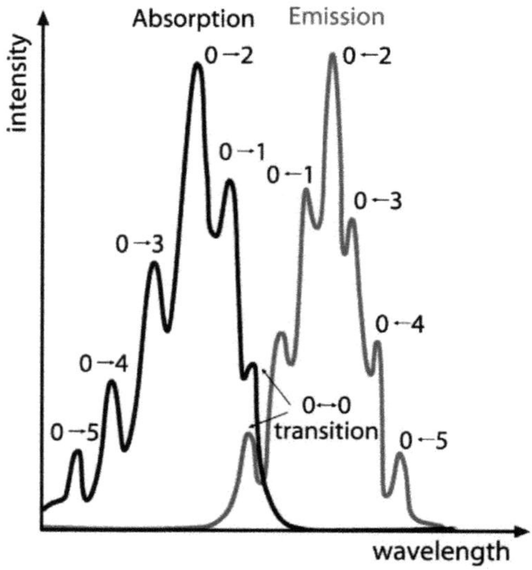

Figure 2.37 Typical absorption and emission bands resolved into fine vibronic structures. The 0 ↔ 0 transition overlaps for the two bands.

It should be also noted that emission spectra are typically almost a mirror image of the corresponding absorption spectra related to the same transition. This is quite obvious when considering that the vibrational progression of the emission process is reversed with respect to the absorption. This empirical observation is often called the *mirror image rule.*

2.3.3.2 Kasha's rule

In polyatomic molecules, many possible excited states, permitted by selection rules, can be accessible to the electron. For diamagnetic organic molecules these states are named with sequential subscripts S_1, S_2, S_3, . . . , S_n, while the ground state is usually indicated as S_0, where the label "S" stands for "singlet." The electronic excited level reached after radiation absorption then depends on the excitation wavelength.

In optically functional molecules, *luminescence generally occurs from the excited state with the lowest energy*, that is, $S_1 \rightarrow S_0$. This statement is known as *Kasha's rule*, first formulated in 1950 by Michael Kasha, and is a general principle of molecular photophysics. Although it is indeed an empirical rule, the reason behind it becomes immediately evident when graphically representing the typical electronic states of a molecule as potential energy (Morse) curves, as depicted in Fig. 2.38. In fact, Morse curves for the electronic excited states are usually so close that their vibrational sublevels overlap. This provides an alternative thermal deactivation pathway via vibrational decay for the excited state, allowing an efficient, that is, fast, radiationless transition from a state to another with the same spin multiplicity, followed by vibrational relaxation. This "cascade" process is called internal conversion (IC) and can occur repeatedly from higher excited states until the electron becomes "trapped" in the lowest vibrational level of the lowest-energy electronic state. Once again, the explanation of this mechanism can be found in the different kinetics of radiative and nonradiative deactivation processes, IC being much faster (10^{-14}–10^{-11} s) than radiative emission (cf. Section 2.3.3.1).

As already pointed out, Kasha's rule has to be considered as a general guideline for typical molecular emitters, not a proper physical principle. Although quite rare, exceptions to this rule exist when the configuration of electronic states is not suitable to allow the mechanism described in Fig. 2.38, for example, when higher S_n states are far enough apart that their vibrational levels to not overlap and the IC between them is not efficient. One typical example is azulene, where the S_1 state is really close in energy to S_0 whereas the gap with S_2 is large. As a result, the azulene emission band unusually appears in between the two main absorption bands, as shown in Fig. 2.39.

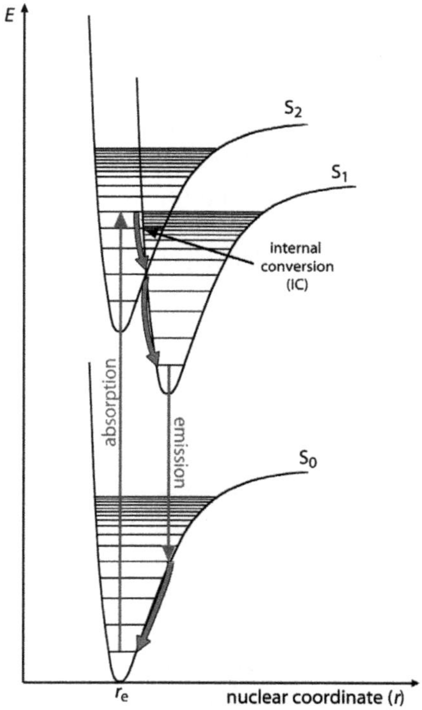

Figure 2.38 Schematic representation of Kasha's rule. Photoexcitation brings the molecule in a superior electronic excited state, S_2. Then nonradiative internal conversion (IC) occurs between states S_2 and S_1, thanks to the overlap of the vibrational sublevels. Emission takes place from the lowest vibrational level of S_1.

Besides exceptions, it must be underlined that Kasha's rule should never be taken as absolute. In fact, it only claims that the most probable emission originates from the lowest excited state, not that radiative deactivations from superior states are forbidden. This means that emission from higher-energy states can actually occur (and it does!) but its intensity is so low that most of the times it cannot be detected.

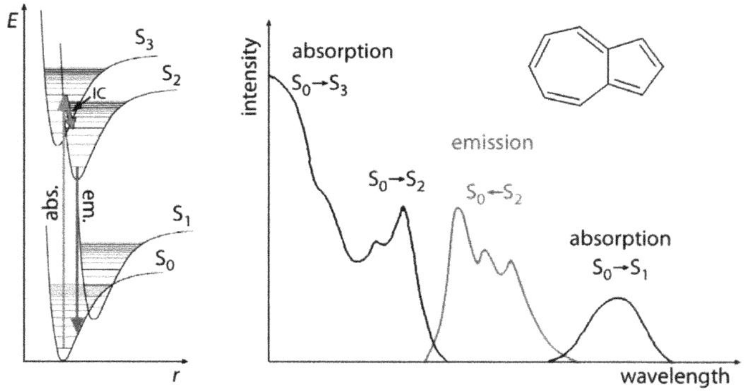

Figure 2.39 Sketch of the photocycle and absorption/emission spectra of the azulene molecule, a typical example of an exception to Kasha's rule.

2.3.3.2.1 *Kasha–Vavilov's rule*

A corollary to Kasha's rule is *Vavilov's rule*, which states that *emission is independent on the excitation wavelength.* More rigorously, the luminescence quantum yield[18] is generally independent of the excitation wavelength. The origin of this statement is quite obvious if we consider that all the excited S_n states will decay to S_0.

2.3.3.3 Fluorescence and phosphorescence

"Luminescence" is the general term to refer to photon emission from an excited species. However, two main kinds of light emission can be recognized that are intrinsically different in nature and have different characteristics. The kind of radiative emission we have taken into consideration so far, which involves states of the same spin multiplicity, is commonly called fluorescence. This process, depicted in Fig. 2.40, is spin-allowed and is usually quite fast, taking place in nanoseconds (ns) to tens of microseconds (μs). On the other hand,

[18] The luminescence quantum yield, indicated by the symbol Φ, is defined as the ratio between the number of emitted photons and the absorbed ones. This parameter is used to indicate the overall efficiency of the emission process, and it will be explained in more detail in Section 2.3.4.

as already pointed out in Section 2.3.2.3, in some circumstances, the heavy atom effect can allow a rearrangement of the spins, leading to the population of a state of different spin multiplicity. Generally, this process, named intersystem crossing (ISC), concerns excited singlet states (S_n), which nonradiatively evolve into lower-energy excited triplet states (T_n), once again thanks to an overlap of the vibrational sublevels, as shown in Fig. 2.40. Radiative emission occurring from an excited state of different spin multiplicity with respect to the ground state is indicated as phosphorescence. The spin-forbidden nature of this process translates into the long survival of the excited state, as if the electron becomes trapped, so phosphorescent emission may occur over a time range of milliseconds up to several seconds or even hours.

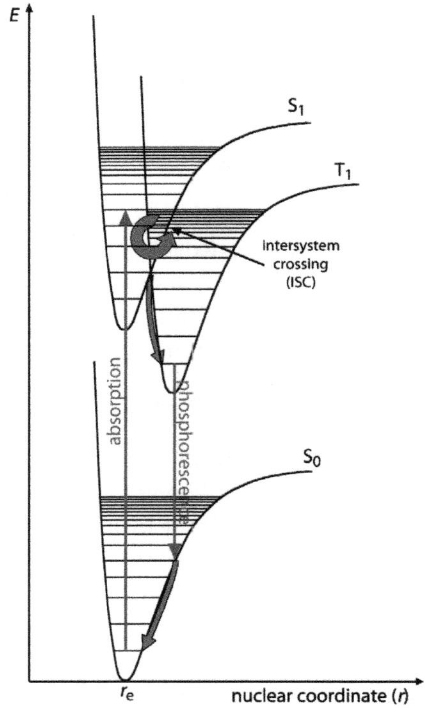

Figure 2.40 General scheme for phosphorescent emission. A molecule is first photoexcited to its first higher singlet state S_1, and then intersystem crossing (ISC) occurs to a lower-energy excited triplet T_1. Eventually, emission occurs for the lowest vibrational level of T_1 to S_0 with $\Delta \mathbf{S} \neq 0$.

Since excited triplets are typically placed at lower energies with respect to singlet states, the entity of the Stokes shift of phosphorescence spectral bands is higher than that of fluorescence bands, that is, phosphorescence is red-shifted with respect to fluorescence.

2.3.3.4 Luminescent and nonluminescent molecules

The reader should have good familiarity with the fact that not all the molecules are luminescent. On the contrary, most of them are not, at least in normal conditions.

Neglecting for a moment the effects of the chemical environment and bearing in mind that IC and ISC can, of course, also occur between the excited state and the ground state, it can be said that an isolated excited molecule has the best probability of decaying radiatively to the ground state when the potential energy curves of the involved electronic states do not overlap. This is, of course, verified when the said energy states are far apart in energy (energy gap law) and place approximately at the same spatial coordinates, as schematically depicted in Fig. 2.41b. In this configuration, the overlap of the energy levels is minimized and radiative decay has the best probability to occur. In fact, if the ground and excited states are too closely separated in energy, the overlap of the vibrational levels favors efficient decay by means of nonradiative IC (Fig. 2.41a). Therefore, the practical meaning of the energy gap law is that *the lower the emission wavelength, the higher is the emission efficiency*. Similarly, the spatial distortion of the excited state translates into a displacement of the corresponding potential energy curve, which can then easily cross that of the ground state, favoring nonradiative deactivation (Fig. 2.41c). For this reason, radiative emission is most often favored in rigid molecules, which have a nonflexible structural arrangement, such as highly aromatic condensed hydrocarbons or heterocyclic rings (azulene, quinoxaline, phenathroline, etc.). A further increase in the rigidity of an organic emitter, which can be achieved at low temperatures, through coordination to a metal ion[19] or by incorporation into a solid matrix (polymers, glasses, etc.), is often a good strategy to enhance its luminescence properties.

[19]To preserve the emission properties of the organic moiety, the coordinated metal ion should be a closed shell in order to avoid metal–ligand MOs mixing, which would dramatically affect the optical features of the molecule (see Section 3.3.2).

However, you must take into account that the rigidity effect in increasing the radiative emission efficiency should be also ascribed to the lowering of thermal deactivation via vibrational modes.

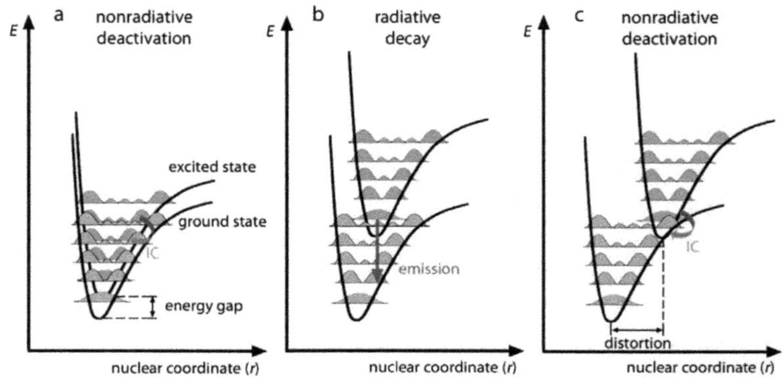

Figure 2.41 Sketch of deactivation processes taking place in molecules where the excited state is at a low-energy gap with respect to the ground state (a), at higher energy and almost vertically aligned with respect to the ground state (b), and spatially displaced due to molecular distortion (c).

2.3.3.5 Jablonski diagram

We have so far represented the electronic states of a molecule through the Morse curve, which actually refers to diatomic species. For polyatomic molecules, the potential energy curve must be adjusted for the increased number of bonds. Sophisticated calculations are usually required to this end and, most of the times, are even unnecessary if the photophysical process under consideration does not involve a photochemical reaction accompanied by molecular rearrangement. Therefore, for convenience' sake, only the energy coordinate is retained while dispensing with the dimensional (r) coordinate. Photophysical processes are then typically described through the Jablonski diagram, shown in Fig. 2.42, in which the spatial orientation of the various electronic states is not specified. Nevertheless, the Franck–Condon principle must be always taken into consideration. Notice that in the Jablonski diagram radiative processes are indicated as vertical solid arrows whereas nonradiative ones are usually represented by dashed (or dotted) or curved arrows.

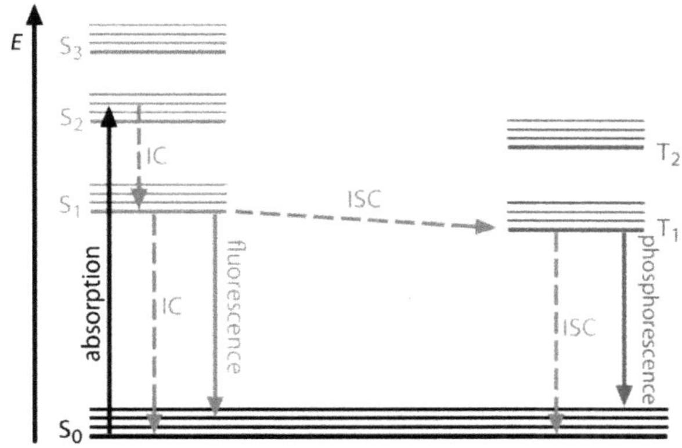

Figure 2.42 Typical Jablonski diagram for a polyatomic molecule showing the energy levels and the main photophysical processes for a molecular emitter. IC, internal conversion; ISC, intersystem crossing.

The main processes taking place in a typical photocycle can be then summarized as follows:

- **Absorption of radiation:** This is the process by which a molecule absorbs one photon of light, causing an electron in the ground state to populate a higher-energy level, depending on the frequency of the incident light. Whenever this process is completely allowed by selection rules (vide supra), it is usually very fast (10^{-15} s).
- **Vibrational relaxation:** This is a radiationless decay process by which a molecule in a superior vibrational level of an electronic state decays through a cascade mechanism to the lowest $v = 0$ level of the same state, losing energy through heat dissipation. This is an efficient process and occurs within a timescale of 10^{-11}–10^{-14} s.
- **IC:** This is a sort of vibrational relaxation that leads to the population of a lower-energy state of the same multiplicity thanks to the overlap of the vibrational sublevels of the involved states. Like intrastate vibrational relaxation, it is also fast (10^{-11}–10^{-14} s).
- **ISC:** This is a nonradiative process involving the population of a lower-lying excited state of a different spin multiplicity from another excited state, most often from singlet to triplet states.

The mechanism is similar to IC, that is, it requires overlap between the states. Being spin-forbidden, ISC is a slow and unlikely process; however, under certain circumstances (heavy atom effect, presence of unpaired electrons, etc.), it can speed up and become competitive with the other processes. A typical timescale for favored ISC processes is of the order of 10^{-8}–10^{-11} s.

- **Fluorescence:** It refers to radiative decay of one electron from an excited state to the ground state, without spin multiplicity change, accompanied by photon emission. It is usually much slower than nonradiative processes and, for allowed transitions, typically occurs in the 10^{-5}–10^{-9} s time range.
- **Phosphorescence**: It is a radiative process related to an electronic transition between the excited and the ground state with a spin multiplicity change (usually triplet to singlet). Being a spin-forbidden process, it is slow in comparison to all the other mechanisms and might endure for a rather long time, from 10^{-3} s up to several hours.

It should be once again underlined that the photophysical processes so far mentioned solely describe changes in the electronic wavefunction of a molecular system, that is, changes in its electron configuration, but they do not necessarily imply a spatial delocalization of the electron undergoing the transition. Although for polynuclear molecular systems electron transitions between different states (either radiative and nonradiative) are most often accompanied by an actual electron transfer to a different molecular site, depending on the shape of the involved MOs (cf. Fig. 2.33), this is, for example, not implied in the case of ISC, where only a change of the electron spin is required. On the basis of these considerations, we should better think of photophysical processes as energy transition processes, not necessarily involving electron transfer.

2.3.4 Emission Efficiency

2.3.4.1 Quantum yield and luminescence quenching

To be able to emit light, a molecular species must first be excited, that is, energy must be provided. As already mentioned briefly, in the field of molecular materials, luminescent species are often employed as molecular light converters, able to absorb light at a certain wavelength and emit light at a different wavelength. With this view, it is then evident how important it is to evaluate the efficiency of such molecular devices in terms of the amount of energy absorbed with respect to the energy recovered in the process. The energy provided to a molecular species upon photoexcitation depends both on the frequency of the incident light and the number of photons absorbed. Therefore, taking for granted the energy dissipation (through heat) related to the change in frequency of the emitted light with respect to the incident radiation (the Stokes effect), the efficiency of the light-converting process can be simply expressed by its quantum yield Φ:

$$\Phi = \frac{\text{Number of emitted photons}}{\text{Number of absorbed photons}} \tag{2.76}$$

It is extremely rare to observe experimental quantum yields having unitary value (100% conversion). As we have already seen, radiative emission in a molecular emitter may be limited by nonradiative IC decay processes. Furthermore, we have so far solely limited the discussion to individual emitting species, neglecting the presence of any external environment. In reality, an emitting species, either dissolved in a solvent or simply exposed to the air, can undergo several nonradiative deactivation phenomena induced by external species. These phenomena are known in molecular photophysics as quenching effects and can occur through several different pathways, as will be explained in detail later in this book. For now, we will simply say that *external quenching*[20] *is a nonradiative deactivation process by which an excited emitter transfers its excitation energy to an external species* (*quencher*). In rough terms, this process is often assimilated to heat dissipation through vibrational relaxation in way similar to the IC mechanism described in Section 2.3.3.2, with the difference of the involvement of external species. For this reason, it is often generically named external conversion (EC) and indicated as a nonradiative process in the Jablonski diagram (cf. Section 2.3.3.5).[21]

[20]"Quenching" actually is a generic term to indicate the depopulation of an excited level by every possible path. Of course, it can also refer to internal ET processes, such as ISC from excited singlet to triplet states. In this specific case, it can be said that the excited singlet gets quenched by ISC.

[21]Obviously, EC could, in principle, also involve excited triplet states, although triplet states are mostly quenched through mechanisms different from vibrational deactivation, involving species of the same spin multiplicity.

Spectroscopic Evaluation of Emission Quantum Yield

In the light of the basic definition of the emission quantum yield simply as the ratio between the amount of emitted light and the absorbed radiation, it becomes evident that Φ depends on two parameters that can be easily experimentally evaluated by means of basic laboratory equipment such as absorption and emission spectrophotometers. The relative method for the spectral determination of the quantum yield then only requires a comparison with an appropriately selected emission reference standard for which the quantum yield is exactly known (directly measured through specific techniques requiring more sophisticated equipment, that is, an integrating sphere).

The equation that can be applied for this purpose, provided that the investigated sample and the reference are measured under the same conditions, is:

$$\Phi = \Phi_R \frac{a_R}{a} \frac{I}{I_R} \frac{n^2}{n_R^2}$$

where a is the absorption factor at excitation wavelength ($a = 1-10^{-A}$, A = absorbance), I is the integrated emission, n is the refractive index of the medium, and the subscript R denotes the reference emitter.

2.3.4.2 Kinetics of photophysical processes

It has already been hinted that a photocycle has to be intended as a kinetic mechanism in which the fastest process has the major probability of occurrence. In general, a molecule can display radiative emission when it has electronic, structural, and chemical environment characteristics suitable for slowing down the nonradiative processes up to the point that radiative emission becomes kinetically competitive.

Each photophysical process taking place in a photocycle is then characterized by a kinetic rate constant k:

$$k = \frac{1}{\tau} \tag{2.77}$$

where τ represents the lifetime of the corresponding excited state and is one of the most important parameters in photophysics. In general, lifetime refers to the average time a chemical species stays in its excited state before decaying to a lower level.

As per these premises, it becomes clear that the quantum yield Φ_r of a radiative process depends on the contribution of all the radiative (k_r) and nonradiative (k_{nr}) rate constants associated with the different steps in the photocycle, as shown in Fig. 2.43. Therefore, the radiative quantum yield can be expressed as follows:

$$\Phi_r = \Phi_{ES} \frac{k_r}{k_r + \sum k_{nr}} \tag{2.78}$$

where Φ_{ES} represents the quantum yield of the formation of the excited state, which is often omitted, given that this step is highly efficient for allowed transitions, so $\Phi_{ES} \approx 1$.

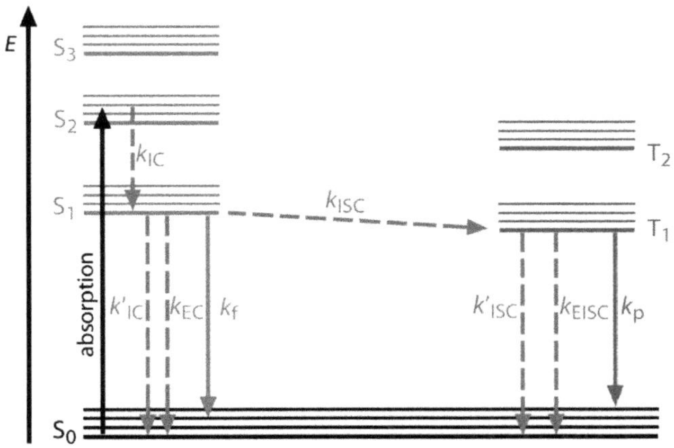

Figure 2.43 Jablonski diagram and rate constants k for the main photophysical processes taking place in a molecular emitter. The solid arrows represent radiative processes, and the dashed ones indicate nonradiative steps. f, fluorescence; p, phosphorescence, IC, internal conversion; EC, external conversion; ISC, intersystem crossing. k_{EC} and k_{EISC} refer to external quenching processes.

When assigning a rate constant for each process in the photocycle, as shown in Fig. 2.43, specific expressions for the quantum yields for radiative fluorescence and phosphorescence can be retrieved. In particular, the equation for the fluorescence quantum yield Φ_f can be simply written by summing up the rate constants for the processes involving the excited and ground singlet states and neglecting all the pathways leading to triplet states:

$$\Phi_f = \frac{k_f}{k_f + k_{IC} + k'_{IC} + k_{EC}} \tag{2.79}$$

For phosphorescence, the quantum yield (Φ_p) equation becomes a little bit more complicated as all the steps preceding the population of the excited triplet must formally be taken into account:

$$\Phi_p = \frac{k_{ISC}}{k_{ISC} + k_f + k'_{IC} + k_{EC}} \times \frac{k_p}{k_p + k'_{ISC} + k_{EISC}} \tag{2.80}$$

However, this expression can be simplified by assuming that, in a system where the spin multiplicity change is favored, the ISC process is faster than the average survival time (lifetime) of the singlet state populating the triplet level in the absence of ISC processes, so $k_{ISC} >> k_f + k'_{IC} + k_{EC}$:

$$\Phi_p = \frac{k_p}{k_p + k'_{ISC} + k_{EISC}} \tag{2.81}$$

Usually, the lifetime of the excited triplet level is so high that nonradiative deactivation is highly competitive ($k_p < k'_{IC} + k_{EC}$) and the Φ_p value is significantly lower than unity.

2.3.4.3 Emission lifetime

The lifetime parameter τ takes on particular relevance when it specifically refers to the overall emission process, as it somehow provides a direct evaluation of the efficiency of the emitter. In this case, it is specifically addressed as emission lifetime. In a general framework, it can in fact be said that the observed emission lifetime, being related to the last step of the photocycle, contains all the information on the preceding steps and so can be written as:

$$\tau = \frac{1}{k_r + \sum k_{nr}} \tag{2.82}$$

and the expression for the quantum yield simply becomes:

$$\Phi_r = k_r \tau \tag{2.83}$$

Here k_r still retains the meaning described in Section 2.3.4.2 and refers to the rate constant of the radiative emission process as if it would have been taken out of the photocycle. In this context, this parameter is called the radiative (or natural) rate constant and is related to the radiative (or natural) lifetime τ_r through the simple relation:

$$\tau_r = \frac{1}{k_r} \tag{2.84}$$

In photophysics, radiative (or natural) lifetime and rate constant refer to the isolated emitter in the absence of any competitive nonradiative process or quenching phenomena. With this view, the quantum yield of an emitting species can be seen as the ratio between the observed emission lifetime, that is, the actual measured decay time, and the radiative or natural lifetime, that is, the emission lifetime under ideal conditions:

$$\Phi_r = \frac{\tau}{\tau_r} \tag{2.85}$$

2.3.4.3.1 *Radiative lifetime*

In the light of the previous considerations, it can be said that the meaning of the radiative time constant is related to the oscillator strength of the emissive transition, that is, the tendency of an excited state to decay radiatively to the ground state. In 1917 Einstein showed that, for a simple two-level system such as the one illustrated in Fig. 2.28, *there exists a fundamental relation between the probability of excitation by the absorption of a photon from the ground to the excited level and the probability of emission of a photon by spontaneous radiative decay* from the upper level. Since then, several different equations have been proposed for a quantitative evaluation of the radiative lifetime in emitting species through easily available spectroscopic data. However, the most used and commonly accepted, in view of its simplicity and reliability, is the Strickler–Berg equation (Strickler and Berg, 1962):

$$\kappa_{\text{rad}} = \tau_{\text{rad}}^{-1} = 8\pi n^2 c \frac{1}{\lambda^3} \frac{g}{g'} \int \frac{\sigma(\lambda)}{\lambda} d\lambda \tag{2.86}$$

where n is the refractive index of the medium; c is the speed of light in vacuum; g and g' refer to the degeneracy of the ground and excited states, respectively; $\langle \lambda^3 \rangle$ refers to the mean value of emission wavelength; and $\sigma(\lambda)$ is the absorption cross section expressed in cm^2.[22]

[22]As a reminder, the following relation between the molar extinction coefficient ε ($M^{-1}cm^{-1}$) and the absorption cross section is valid: $\sigma(\lambda) = 2303\varepsilon(\lambda)/N_A$, where N_A is Avogadro's number.

It has to be, however, remarked that in the light of the Einstein's postulate mentioned above, the Strickler–Berg equation only relates absorption and emission transitions involving the same energy levels. This means that this equation can only be applied if the absorption spectrum to the emitting level is accurately known (which is not always possible due to the spectral superposition of higher energy transitions) and if the emission decay terminates on the ground level. Therefore, it cannot, for example, be applied to estimate phosphorescence radiative rate constant.

2.3.4.4 Dynamics of radiative decay

The origin of the emission rate constant k is, of course, related to a kinetic law for a process in which an excited state of population $N(t)$ at time t evolves into a ground state with concomitant photon emission within a certain period of time. Therefore, the common equations of kinetics can be applied to describe the temporal evolution of this mechanism, and the depletion of the excited-state population over time is:

$$\frac{\partial N(t)}{\partial t} = -kN(t) \tag{2.87}$$

The solution of this differential equation is an exponential function:

$$N(t) = N(t_0)e^{-k\tau} \tag{2.88}$$

where $N(t_0)$ is the initial population of the excited state, τ is the decay time constant or lifetime, and k is the decay rate constant of the transition. $N(t)$ thus decays exponentially with time and after one lifetime ($\tau = 1/k$) the number of excited states decays to 36.8% of its original value (1/e). Since $N(t)$ is directly proportional to the intensity $I(t)$ of the emission (Grotthus's law), which can be directly measured, the above equation can be written as:

$$I(t) = I(t_0)e^{-k\tau} \tag{2.89}$$

The temporal evolution (decay dynamics) of an excited state giving rise to radiative emission can be measured by means of time-resolved photoluminescence experiments. Unlike spectral analysis, which is performed with continuous-wave (cw) light sources so that the population of excited states accumulates (steady state emission), time-resolved measurements are conducted with pulsed excitation irradiation with duration shorter than the expected emission lifetime. t_0 in this case represents the time at which irradiation is switched off. The typical plot of time-dependent emission intensity for fluorescence and phosphorescence mechanisms is shown in Fig. 2.44.

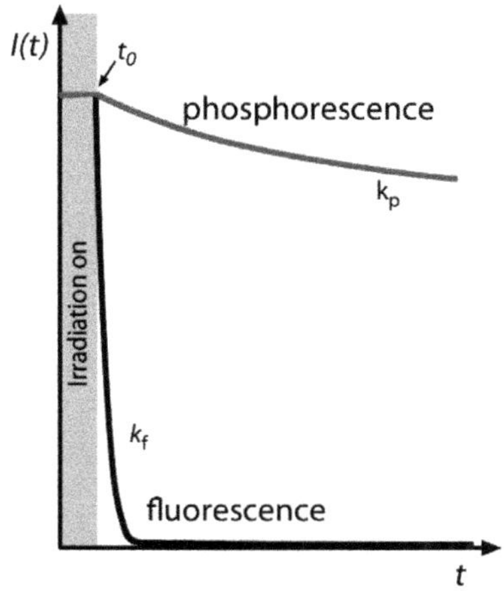

Figure 2.44 Typical time-resolved exponential decays for fluorescence and phosphorescence with rate constants k_f and k_p, respectively. Since phosphorescence is usually much slower than fluorescence, $k_f >> k_p$. The gray-shaded zone represents the time of the irradiation pulse.

It should be underlined that the above equation refers to a single emissive process of an isolated species or an ensemble of emitters decaying through the same deactivation channel. However, outside ideal conditions, it often happens that not all the excited emitters constituting a material decay through the same pathway, in particular because of sample inhomogeneity and differences in the chemical environment. In such an instance, the observed emission decay becomes multiexponential as a consequence of the existence of different populations of emitters associated with different emission mechanisms:

$$I(t) = I(t_0)\sum_i a_i e^{-k_i \tau} \tag{2.90}$$

In this equation, the coefficients a_i represent the weight of each exponential decay. The ratio between the excited-state populations decaying through different deactivation channels is called the branching ratio.

2.3.5 Energy Transfer

In photophysics, energy transfer (ET) is a nonradiative phenomenon by which an acceptor species in its ground state becomes excited at the expense of a donor species initially in an excited state without photon emission:[23]

$$\text{Donor}^* + \text{Acceptor} \rightarrow \text{Donor} + \text{Acceptor}^*$$

ET is one of the most important fundamental concepts in the field of luminescent molecular materials as it governs the nature and efficiency of the resulting optical functionality depending on the nature of the acceptor. In fact, energy can be transferred from an excited entity to either a nonluminescent species, causing quenching phenomena, or an entity that becomes consequently able to emit light. In this latter case the ET process takes the name of sensitization and is of particular interest as it allows significant control of light conversion between the radiation absorbed by the donor and that emitted by the acceptor. The extra energy acquired by a molecular species through photon absorption can be transferred either within different sites of the molecule itself (intramolecular ET) or to a second individual species (intermolecular ET).

In light of these considerations, the photophysical properties of molecular materials are mostly interpreted on the basis of ET processes that determine both nonradiative deactivation phenomena, usually detrimental, and the sensitization of emitting species, as schematically illustrated in Fig. 2.45.

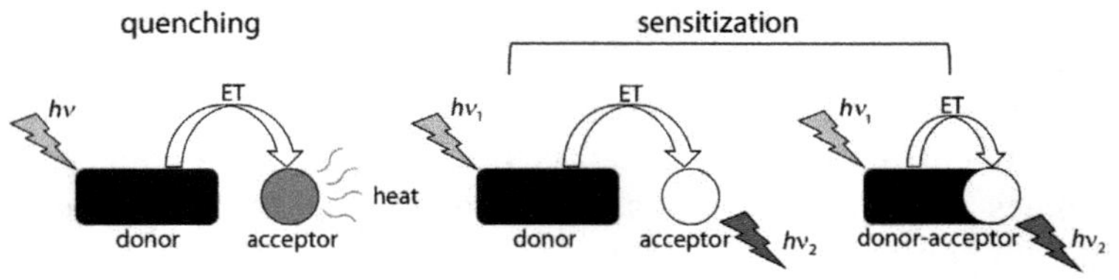

Figure 2.45 Schematic depiction of quenching (left) and sensitization (center and right) phenomena through energy transfer from a donor to an acceptor. Sensitization could occur either between individual entities (center) or at different sites of the same species (right).

---

[23]Sometimes, the term "radiative energy transfer" is used to indicate the trivial phenomenon by which a chemical species becomes excited by absorbing the light emitted by another species. This phenomenon occurs when the emitter and the absorber are at a distance longer than the radiation wavelength.

ET from a donor to an acceptor is only allowed if the resonance condition is satisfied, that is, if there is overlap between the emission band of the donor and the absorption band of the acceptor, as illustrated in Fig. 2.46.

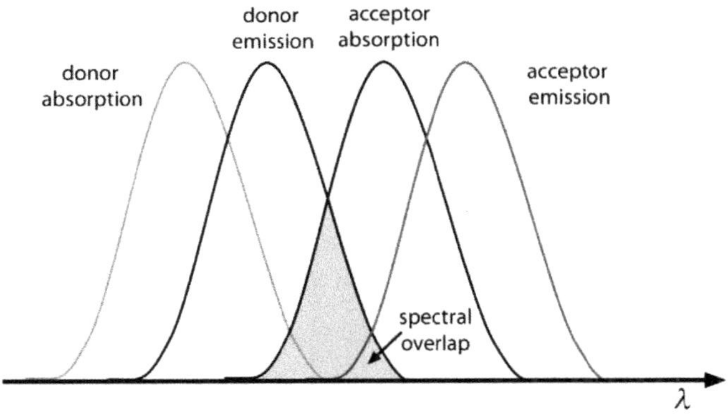

Figure 2.46 The resonance condition for donor–acceptor ET between a donor and an emitting acceptor species. Notice the large Stokes shift between the absorbed and the emitted radiation.

The entity of the spectral overlap between the donor (D) emission band and the acceptor (A) absorption band determines the efficiency of the ET and can be calculated from normalized spectral data through the spectral overlap integral J:

$$J = \int F_{\mathrm{D}}(\lambda)\sigma_{\mathrm{A}}(\lambda)\lambda^4 d\lambda \qquad (2.91)$$

where F_{D} and σ_{A} are the donor emission and the acceptor absorption cross sections, respectively.

ET is usually described according to two main different model mechanisms, both developed in the 1950s. These two models share important similarities in that they are both dependent on the donor–acceptor spatial separation and on the resonance condition. However, they are intrinsically different and have proved to be reliable and appropriate in the description of ET mechanisms under different circumstances. The main distinction between them relies on the fact that one model, that is, Dexter's or exchange mechanism (Dexter, 1953), involves electron exchange between the donor and the acceptor whereas the second, Förster's or dipolar mechanism

(Förster, 1959), only takes into consideration coulombic interactions. These models will be described in more detail in the following paragraphs.

2.3.5.1 Dexter's energy transfer

As mentioned, Dexter's mechanism of ET involves the exchange of one electron in an excited energy state of the donor (D) to an upper energy level of the acceptor (A), accompanied by the simultaneous transfer of a ground state electron of the acceptor to the ground state of the donor. This mechanism will leave the initially excited donor in the ground state and the acceptor in an excited state and could either involve singlet states only or lead to singlet-triplet conversion of the acceptor according to the scheme in Fig. 2.47.

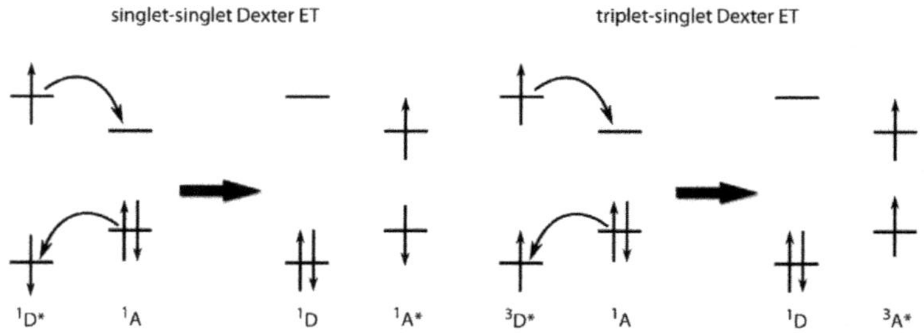

Figure 2.47 Scheme of Dexter's mechanism of energy transfer.

Since Dexter's mechanism requires an exchange of electrons, it is quite obvious that it implies a superposition of donor and acceptor wavefunctions, that is, orbital overlap must exist other than spectral overlap. For this reason, Dexter's mechanism is a short-range phenomenon that is usually efficient only for intramolecular ET or in the case of donor–acceptor species coming in close contact as in collisional processes. The rate constant k_{DET} of this process is then strongly distance dependent according to an exponential relation:

$$k_{\text{DET}} = Je^{\frac{-2r}{L}} \tag{2.92}$$

where r is the donor–acceptor distance, L is the sum of van der Waals radii, and J is the spectral overlap integral as defined by Eq. 2.91.

2.3.5.2 Förster's energy transfer

Förster's or dipolar mechanism of ET is commonly referred to as resonance energy transfer (RET) or Förster's resonance energy transfer (FRET), and it often takes on more importance with respect to Dexter's model since it can account for observed ET phenomena between species that do not have orbital overlap and are not in close contact. RET is a radiationless transmission of an energy quantum from a donor, previously excited, to an acceptor by dipolar through-space interaction without conversion to thermal energy, and without the donor and acceptor coming into kinetic collision, as schematically illustrated in Fig. 2.48.

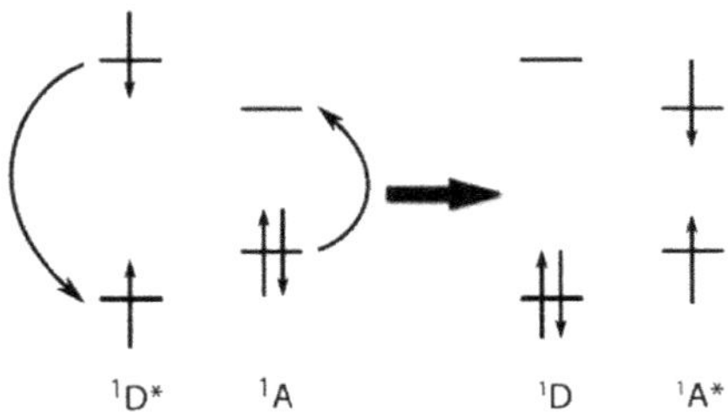

Figure 2.48 Schematic representation of Förster's resonance energy transfer.

Unlike Dexter's model, Förster's model only accounts for singlet-singlet mechanisms. Being a dipolar through-space process, which does not require a physical exchange of electrons, FRET is usually a very fast process, occurring in the picosecond timescale.

From a classic viewpoint the FRET process could be summarized in three basic steps: (i) the excited donor acts as an oscillating electric dipole, (ii) the acceptor species interacts with this oscillating field, and (iii) excitation transfer occurs from donor to acceptor without photon emission through a dipolar–dipolar interaction[24] through space. Förster's mechanism is then a long-range mechanism, and the rate constant k_{FRET} is inversely proportional to the sixth power of the donor–acceptor distance r according to the equation:

$$k_{\mathrm{FRET}} = \frac{\Phi_D \kappa^2}{\tau_r r^6} \frac{9 \ln 10}{128 \pi^2 N_A n^4} J \qquad (2.93)$$

[24]Actually, all electric field interactions, including quadrupolar or multipolar, can determine FRET.

where N_A is Avogadro's number; τ_r and Φ_D are, respectively, the radiative lifetime and the quantum yield of the donor in the absence of the acceptor (see Section 2.3.4.3); n is the refractive index of the medium; J is, once again, the spectral overlap integral defined by Eq. 2.91; and κ^2 is a geometrical factor that takes into account the relative orientation of transition dipole moments of the donor and the acceptor, as we shall discuss a little later in the text.

Equation 2.93 can be written in a simplified way by defining Förster's radius R_0 as:

$$R_0^6 = \frac{\Phi_D \kappa^2 9\ln 10}{128\pi^2 N_A n^4} J \tag{2.94}$$

so Eq. 2.93 becomes:

$$k_{FRET} = \frac{1}{\tau_r}\left(\frac{R_0}{r}\right)^6 \tag{2.95}$$

R_0 takes on an important meaning as it represents the shorter donor–acceptor distance at which RET becomes predominant over the radiative decay of the donor.

As mentioned before, the efficiency of the FRET depends on geometrical considerations beyond the donor–acceptor distance. In particular, the orientation factor κ^2 accounts for the fact that dipolar ET is more or less effective for different relative orientations of the dipoles of the donor and the acceptor.

Geometrically, κ^2 is defined as:

$$\kappa^2 = \left(\cos\theta_T - 3\cos\theta_D \cos\theta_A\right)^2 \tag{2.96}$$

where θ_T is the angle between the transition dipole moments of the donor D and the acceptor A and θ_D and θ_A are, respectively, the angles between the donor transition moment and the connecting vector **r** and between the latter and the acceptor transition moment, as illustrated in Fig. 2.49. The maximum value of κ^2 is obtained for collinear transition dipole moments (κ^2 = 4), whereas for parallel transition dipole moments κ^2 = 1. If dipole moments are oriented perpendicular to each other κ^2 takes the minimum value 0. In Fig. 2.49 examples of relative orientation of donor and acceptor dipole transition moments are shown. Normally, if the donor and acceptor species are independently free to rotate and translate in space, the number of the possible relative orientations between the corresponding dipoles is averaged. This is the typical case, and κ^2 takes the value 2/3. However, if the donor and the acceptor are not independent, for example, they lie in the same molecular architecture, a careful consideration of the orientation factor is clearly necessary.

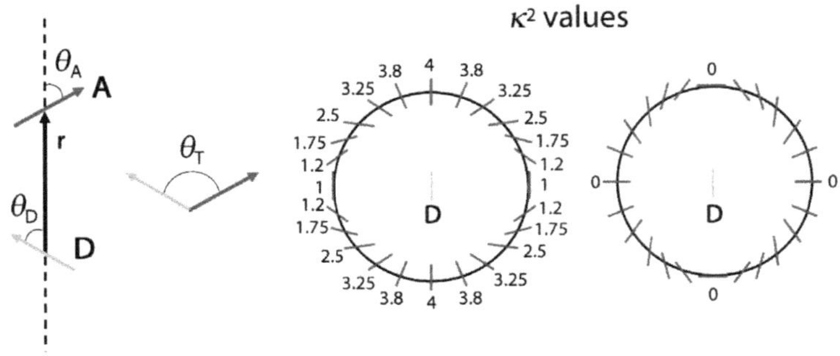

Figure 2.49 Illustration of the angles θ_D, θ_A, and θ_T (left). **D** and **A** are, respectively, the transition dipole moments of the donor and the acceptor, and **r** is the vector connecting their centers. κ^2 values for different orientations of the donor and acceptor dipole moments (right).

In the classical depiction of Förster's mechanism illustrated in Fig. 2.48, the acceptor becomes electronically excited at the expense of an initially electronically excited donor as a result of the interaction between the electric transition dipoles related to the two species. On the other hand, dipolar ET can also take place through a different mechanism between an electronically excited species and a vibrational oscillator, that is, a chemical entity possessing vibrational modes, such as, for example, water molecules undergoing stretching and bending vibrations. In this instance, the dipole moment associated with the vibrational mode of the acceptor is responsible for the ET in the condition that the frequency of the vibration is resonant with the frequency of the donor oscillator (cf. footnote 15).[25] Eventually, the acceptor becomes vibrationally excited as a result of energy absorption. This latter phenomenon

[25]This is, in practice, a derivation of the Bohr frequency condition described in Section 2.3.2.2. In this case, the source of excitation is the electronically excited donor, which is regarded as an emitting oscillator, and the species able to absorb energy is the vibrating chemical entity.

is of particular importance since it is often invoked to account for quenching mechanisms that are said to occur via vibrational (or phonon) deactivation.

2.3.6 Quenching

Quenching is, in general, a phenomenon by which an excited energy state of an emissive species becomes depopulated by means of nonradiative ET to another entity, would it be part of the same emitter or an individual species (cf. Section 2.3.4.1). Although this term has, strictly speaking, the general meaning described earlier, it is most often used to indicate *the decrease of emission intensity resulting from nonradiative deactivation induced by a nonemissive quencher*, and it will have this connotation in this book from now on, also with the view to distinguish it from *sensitization* (cf. Section 2.3.5).

2.3.6.1 Dynamic and static quenching

The nonradiative deactivation of excited states can occur through a variety of paths, involving electron transfer or ET. However, without entering into the details of quenching mechanisms, two general scenarios can be pictured.

Dynamic quenching is, in general, a collisional process occurring when the excited emitter M* interacts with the quencher Q, diffusing through the medium and providing an alternative nonradiative deactivation channel competing with radiative decay. This process results in the transient formation of a quencher in a "hot" energy level "Q," which subsequently dissipates the extra energy through heat, restoring the initial conditions with both M and Q in their ground states. Conversely, static quenching occurs when the quencher Q associates with the emitter M to form a nonluminescent species MQ before irradiation. Although both types of processes require the emitter M and the quencher Q to come into close contact, only static quenching necessitates some kind of chemical association between the two species. Dynamic and static quenching mechanisms are schematically illustrated in Fig. 2.50.

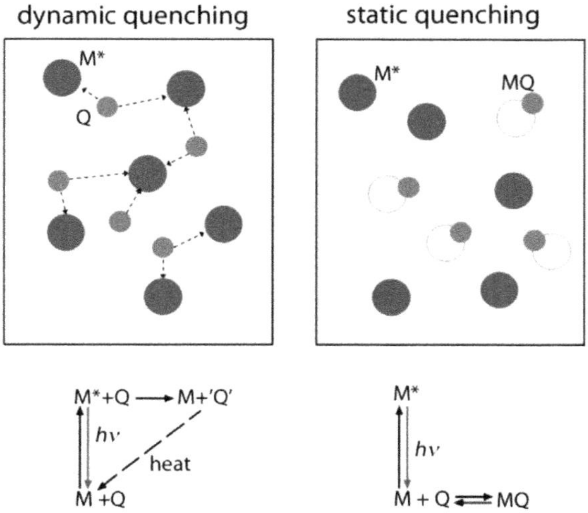

Figure 2.50 Schematic depiction of dynamic (left) and static (right) quenching.

2.3.6.1.1 *Dynamic quenching*

As already pointed out, quenching phenomena affect the efficiency of the emission process, and this would result in a decrease in the quantum yield. Being a process typically governed by diffusion and collisional events, dynamic quenching is usually very fast and can be associated with a kinetic rate constant, in turn, dependent on the concentration of the quencher Q, $k_q[Q]$. Therefore, recalling Eqs. 2.78 and 2.82, it is possible to write the expression for the quantum yield in the case of dynamic quenching by the species Q:

$$\Phi_r = \frac{k_r}{k_r + \sum k_{nr} + k_q[Q]} = \frac{k_r}{\frac{1}{\tau} + k_q[Q]} \tag{2.97}$$

Then, by denoting with the superscript 0 the quantum yield of the emitter in the absence of the quencher and dividing Eq. 2.83 by Eq. 2.97, we get:

$$\frac{\Phi_r^0}{\Phi_r} = 1 + k_q \tau [Q] \tag{2.98}$$

Since quantum yield is proportional to lifetime (Eq. 2.83), Eq. 2.98 could be written as:

$$\frac{\tau}{\tau_q} = 1 + k_q \tau [Q] \qquad (2.99)$$

where τ_q represents the observed lifetime in the presence of the quencher. Equation 2.99 clearly implies that dynamic quenching affects the efficiency of the emission by directly causing a shortening of the observed emission lifetime. A plot of the ratio of the observed lifetimes in the absence and in the presence of the quencher against the quencher's concentration will give a straight line of slope $k_q\tau$ and intercept 1, allowing for very easy evaluation of the quenching rate constant k_q. However, for practical reasons, since equipment for time-resolved spectroscopy is not always available, it is often preferred that data be reported as a function of intensity variation, taking into account that the emission intensity has a proportional relation to lifetime. In such an instance, the product $k_q\tau$ in Eq. 2.99 is conveniently replaced by the Stern–Volmer constant K_{SV} and the Stern–Volmer equation can be written as follows:

$$\frac{I^0}{I} = 1 + K_{SV}(Q) \qquad (2.100)$$

where I^0 and I are the intensities at the band maximum in the absence and the presence of the quencher and $K_{SV} = k_q\tau$ represents the slope of the plot I^0/I versus [Q], as reported in Fig. 2.51.

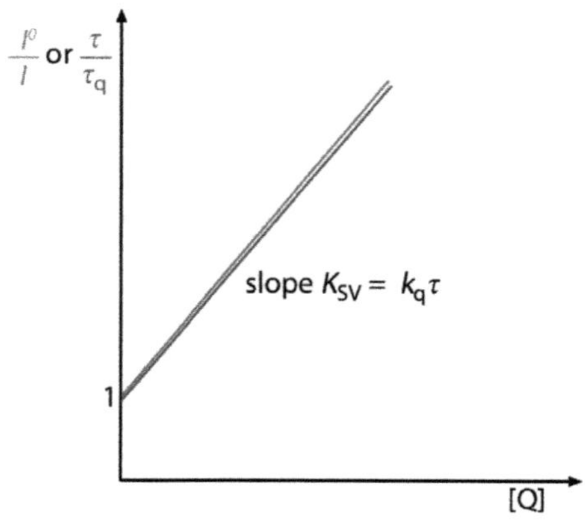

Figure 2.51 Typical plot for the Stern–Volmer equation.

It is evident that in the absence of time-resolved experiments, no direct information on the dynamics of the quenching process, in terms of the kinetic rate constants related to the emitter deactivation, can be obtained. Equation 2.100 and the related plot are nonetheless useful to quantitatively compare the efficiencies of different quenchers through the Stern–Volmer constant parameter K_{SV}.

2.3.6.1.2 *Static quenching*

As already pointed out in Section 2.3.6, static quenching occurs when the emitting species M and the quencher Q associate through some kind of chemical interaction, either through covalent or noncovalent bonding, forming a new species MQ which is no more able to emit light, according to the reaction shown in Fig. 2.50. Therefore, differently from dynamic quenching, which is a kinetic diffusional phenomenon, static quenching is instead governed by a chemical equilibrium whose equilibrium constant is:

$$K_Q = \frac{[MQ]}{[M][Q]} \tag{2.101}$$

If we take into account that the initial concentration of the emitter (in the absence of the quencher) equals the sum [MQ] + [M], and considering that the intensity of emission is directly proportional to the concentration of the emitter, it is then possible to derive the equation:

$$\frac{I^0}{I} = 1 + K_Q[Q] \tag{2.102}$$

which is formally identical to the Stern–Volmer equation, so a plot of I^0/I versus [Q] would give a straight line of intercept 1 and slope K_Q, as reported in Fig. 2.52. Unlike dynamic quenching, however, static quenching does not affect emission lifetimes, as the fraction of free emitters is not influenced by the presence of nonemitting species and is able to luminesce as if there is no quenching species present. Therefore, in this instance, τ/τ_q would be a constant with respect to the quencher concentration.

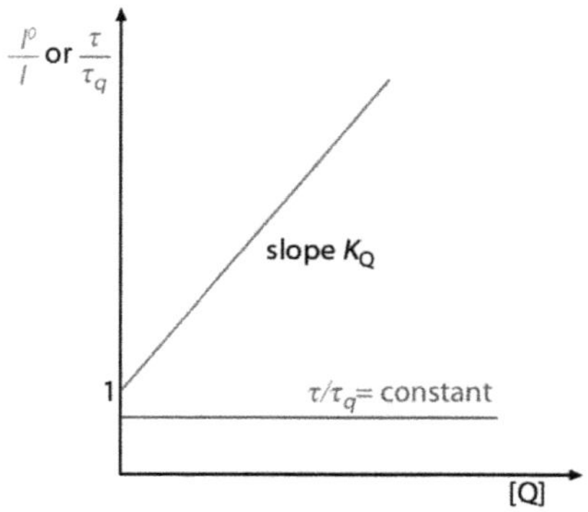

Figure 2.52 Typical plot of intensity ratio versus quencher concentration in the case of static quenching.

2.3.6.1.3 *Dynamic or static quenching and energy transfer*

The similarity of Eqs. 2.100 and 2.102 does not allow easy discrimination between the type of quenching mechanism, either dynamic or static, involved in the studied system. In principle, only time-resolved studies would unravel the actual mechanism. However, another basic physical parameter could be of help in solving this issue, that is, the temperature dependence of emission intensity. It is in fact evident that an increase in the temperature will favor collisional processes, leading to an increase in the rate of dynamic quenching. On the contrary, the binding between the emitter and the quencher tends to become weaker on raising the temperature, leading to an increase in emission intensity. On this basis, it would be expected that the slope of the I^0/I versus [Q] plot will increase in the case of dynamic quenching and, conversely, decrease for the static quenching, as illustrated in Fig. 2.53.

Sometimes quenching phenomena result from a combination of dynamic and static quenching mechanisms. In such an instance, deviations from the straight-line trend of intensity ratio against quencher concentration are observed.

There is another important mechanism of quenching that cannot be classified either as dynamic or as static, that is, Förster-type quenching via ET, already described in Section 2.3.5.2. In fact, while FRET indeed induces a shortening of the emission lifetime, characteristic of dynamic quenching, it does not necessarily involve collisional processes. Actually, FRET often occurs between different sites of the same molecular architecture, which are therefore associated as in static quenching. A typical case is vibrational quenching through chemical groups of the ligands in a molecular complex, as we shall see through actual examples in Chapter 3. Similar considerations can be also made for Dexter-type ET processes, which are, however, less often invoked to account for quenching phenomena.

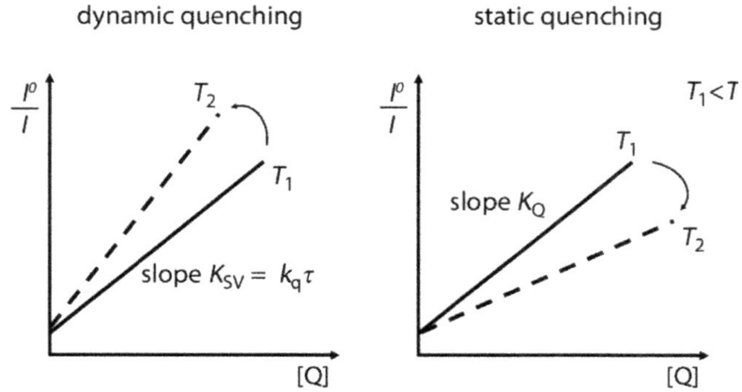

Figure 2.53 Effect of temperature increase on dynamic and static quenching.

2.3.7 Metal-Centered Electronic Transitions

We have so far taken into consideration general photophysical processes mostly related to organic molecules. Before entering a general description of the types of electronic transitions commonly encountered in the field of molecular materials, which will be dealt with in Chapter 3, it is worth providing some basic notions of electronic transitions involving the metal counterpart of coordination complexes with organic ligands.

On the basis of the selection rules discussed in Section 2.3.2.3, metal-centered intrashell d–d and f–f electronic transitions are, in principle, forbidden by parity (or Laporte's) rules. However, we have also seen that, under certain conditions, selection rules can be somewhat relaxed, as for example, by vibronic coupling. In molecular complexes, the effect of the ligand orbitals' interaction with the atomic orbitals (AOs) of the coordinated metal induces the mixing of orbital wavefunctions and allows metal-centered transitions to become observable up to the point that they determine the properties of a molecular material.

Unlike the discussion made for polyatomic organic molecules, metal-centered transitions are dealt with by, first, taking into consideration the model for free atoms and subsequently introducing the effects of the chemical environment as perturbations of the said model. In Section 2.1.6.1 we have seen that Hund's rules are used to determine the lowest-energy level of an atomic electronic configuration according to the Russell–Saunders coupling scheme. Indeed, many different terms are permitted for each electronic configuration depending on the arrangement of the electrons, and interelectronic repulsion determines their energy separation (term splitting). Each term consists of a set of multiple microstates, and its degeneracy is equal to $(2\mathbf{S} + 1)(2\mathbf{L} + 1)$, whereas additional $2\mathbf{J} + 1$ m_J sublevels must be taken into account where spin–orbit coupling becomes relevant (in heavy ions). In metal complexes, the ligand/crystal field effects induce a removal of the degeneracy of the level manifolds, resulting in a wealth of energy levels (Stark splitting). It is then possible to envisage three main cases where the magnitude of each of the above-mentioned forces determines the sequence and entity of the separation of the spectroscopic terms:

- Crystal/ligand field > term splitting > spin–orbit coupling
- Term splitting > crystal/ligand field > spin–orbit coupling
- Term splitting > spin–orbit coupling > crystal/ligand field

The first two cases are typical of light transition metal ions where the spin–orbit coupling is almost negligible or much lower than coulombic interactions, as a result of which, the splitting of the energy levels actually depends on the strength of the crystal/ligand field. The third case, instead, is typical of lanthanide ions, as will be better clarified in Section 2.3.7.2.

The determination, calculation, and assignation of energy levels in transition metal–containing compounds is indeed a complicated and perhaps tedious task, which was nevertheless first undertaken in the 1940s, with the aim to properly describe electronic transitions in such compounds. Fortunately, the availability of calculated energy level diagrams for transition metal ions nowadays saves us a considerable amount of work and allows us to predict, understand, and interpret the optical properties of metal compounds.

2.3.7.1 d–d transitions

d orbitals are strongly directional along a spatial reference system made by the three Cartesian coordinates, and the effect of the ligand field is in general remarkable. So, as the metal ion interacts with an external matrix, the degeneracy of the orbitals is removed and their energy sequenced according to the symmetry of the chemical environment. For example, in purely octahedral complexes (O_h symmetry point group), d orbitals are split into two sets of different energies: three stabilized t_{2g} orbitals (d_{xy}, d_{xz}, and d_{yz}) and two destabilized e_g orbitals (d_{z^2} and $d_{x^2-y^2}$). This would allow electronic transitions to occur between different, energetically separated, d orbitals,[26] as shown in Fig. 2.54 for a metal of d^1 configuration in an octahedral field as an example. The entity of such splitting is of the order 10^3–10^4 cm^{-1}, therefore d–d transitions usually spectroscopically fall in the visible region.

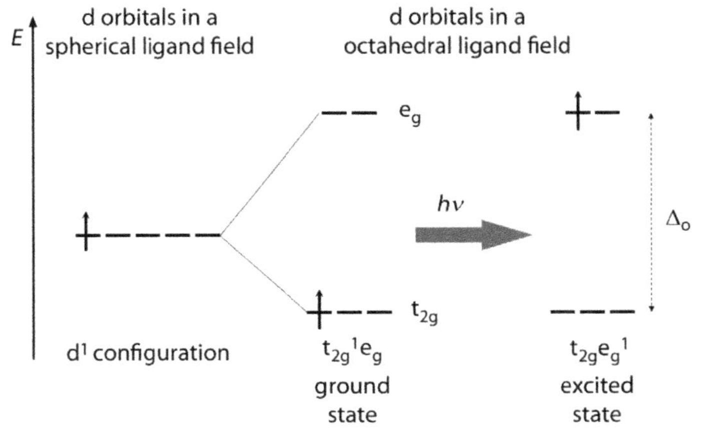

Figure 2.54 Electronic transition for a d^1 metal ion in an octahedral ligand field. The energy gap between t_{2g} and e_g orbitals is indicated as Δ_O.

[26] To be more precise, it should, however, be pointed out that, as already discussed in Section 2.3.3.5 for the general case of organic molecules, electronic transitions in d metal ions do not necessarily imply an electron transfer between two different orbitals, but states simply differing for spin multiplicity can be involved. This point will be better clarified in Chapter 3.

As already stated in Section 2.3.2.3, either vibronic or static distortion induced by the interaction of the vibrating ligands with the transition metal d orbitals results in a small symmetry disruption and p-d orbital mixing, allowing for partially permitted intrashell d–d transitions. This is, of course, entirely true for complexes of noncentrosymmetric point groups such as tetrahedral geometries. However, among the possible symmetries of transition d-metal complexes, the octahedral geometry, which is perhaps the most common, is by far the most studied in regard to optical properties. The reason for this stems from the work of Giulio Racah in the 1940s (Racah, 1942), followed by that of Yukito Tanabe and Satoru Sugano in the 1950s (Tanabe and Sugano, 1954a, 1954b, and 1956), which made a fundamental contribution to the study of the optical properties of octahedral complexes, which will be briefly recalled herein.

The derivation of the term symbols according to the Russell–Saunders coupling scheme for the monoelectronic d^1 configuration, shown in Fig. 2.54, as well as for the "monohole" d^9, is trivial, as only one term, ^2D, can be assigned. For the other d^2–d^8 configurations, of course, more than one spectroscopic term exists, as summarized in Table 2.5.

Table 2.5 Term symbols for d metal ions as free ions

d^n	Free-ion term
d^0, d^{10}	^1S
d^1, d^9	^2D
d^2, d^8	^1S, ^1D, ^1G, ^3P, ^3F
d^3, d^7	^2P, $^2D(2)$, ^2F, ^2G, ^2H, ^4P, ^4F
d^4, d^6	$^1S(2)$, $^1D(2)$, ^1F, $^1G(2)$, ^1I, $^3P(2)$, ^3D, $^3F(2)$, ^3G, ^3H, ^5D
d^5	^2S, ^2P, $^2D(3)$, $^2F(2)$, $^2G(2)$, ^2H, ^2I, ^4P, ^4D, ^4F, ^4G, ^6S

In an octahedral symmetry, the free-ion terms are further split due to electrostatic interactions with the ligand field. For the d^1 system (as well as for the d^9), two terms, that is, the triply degenerate T_{2g} and the doubly degenerate E_g, originate from the term ^2D. For the rest of the terms related to the other electronic configurations

(Table 2.5), the corresponding irreducible representations in the symmetry point group O_h are summarized in Table 2.6.[27]

Table 2.6 Splitting of terms of d metal ions in O_h symmetry

Free-ion term	Terms in O_h point group
S	A_{1g}
P	T_{1g}
D	$E_g + T_{2g}$
F	$A_{2g} + T_{1g} + T_{2g}$
G	$A_{1g} + E_g + T_{1g} + T_{2g}$
H	$E_g + 2T_{1g} + T_{2g}$
I	$A_{1g} + A_{2g} + E_g + T_{1g} + 2T_{2g}$

Tanabe and Sugano (1954a, 1954b, 1956) constructed useful diagrams correlating the energy of each microstate, arising from the splitting of the free-ion terms in an octahedral symmetry, with the strength of the ligand field, which determines the energy separation between the two sets of d orbitals (Δ_O, cf. Fig. 2.54). Both these parameters are expressed in terms of Racah parameter *B*, which takes into account the interelectronic repulsion and the electron cloud delocalization on the ligand (nephelauxetic effects). Where appropriate, the Tanabe–Sugano diagrams also account for the change in energy of the spectroscopic terms when passing from a low-spin (ls) to a high-spin (hs) configuration. This results in an abrupt discontinuity in the slope of the energy curve, as seen, for example, in the case of the d^6 configuration whose diagram is reported in Fig. 2.55.[28]

It is then evident that while for the d^1 and d^9 configurations only one absorption transition is possible, that is, ${}^2T_{2g} \rightarrow {}^2E_g$, which is spin-allowed, for the other configurations, transitions may occur between ground and excited states of the same or of different spin

[27]For other molecular geometries, the terms in the corresponding symmetry point group can be retrieved by finding the reducible representations of the electronic configurations through the direct product of the single monoelectronic functions and subsequently retrieving the corresponding irreducible representations.

[28]Complete Tanabe–Sugano diagrams can be found in many advanced inorganic chemistry textbooks, such as *Molecular Symmetry and Group Theory* (Carter, 1998).

multiplicity. In the case of the d^5 configuration, all the possible transitions would involve a change in the spin, implying very weak absorption bands.

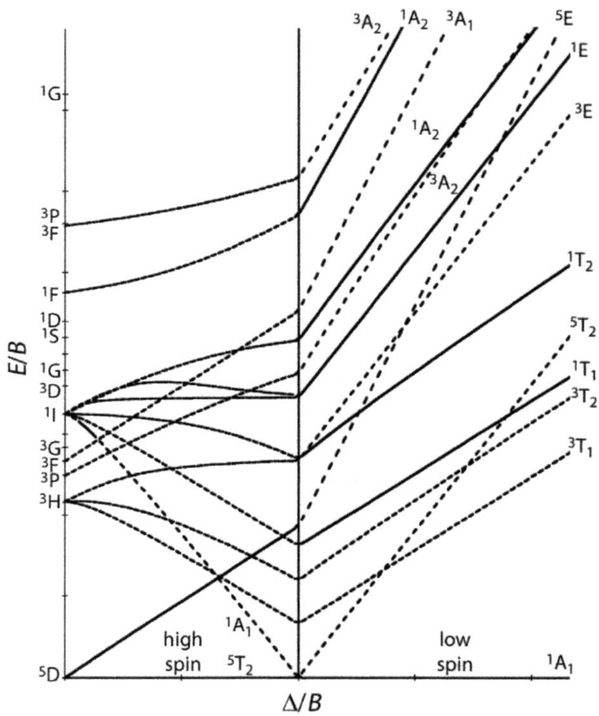

Figure 2.55 Tanabe–Sugano diagram for a d^6 configuration in an octahedral field. The subscript "g" was removed as all the spectroscopic terms are understood to be *gerade*. The solid lines represent accessible states for electronic transitions with $\Delta S = 0$, whereas the dotted curves refer to high-energy states forbidden by selection rules.

2.3.7.2 f–f transitions

Before beginning the discussion on the optical transitions typical of the lanthanides, it is important to point out that, whereas transition metals often display d configurations in a bunch of possible oxidation numbers, the chemistry and spectroscopy of f elements are in practice solely dominated by the trivalent state, corresponding to a

$4f^n$ configuration.[29] Therefore the following discussion will strictly refer to triply charged lanthanide ions.

Contrary to the case of transition metals, where d orbitals significantly interact with the crystal/ligand field, partially allowing intrashell transitions that are largely influenced by the chemical environment, f–f transitions typical of lanthanide ions are hardly affected by the external matrix. This is due to the fact that the outer 5s and 5p shells of lanthanide ions have a larger radial dispersion than 4f orbitals, providing an effective shield against external interactions. Thus, lanthanide ions retain to a great extent the spectroscopic and magnetic properties of the free ions and the electronic spectra of lanthanide compounds feature a line-like nature. It also becomes evident that Laporte's rule is hardly relaxed for lanthanide compounds and f–f transitions are usually characterized by weak intensities and long emission lifetimes, up to several milliseconds.

In principle, the energy levels of lanthanide ions would be appropriately described on the basis of the j–j coupling scheme (cf, Section 2.1.6), which takes into greater account the spin–orbit interaction, which is particularly relevant in these elements, rather than the interelectronic repulsion. Even better, the intermediate coupling scheme, in which coulombic and spin–orbit interactions have the same importance and terms with the same **J** but different **L** and **S** could mix, could be used. However, for convenience, the less sophisticated Russell–Saunders scheme is most often adopted as a convenient way to label the energy levels of a lanthanide ion. According to Hund's rule, the ground term for the free ion is derived from the largest values of **L** and **S** quantum numbers, while the spin-orbit coupling quantum number, **J**, ranges from $\mathbf{L} - \mathbf{S} \leq \mathbf{J} \leq \mathbf{L} + \mathbf{S}$ and takes, for the fundamental level, the value $\mathbf{J} = \mathbf{L} - \mathbf{S}$ for $n < 7$ (La–Eu) and $\mathbf{J} = \mathbf{L} + \mathbf{S}$ for $n \geq 7$ (Gd–Lu). The energy separation between the $^{(2S+1)}\Gamma$ terms induced by the coulombic effect is about 10^5–10^4 cm^{-1}, while the energy gap between the **J** levels induced by spin–orbit coupling in the heavy lanthanide ions is relevant and of the order of 10^3 cm^{-1}. Therefore, since this energy gap is much greater than the

[29]Deviations from the +3 oxidation state can be only observed in special cases where the empty $4f^0$ (Ce, Pr, Nd), half-filled $4f^7$ (Tb, Dy, Sm, Eu), or filled $4f^{14}$ (Yb) configurations can be reached. Nonetheless +4 and +2 charged ions are seldom thermodynamically stable, with the exception of Ce^{IV}.

thermal energy at ambient temperature ($k_B T \simeq 200$ cm^{-1}), **J** levels other than the fundamental level are hardly populated (although exceptions exist). The entity of further splitting induced by crystal/ligand field effects (Stark splitting) into 2**J** + 1 m_J sublevels is very small (~10^2 cm^{-1}) in comparison to transition d metal ions, and this typically translates barely into the resolution of lanthanide ions' spectra into a fine structure,[30] while the positions of the absorption and emission bands do not change significantly upon coordination. A schematic example of the energy-level splitting induced by different forces (interelectronic repulsion, spin–orbit coupling, and crystal/ligand field) for the case of Eu^{III} (4f^6) is shown in Fig. 2.56.

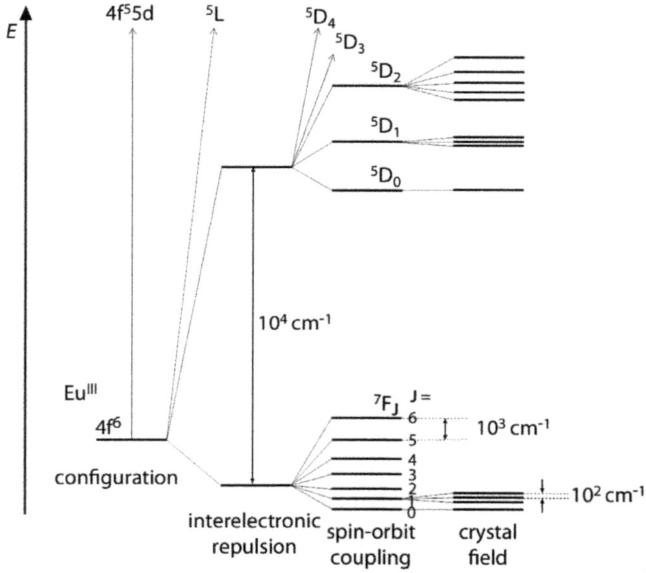

Figure 2.56 Typical splitting of energy levels for Eu^{III}. The interelectronic repulsion separates the energy levels of the ion's configuration into terms, whereas relevant spin–orbit coupling splits them further into *levels*. Weak crystal/ligand field effects induce a removal of the degeneration of each level into 2**J** + 1 m_J sublevels. Reprinted from BÜnzli (1987), Copyright (1987), with permission from Elsevier.

[30]Actually, the extent of the separation between m_J sublevels is of utmost importance to observe peculiar magnetic behavior in specifically designed lanthanide complexes having a symmetrical arrangement along an easily recognizable molecular axis, as will be highlighted in Section 3.1.7.2.

The sequence of the first energy levels for each lanthanide ion is reported in the Dieke diagram, shown in Fig. 2.57 for some significant examples.

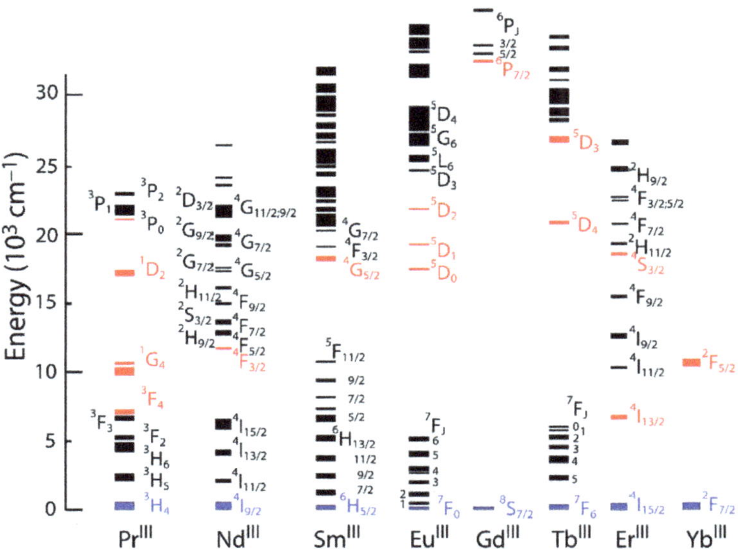

Figure 2.57 Simplified Dieke diagram showing the energy levels of some luminescent lanthanide ions. Emissive levels are highlighted in red, and ground states are depicted in blue.

It can be noticed that the energy levels of lanthanide ions are often markedly subdivided into two main groups consisting of closely spaced levels. This energy gap is the reason why many lanthanide ions give rise to intra-atomic emission arising from decay transitions usually occurring from the lowest level of the higher multiplet, depicted in red in Fig. 2.57, populated by nonradiative cascade decay from the excited upper levels, to the ground manifolds, represented in blue in Fig. 2.57. Of course, the entity of the energy gap between the level manifolds determines the spectral region of the emitted light. Differently to d metal ions, whose electronic transitions usually fall in the visible region, lanthanide emission can span the entire electromagnetic spectrum, depending on the nature of the metal. In general, two classes of 4f transitions giving rise to radiative emission can be identified:

- Transitions between states with different spin multiplicity (spin-forbidden), which usually give rise to energy-level-splitting UV and visible emissions (e.g., $^5D_{J'} \rightarrow {}^7F_J$ for Eu^{III} and Tb^{III})
- Transitions between levels that differ only in the value of the total spin–orbit coupling constant **J**, which correspond to NIR emission (e.g., $^4I_{13/2} \rightarrow {}^4I_{15/2}$ for Er^{III})

In principle, according to the International Union of Pure and Applied Chemistry's (IUPAC's) rules, one can refer to the first class of emissive transitions by using the term "phosphorescence" and to the second with the term "fluorescence." However, since emission arises from f orbitals of heavy lanthanide ions with high spin–orbit effects (**L** and **S** quantum numbers somehow lose their meanings),[31] some authors prefer to use the general term "luminescence" to indicate the radiative emission from these ions, so this definition will be adopted in this book from now on. Lanthanide ions with completely filled (Lu^{III}) or empty (La^{III}) 4f orbitals are not luminescent, whereas for the other ions emissions range from the UV to the NIR regions. Table 2.7 summarizes the most important emission lines of luminescent lanthanide ions, although it has to be remarked that this list is far than exhaustive as many ions have complicated spectra and undergo several radiative transitions corresponding to different spectral windows.

Table 2.7 Main luminescent transitions of lanthanide ions

Lanthanide ion	Transition	Emission wavelength (nm)	Emission color
Pr^{III}	$^1G_4 \rightarrow {}^3H_J$	1300	NIR
	$^1D_2 \rightarrow {}^3F_J$	890, 1060	NIR
	$^3P_0 \rightarrow {}^3H_J$	525-680	Orange

[31]Since for lanthanide ions the spin–orbit coupling **J** takes on a much greater importance over **L** and **S**, the spin rule as well as the Δ**L** rule are greatly lifted. This is even more convincing if we keep in mind that the large spin–orbit coupling in these ions actually induces a mix in the **L** and **S** states, as best described with the intermediate coupling scheme, and that Russell–Saunders terms are an oversimplified approximation. In principle, the Δ**J** rule, $\mathbf{J} = 0 \leftrightarrow \mathbf{J}' = 0$ forbidden, applies to intraconfigurational lanthanide transitions, but it may be nonetheless relaxed due to the mixing of closely spaced energy states.

Lanthanide ion	Transition	Emission wavelength (nm)	Emission color
Nd^{III}	$^4F_{3/2} \rightarrow {}^4I_J$	890, 1060, 1330	NIR
Sm^{III}	$^4G_{5/2} \rightarrow {}^6H_J$	590	Orange
Eu^{III}	$^5D_0 \rightarrow {}^7F_J$	620	Red
Gd^{III}	$^6P_{7/2} \rightarrow {}^8S_{7/2}$	312	UV
Tb^{III}	$^5D_4 \rightarrow {}^7F_J$	550	Green
Dy^{III}	$^4F_{9/2} \rightarrow {}^6H_J$	570	Green-orange
Ho^{III}	$^5F_5 \rightarrow {}^5I_J$	970, 1450	NIR
	$^5S_2 \rightarrow {}^5I_J$	540	Green
Er^{III}	$^4I_{13/2} \rightarrow {}^4I_{15/2}$	1530	NIR
Tm^{III}	$^1G_4 \rightarrow {}^3H_J$	475	Blue
Yb^{III}	$^4F_{5/2} \rightarrow {}^4F_{7/2}$	980	NIR

2.3.7.2.1 *The nature of lanthanide ions' transitions*

For a more complete picture of lanthanide ions' optical properties, it should be mentioned that, apart from crystal-/ligand-field-induced electric dipolar (ED) transitions, magnetic dipolar (MD) transitions are often recognizable in the spectra of these ions. Differently to ED transitions, MD transitions originate from the interaction of the 4f electrons with the oscillating magnetic vector of the incident light and are promoted by the even-parity magnetic dipole moment operator (cf. 2.3.2.3). As a consequence, MD transitions do not obey the same selection rules as ED transitions and are fully parity allowed. Similarly, electric quadrupolar (EQ) transitions are also allowed by parity. Although MD and EQ spectral bands are generally very weak, the existence of such transitions accounts for the appearance of the spectral lines of lanthanide ions despite the fact that ED transitions are, in principle, prohibited. In fact, back in 1937, Van Vleck (1937) concluded that the observed spectral features arise from a combination of ED, MD, and EQ transitions. Following the work of Judd (1962) and Ofelt (1962), who developed a theoretical model to predict the intensity of 4f transitions for lanthanide ions subjected to the influence of a ligand/crystal field, a revision of the main selection rules for lanthanide ions' transitions in molecular complexes was made, which is summarized in Table 2.8.

Table 2.8 Selection rules for lanthanide ions' transitions

Transition	S	L	J (no 0 ↔0)	Parity
ED	Δ**S** = 0	Δ**L** ≤ 6	Δ**J** ≤ 6 Δ**J** ≤ 2, 4, 6 (if J or **J**′ = 0)	Opposite
MD	Δ**S** = 0	Δ**L** = 0	Δ**J** = 0, ±1	Same
EQ	Δ**S** = 0	Δ**L** = 0, ±1, ±2	Δ**J** = 0, ±1, ±2	Opposite

Without entering into the details of *Judd–Ofelt theory*, for which the reader is referred to the original publications or to the numerous reviews and books on this topic, it is important to highlight that this model takes into account the admixing of states of opposite parity wavefunctions, including $4f^n4f^{n-1}5d$ configurations, to account for the intensity of lanthanide transitions. In other words, under the influence of a ligand/crystal field, some d orbital character can mix with f orbitals so that parity rules are greatly lifted and intrashell 4f transitions are more easily allowed. For this reason many authors prefer to include the term "forced" or "induced" when referring to ED transitions in lanthanide ions. It must be nonetheless underlined that, although this model works quite well for the interpretation of the optical spectra of a number of lanthanide compounds, it is, however, quite a simplified approximation that is grounded in some rough assumptions such as the degeneracy of the 2**J** + 1 m_J sublevels that are considered as equally populated, and several exceptions to the above-reported selection rules can be found.

Though weak, detectable pure MD lines are often very useful as internal spectral references as they are not influenced by coulombic interactions within the chemical environment, unlike ED bands, whose intensity and shape can be noticeably affected by the ligand/crystal field. In this regard, a third type of important transitions in lanthanide complexes is the so-called hypersensitive transitions, whose oscillator strength appears to be particularly sensitive to the nature of the coordination sphere. Their intensity is significantly increased as the symmetry of the complex is lowered or the polarizability of the ligands increased, making them a sort of probe of the metal surroundings. These transitions are induced ED transitions that obey the same selection rules as EQ transitions. Therefore they are often referred to as "pseudo-quadrupolar." They occur through a concerted mechanism where the ED component of the radiation induces a set of transient electric dipoles in the ligand environment, which couples to the 4f electron distribution via electrostatic quadrupolar (metal)–dipolar (ligand) interactions. The correct recognition of the nature of absorption and emission lines in lanthanide compounds is of crucial importance for correlating the structural/chemical features to the luminescence properties, as will be discussed with practical examples in Chapter 3.

2.3.8 Nonlinear Molecular Optics

Nonlinear optics (NLO) deals with the nonlinear optical response of matter upon interaction with light of high intensity. This is indeed a field of high complexity, and many nonlinear optical phenomena have been discovered and studied so far. However, a rigorous description of NLO is beyond the scope of this book, and the discussion will be limited to the main principles and phenomena of importance in the field of molecular materials, such as second and third harmonic generation (SHG and THG, vide infra) and two-photon (or multiphoton) absorption (TPA, vide infra), whereas the reader can find many other exhaustive and more specialized texts for deeper knowledge on this topic.

We have already pointed out at the very beginning of this section (Section 2.3.1) that the oscillating electric field of the electromagnetic radiation can interact with the electric field of atomic and molecular species as carriers of electron charge distribution, inducing a perturbation of the electron cloud. In the case of linear optical phenomena, as the ones that we have discussed so far, the electromagnetic radiation induces an instantaneous displacement (polarization) of the electron density, resulting in an induced dipole. Whenever the response is instantaneous, the induced polarization will have the same frequency and phase as those of the oscillating electric field of the incident radiation. If the intensity of the applied field is high enough, that is, when using intense power lasers, the response of the matter to the electromagnetic beam, that is, the induced oscillation of the electron cloud, goes beyond the linear regime and nonlinear optical effects become significant.[32]

[32]To envisage this concept in simple terms, one may recall the drag force in fluids, which has a linear dependence on the velocity in a laminar flow regime but is instead proportional to the squared velocity, that is, a nonlinear relationship, in the case of turbulent flow.

NLO phenomena were not experimentally evidenced until the development of efficient laser beams in the early 1960s, in particular by Kaiser and Garett (1961) for TPA and by Franken et al. (1961) for SHG.

From the point of view of the quantum theory, the main conceptual difference between LO and NLO basically relies on the fact that whereas linear processes involve real quantized excited states as a result of photon absorption, nonlinear phenomena are rather described to occur through "virtual" high-energy states, resulting from the perturbation of the electron distribution of polarizable atomic or molecular species and the consequent modification of the ground state by the strong electric field associated with a very intense laser beam (Fig. 2.58). In other words, the irradiated species reaches a higher-energy state, which corresponds to the combination of the unperturbed molecule with the energy of the absorbed photons. The as-generated, instantaneously polarized wave can be considered as arising from the mixing of the dipole moments of the ground and excited states of the system with the transition moment, and the related "virtual state" can be described by a linear combination of the corresponding wavefunctions. Since, in general, excited states have a different electronic distribution with respect to the ground state, their mixing ensures the occurrence of a variation in the electric dipole moment between the initial and the final state that is the fundamental requirement for photon absorption (cf. Section 2.3.2.2)

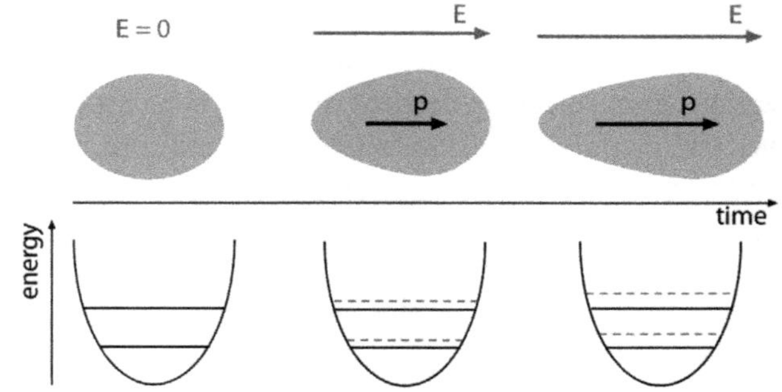

Figure 2.58 Schematic representation of electron density perturbation accompanied by polarization change and modification of energy levels (red dashed lines) for a molecular species under an external oscillating electric field, **E**.

From the above considerations it becomes also evident that, for observing NLO phenomena, the Bohr frequency condition (cf. Section 2.3.2.2), required for linear processes, no longer holds. In principle, light of any frequency can induce NLO processes as a consequence of the perturbation of the electron density of an unexcited entity. However, a certain dependence of observable NLO phenomena on the wavelength of the incident light still applies and, in the field of molecular materials, two main classes can be envisaged:

- **Nonresonant NLO phenomena**: These are mainly second and third harmonic generation processes (vide infra) that can be simply described as arising from the perturbation of the ground state of molecular or atomic species and can in principle occur for any irradiation wavelength.
- **Resonant NLO phenomena**: Multiphoton absorption processes that require an excitation radiation of multiple wavelengths with respect to that corresponding to the energy gap between the ground and the excited state of an atom or molecule belong to this class.

Both these mechanisms imply the conversion of two or more photons into a single photon of a different frequency. However, whereas nonresonant processes only involve virtual high-energy states as perturbation of the ground state, in the case of multiphoton absorption the irradiated species reaches its excited state and the resulting optical response is emitted light with the same characteristics as those of the linear processes (two-photon induced fluorescence [TPIF]). These two classes of mechanisms are schematically illustrated in Fig. 2.59.

Despite the existence of different NLO phenomena, they all have the common characteristic of a nonlinear (power) dependence on the intensity of the applied field (i.e., the intensity of the light source). This concept can be better expressed through simple mathematical relations. In the linear optical regime, that is, at low irradiation intensities, the induced polarization **p**, that is, the displacement of the electron density of an atomic or molecular species under light irradiation, is linearly proportional to the intensity of the applied electric field **E** through the linear polarizability α according to the scalar relation $\mathbf{p} = \alpha\mathbf{E}$. However, when high-power laser sources are used, the induced polarization is no more linearly dependent on the intensity of the applied field but can be expressed in the form of a Taylor series, where quadratic and cubic terms appear:

$$\mathbf{p}(\mathbf{E}) = \alpha\mathbf{E} + \beta\mathbf{E}^2 + \gamma\mathbf{E}^3 + \cdots \quad (2.103)$$

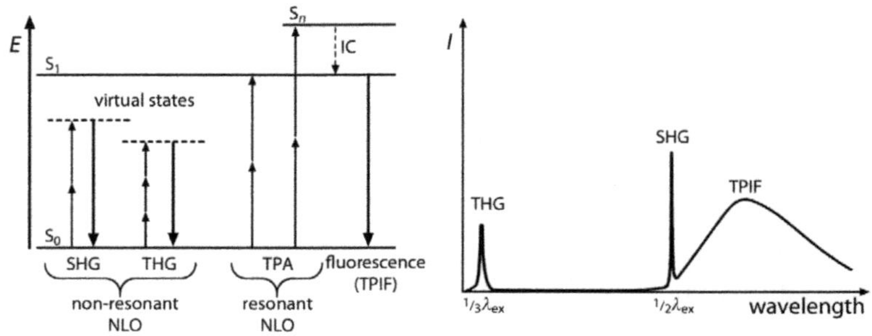

Figure 2.59 Simplified Jablonski diagram depicting nonresonant SHG and THG NLO phenomena, resonant TPA, and two-photon induced fluorescence (TPIF) (left). A typical spectrum representing resonant and nonresonant NLO phenomena in an organic molecule as a result of irradiation with a light source of wavelength λ_{ex} (right). Notice that, whereas SHG and THG signals are almost monochromatic, TPIF has the typical shape of a fluorescent band for an organic molecule.

In this equation β and γ are, respectively, the first (or quadratic) and second (or cubic) hyperpolarizabilities and account for second harmonic generation (SHG) and third harmonic generation (THG).[33] γ is also representative of TPA phenomena. Since their values are typically much smaller than α, NLO properties can be observable only for high values of **E**.

Figure 2.59 also helps explain why NLO processes are often exploited to achieve frequency doubling, one of the phenomena of utmost importance in photonics. This can be also very simply explained mathematically by considering that the electric field of a plane light wave of frequency ω is $\mathbf{E} = \mathbf{E}_0\cos(\omega t)$. Therefore, Eq. 2.103 can be rewritten as:

$$p(\mathbf{E}) = \alpha\mathbf{E}_0\cos(\omega t) + \beta\mathbf{E}_0^2\cos^2(\omega t) + \gamma\mathbf{E}_0^3\cos^3(\omega t)\cdots \quad (2.104)$$

Neglecting for a moment third-order terms and recalling that trigonometry tells us that $\cos^2(\omega t) = ½ + ½\cos(2\omega t)$, Eq. 2.104 becomes:

[33]More precisely, α, β, and, γ are polarizability tensors of the first, second, and third order, respectively. The tensor attribute of these variables accounts for the anisotropy of the electron cloud distribution in atoms and molecules.

$$p(\mathbf{E}) = \tfrac{1}{2}\beta\mathbf{E}_0^2 + \alpha\mathbf{E}_0\cos(\omega t) + \tfrac{1}{2}\beta\mathbf{E}_0^2\cos(2\omega t) \quad (2.105)$$

Thus, other than the frequency-independent first term, a new component of the radiation of frequency 2ω, doubled with respect to that of the incident beam ω, appears (see Fig. 2.60). Similarly, for third-order nonlinearities, it is easy to demonstrate, by applying trigonometry rules, that a term function of 3ω is included in the equation.[34]

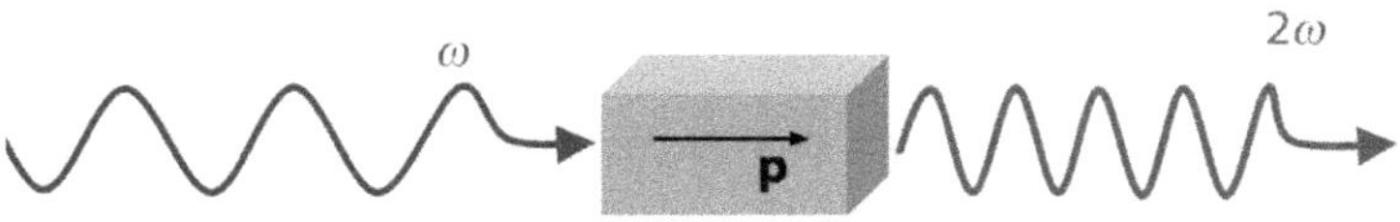

Figure 2.60 Second harmonic generation in a molecular material under irradiation of a laser beam of frequency ω.

However, not all systems can display second-order NLO properties, irrespective of the light source intensity. In fact, *the mandatory requirement for observing SHG is that the system must be noncentrosymmetric*. This can be easily demonstrated by taking into account that the symmetry inversion operation is an odd function[35] (cf. Section 2.3.2.3), so if applied to the polarization field (Eq. 2.103):

$$-\mathbf{p}(-\mathbf{E}) = -\alpha(-\mathbf{E}) - \beta(-\mathbf{E})^2 - \gamma(-\mathbf{E})^3 = \alpha\mathbf{E} - \beta\mathbf{E}^2 + \gamma\mathbf{E}^3 \quad (2.106)$$

It is then clear that Eqs. 2.103 and 2.106 hold together, that is, $\mathbf{p}(\mathbf{E}) = -\mathbf{p}(-\mathbf{E})$, only for $\beta = 0$, so second-order hyperpolarizability terms are null for centrosymmetric systems. On the other hand, no formal restrictions apply for third-order terms and linear processes follow the symmetry selection rules already described in Section 2.3.2.3.

---

[34]SHG is actually a special case of sum frequency generation (SFG) for which two photons of frequencies ω_1 and ω_2 sum up to generate a photon of frequency $\omega_3 = \omega_1 + \omega_2$. Similarly, difference harmonic generation (DHG) arises from a combination of photons that results in light with a frequency that is the difference between two frequencies of incident light: $\omega_3 = \omega_1 - \omega_2$. These phenomena can be only induced by irradiation with two different laser sources, whereas for SHG $\omega_1 = \omega_2$ only one input light beam is required. For this reason, SHG is more conveniently generated and is, by far, the most-studied NLO process in molecules.

[35]For spatial symmetry inversion operation in the three Cartesian coordinates: $g(x,y,z) = -g(-x,-y,-z)$.

The Molecular Origin of the Noncentrosymmetry Requirement in SHG

While the simple mathematical explanation provided through Eqs. 2.103 and 2.106 can be an immediate demonstration of the requirement of noncentrosymmetry for molecular systems to display SHG, it could nonetheless be useful to clarify the microscopic molecular origin of this rule from the chemist's point of view.

Suppose there is a system with a symmetric distribution of electron density under the effect of the symmetric oscillating electric field of the radiation. Charges will then start to be displaced by the same extent, $-x$ and $+x$, around their equilibrium position in an alternate motion in phase with the incident radiation against the restoring force of the positive nuclei, which has the form of Hooke's law $\mathbf{F} = -\mathbf{kx}$. This means that electrons will experience a harmonic potential around the equilibrium position and that the induced polarization will have a linear dependence on the applied electric field. On the other hand, whenever the initial distribution of electron density is asymmetric, which happens when there are atoms or molecular moieties with electron-withdrawing properties, charges will then be subjected to an anharmonic potential. The related diagram as a function of the electron displacement will then resemble that of the Morse curve on dependence of the nuclear coordinates and it will account for the asymmetry of the system. In other words, one would expect that, under the effect of the symmetric oscillating electric field of the radiation, electrons will be more easily displaced toward the most electron-withdrawing molecular moiety. Therefore, the response of the induced polarization wave will be asymmetric and no longer linear with the applied field. These concepts can be probably made more intuitive by visualizing them as in the figure given. The as-generated polarization wave, depicted in red in the lower-left corner of the figure, can be decomposed by a Fourier transform into a zero-frequency component and two other components at the fundamental and second harmonic frequency, corresponding to the three terms of the right side of Eq. 2.105, respectively.

In light of these considerations, it is then possible to conclude that to achieve asymmetry in the polarization response, asymmetry at the molecular level must exist, that is, the molecule must be noncentrosymmetric.

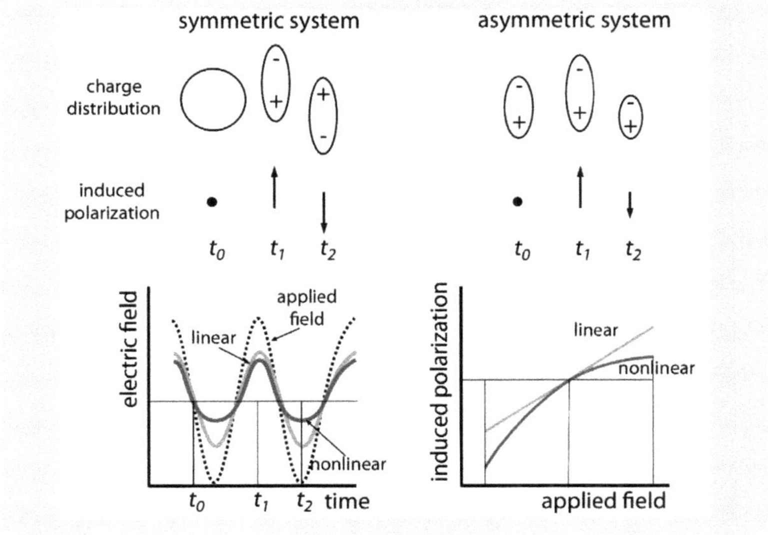

As a matter of fact, many other requirements must be fulfilled by molecular systems to display significant NLO properties. These points will be more specifically addressed when discussing related molecular materials in Chapter 3.

Chapter 3

Functional Molecular Materials

This chapter provides an overview of the most encountered types of functional molecular materials, presenting them as a function of their physical properties. They will be described taking into account the basic concepts summarized in Chapter 2, with the aim to point out the main chemical, structural, or electronic features that a certain system should have to show the desired properties. Only after that, for each property, will we present general archetype examples of functional molecular materials and a thorough discussion on the correlation of their features with the observed functionality. The chapter is organized into three main sections that describe the physical properties and the related archetype examples of molecular materials showing magnetic, conducting, and optical properties.

3.1 Magnetic Molecular Materials

The field of magnetic molecular materials comprises a wealth of molecular systems, ranging from mononuclear transition metals and lanthanide complexes to coordination polymers of various dimensionalities, polynuclear assemblies, organic radicals, and supramolecular architectures. Such chemical and structural variety is essentially related to the wide range of magnetic properties that they can exhibit. Indeed, the observed magnetic behavior results from specific electronic and structural features, which can

Functional Molecular Materials: An Introductory Textbook
Matteo Atzori and Flavia Artizzu

ISBN 978-981-4774-76-5 (Hardcover), 978-1-351-23366-8 (eBook)
www.panstanford.com

widely diversify across the range of magnetic molecular materials, making this field particularly varied in terms of classes of chemical compounds.

The interest in molecular materials showing magnetic properties is boosted by their potential practical uses, for example, as permanent magnets for transformer applications, magnetic switches activated by external stimuli (temperature, light irradiation, pressure, etc.), and high-density magnetic memories. Some properties shown by molecular materials are similar to those observed in classical inorganic compounds, with their own advantages and disadvantages, but more interestingly, other properties are completely different and observable only in molecular systems.

The most encountered magnetic properties and peculiar behaviors shown by magnetic molecular materials can be summarized as follows:

- Paramagnetism
- Ferromagnetism
- Antiferromagnetism
- Ferrimagnetism
- Spin crossover (SCO)
- Valence tautomerism (VT)
- Slow magnetic relaxation

These magnetism-related properties will be separately addressed in this section by correlating them to the specific electronic and structural features needed for a molecular compound to display such properties, through the presentation of archetype molecular materials.

3.1.1 Paramagnetism

Paramagnetism is the property of matter that represents the ability of a certain material to be attracted by a magnetic field. This property is observed in all materials having chemical species with unpaired electrons, which results in a permanent magnetic moment. The magnetic moments of a paramagnetic material are randomly orientated as a result of their thermal energy (Fig. 3.1) but they can be aligned by an external magnetic field. This means that paramagnetic materials show an induced magnetization in the presence of a field but do not retain magnetization once the field is removed.

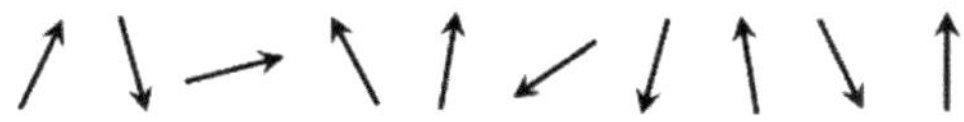

Figure 3.1 Schematic representation of the randomized orientation of the magnetic moments of a paramagnet caused by thermal energy.

Paramagnetism can be considered the simplest magnetic property of matter, and even if paramagnetic materials do not find as many applications as ferro- and ferrimagnets (vide infra), it is nevertheless at the basis of the most intriguing and more complex magnetic properties. The description of paramagnetism is easier with respect to other magnetic properties, and this is related to the fact that it is a property of isolated magnetic moments and does not involve any directional interactions between them.

Molecular materials exhibiting paramagnetism are molecules with an odd number of electrons so there is at least one unpaired electron, which gives rise to a net magnetic moment. This includes metal complexes of transition metals or lanthanide ions with partially filled d and f inner orbitals, respectively, and organic molecules with unpaired p electrons (usually one per molecule). Whereas transition metal or lanthanide complexes usually have localized unpaired electrons, since they reside in d or f orbitals of the metal ions,[1] paramagnetic organic molecules (organic radicals) present unpaired electrons with a high degree of electron delocalization. The stability of an organic radical is strongly correlated to the ability of the molecular structure to efficiently delocalize the unpaired electron. Those systems that do not have a structure suited for electron delocalization are in fact highly reactive species with short lifetimes that rapidly undergo dimerization or redox phenomena, such as polymerization reactions, photo-oxidations, and ozone depletion events.

While organic radicals usually have only one unpaired electron within a molecular structure, that is, they are $\mathbf{S} = 1/2$ species,

[1]In some metal complexes with non-innocent ligands, such as in d^8 metal dithiolenes, the unpaired electron can be delocalized between the metal core and the organic ligand moiety, thus representing an intermediate case between the localized case of most metal complexes and the delocalized case of organic radicals.

transition metals and lanthanide ions can bear up to five (**S** = 5/2) or seven (**S** = 7/2) unpaired electrons, respectively, thus providing higher magnetic moments. The total magnetic moment of a lanthanide ion can be even higher due to the presence of a non-negligible orbital contribution (cf. Section 2.1.6) and can be also associated with a high degree of magnetic anisotropy. The higher magnetic moment per ion and the stability of the odd electronic configurations of transition metal and lanthanide ions with respect to those of organic radicals further increased the interest in using these systems for the preparation of molecular materials.

Figure 3.2 shows three archetypes of paramagnetic building blocks:

- A molecular paramagnet based on a metal complex of a transition metal ion
- A molecular paramagnet based on a metal complex of a lanthanide ion
- A molecular paramagnet based on a stable organic radical

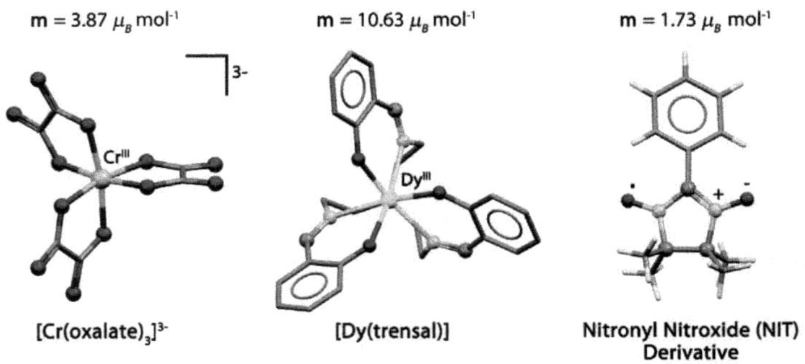

Figure 3.2 Three archetype examples of molecular paramagnets, with their calculated magnetic moment per molecule: (left) a transition metal complex, (middle) a lanthanide complex, and (right) an organic radical.

These three classes of compounds represent the fundamental building units for the realization of paramagnetic molecular materials. Any other system that satisfies these structural and electronic criteria can be, in principle, categorized under these classes.

The properties of paramagnets have been already described in Chapter 2 as they represent the fundamental basis of molecular magnetism. However, it can be useful to summarize here the most salient features. Molecular paramagnets with localized magnetic moments show magnetic properties that are unaffected by temperature variations, unless any other more complicated phenomena like ferromagnetic or antiferromagnetic interactions or phase transitions are involved. These latter phenomena, if present, are usually observed at low temperature (indicatively $T < 150$ K). Thus, in a high-temperature regime ($T > 150$ K), a pure paramagnetic behavior can be easily observed. Temperature-independent magnetic properties are most often encountered in transition metal complexes and organic radicals but are not commonly observed in the case of lanthanide compounds. In the latter case the χT value shows strong deviations as the temperature decreases. This is due to the fact that the thermal depopulation of low-energy excited states (easily accessible for lanthanides) induces a variation of the expected value of the paramagnetic susceptibility in the low-temperature regime (Fig. 3.3).

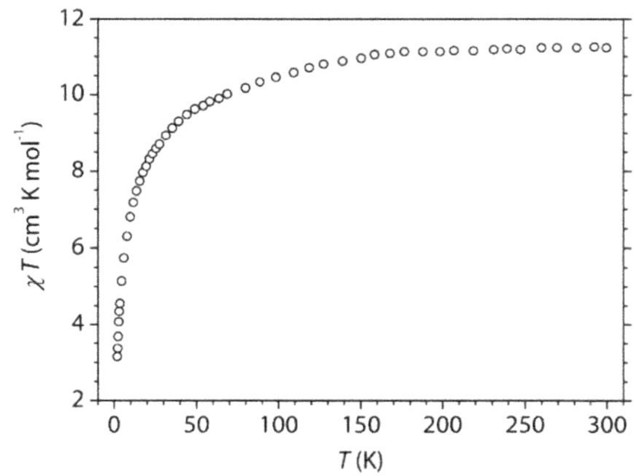

Figure 3.3 Thermal variation of the χT product for a generic lanthanide-based complex with $\mathbf{L} \neq 0$. The thermal depopulation of the high-temperature populated states results in strong deviations from the constant value expected in the case of $\mathbf{L} = 0$ already below 150 K. Note the drop from ca. 11.5 cm^3 K mol^{-1} at 300 K to ca. 3.0 cm^3 K mol^{-1} at 2.0 K.

The magnetic field dependence for pure paramagnets is well described by Eq. 2.53 (Section 2.1.10), where the expression of the Brillouin function is written in terms of atomic spin angular momentum **S** in the case of spin-only approximation (first-row transition metal ions and organic radicals) and in terms of atomic total angular momentum **J** in the case of not negligible spin–orbit coupling (second- and third-row transition metal ions and lanthanides).

As far as the application of paramagnetic building blocks for the preparation of functional molecular materials is concerned, we can say that whereas lanthanide-based complexes are better suited for single-molecule properties such as slow magnetic relaxation (cf. Section 3.1.7) as a consequence of the large ground spin state and the magnetic anisotropy resulting from significant spin–orbit coupling, transition metal-based complexes can be exploited as molecular building blocks for the preparation of materials showing cooperative magnetic properties. This originates from the more extended nature of the d orbitals with respect to the f ones, which favors magnetic exchange interactions throughout proper bridging ligands (superexchange). This aspect is fundamental to the design of molecular materials showing cooperative magnetic properties and will be encountered often throughout this book. Organic radicals can also be exploited for the preparation of magnetic materials with cooperative magnetic properties, thanks to the high mobility of the unpaired electron over the molecular structure, which can favor superexchange interactions (cf. Section 2.1.12). In fact, when they are used as bridging ligands between paramagnetic metal centers, their effectiveness in mediating superexchange magnetic interactions is higher than that of bridging ligands having an even electron configuration. However, these species can have the disadvantage of easily undergoing dimerization phenomena, with concomitant electron pairing and loss of magnetic properties. Moreover, the relatively low number of stable organic radicals toward oxidation or reduction makes them a restricted class of compounds with respect to paramagnetic metal-based complexes.

Even if the magnetic properties of metal complexes, either transition metals or lanthanides based, are mostly determined by the nature of the involved metal ion, that is, its electronic configuration, subtle changes can be nonetheless achieved by playing with the nature of the coordinating ligands. Several ligand field effects may strongly influence the paramagnetic properties of a coordination complex, for example:

- The determination of the high-spin (*hs*) or low-spin (*ls*) state of the metal ion[2]
- Electron transfer from a redox-active ligand to the metal or vice versa in systems exhibiting VT (cf. Section 3.1.6)
- Electron density modulation of the metal ion as a function of the electron-withdrawing or electron-donating nature of the ligands and different degrees of orbital mixing between the metal and the ligands
- Specific metal coordination geometries induced by multidentate ligands having a rigid structure, which induce a variation of the d-orbital splitting and the resulting electron configuration of the coordinated metal ion

Some of these effects can be so significant that important changes in the physical properties can result. Materials that show such properties are not considered simple paramagnets anymore and are categorized under specific classes. This is the case, for example, for those systems exhibiting SCO and VT that will be described in detail a little later in this chapter.

3.1.2 Ferromagnetism

Ferromagnetism is the property of matter that represents the ability of a certain material to undergo spontaneous magnetization below a critical temperature. For temperatures below the Curie temperature, the magnetic moments of a ferromagnetic material are parallelly aligned within microscopic portions of its magnetic structure called magnetic domains (Fig. 3.4), giving rise to what is commonly called ferromagnetic ordering. This results from a magnetic exchange energy between them, which favors alignment of the magnetic moments with respect to the thermal randomization typical of paramagnetism.

Figure 3.4 Schematic representation of a parallel-aligned orientation of magnetic moments within a ferromagnet as a result of magnetic exchange interactions.

[2]The most dramatic manifestation of this ligand field effect are the *hs*–*ls* transitions shown by those systems that are called SCO complexes (cf. Section 3.1.5).

This behavior can be explained in a phenomenological way by the Weiss field interaction, which was originally suggested to explain the temperature dependence of the paramagnetic susceptibility of certain materials (Jiles, 1998). As the temperature of a ferromagnet is increased, the thermal energy increases while the magnetic exchange interaction energy between magnetic moments remains unaffected. At a critical temperature T_c, the randomizing effect of the thermal energy overcomes the aligning effect of the interaction energy, and above this temperature the magnetic state becomes disordered. If the magnetic moments are localized to the atomic or ionic sites, the magnetic interaction between them can be described in the same manner as that discussed in Section 2.1.13 for the simplest case of two interacting magnetic ions. In this case the magnetic superexchange interaction is extended to an ideally infinite number of atoms arranged in a 3D or 2D lattice instead of a discrete and isolated system (Fig. 3.5).

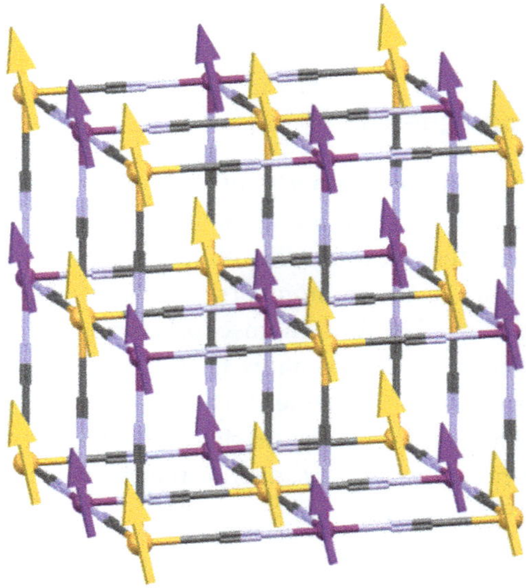

Figure 3.5 Schematic representation of a portion of a 3D molecular lattice where the magnetic moments (indicated with arrows) of each paramagnetic metal ion (yellow and purple balls) are parallelly aligned toward the same direction. Superexchange interactions are mediated by bridging ligands (blue and gray sticks) along the three space dimensions.

The number of atoms interested in these interactions, and consequently the diverse magnetic structural dimensionality, clearly defines the difference between ferromagnetic (or antiferromagnetic) ordering and coupling. The former are long-range interactions extending over space and encompassing a large number of units while the latter are short-range interactions most often encountered in discrete molecular units.

The presence of relevant exchange interactions between magnetic moments leads to the introduction of the Weiss constant θ to take into account that thermal energy is not enough to cause the random paramagnetic orientation of the magnetic moments for temperatures below the critical temperature T_c. The analytical expression that quantifies this behavior is the modification of the Curie law in the Curie–Weiss model described in Section 2.1.8.

The thermal variation of the susceptibility for a ferromagnet is already reported in Fig. 2.4, where it is highlighted that its thermal dependence below T_c varies in a more complicated way than that of a paramagnet. While for paramagnets it is known that the thermal variation of χ varies with T^{-1} according to the Curie law, the χ dependence on T for a ferromagnet is not usually described through an analytical expression since ferromagnetic susceptibility can also assume different values at a given temperature (vide infra).

Ferromagnets and ferrimagnets (cf. Section 3.1.4) represent to date the most important classes of magnetic materials for technological applications. In engineering, ferromagnets are used in view of their high permeability, which enables high magnetic inductions on applying modest magnetic fields, their ability to retain magnetization and thereby act as a field source, *etc.*

The most important property of ferromagnets is their high relative permeability. The permeability of a ferromagnet is not constant as a function of the magnetic field in the way it is for paramagnets. For this reason, in order to provide a full account of the magnetic properties of a ferromagnet, it is necessary to measure the magnetization **M** as a function of the magnetic field **H** over a continuous range of values to obtain a hysteresis curve (Fig. 3.6).

From magnetization measurements as a function of the applied field it can be seen that a ferromagnet in its initial state is not magnetized. The application of an external magnetic field **H** causes magnetization in the field direction. If **H** is increased indefinitely, the magnetization reaches saturation up to a value designated as $\mathbf{M}_S$. This value represents a state where all the magnetic moments within the material are aligned in the direction of the magnetic field. The saturation magnetization depends only on the magnitude of the atomic magnetic moments **m** and the number of atoms per unit volume or per mole as described in Section 2.1.11 and as stated by Eq. 2.53. Once magnetized, if a ferromagnet is subjected to the action of an external magnetic field that is in the opposite direction, the value of the magnetization does not assume the initial zero value at zero field. This nonzero value represents the remanent magnetization $\mathbf{M}_r$. To reduce the magnetization of a ferromagnet to zero it is necessary to apply a reverse magnetic field of strength $\mathbf{H}_C$, called coercive field, which quantifies the tendency of the material to retain its magnetization (coercivity). Materials characterized by high coercive fields are called hard magnets, while those exhibiting low coercive fields are soft magnets.

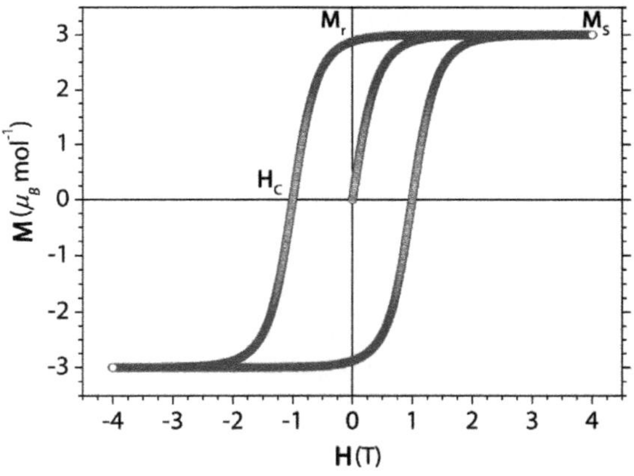

Figure 3.6 A typical hysteresis loop for a ferromagnetic material: $\mathbf{M}_S$ = saturation magnetization, $\mathbf{M}_r$ = remanent magnetization, and $\mathbf{H}_C$ = coercive field. Note that experimental hysteresis plots for real materials can show deviations from this idealized picture obtained through data simulation.

The suitability of a ferromagnetic material for technological applications is essentially evaluated from the information emerging from its hysteresis loop. For example, transformer applications, that is, increase or decrease of the alternating voltages in electric power applications, require materials showing high permeability and low hysteresis losses for efficient energy conversion. Instead, permanent magnets require high remanence and coercivity in order to retain the magnetization as long as possible.

From the considerations reported so far, it emerges that the molecular materials showing ferromagnetic ordering below a certain critical temperature T_c are those systems that satisfy the following electronic and structural features:

- A net magnetic moment per atom resulting from a noninteger electron configuration (as described for paramagnetic materials)
- An extended 2D or 3D structure required to give rise to those portions of the magnetic structure, the magnetic domains, where the magnetic moments can align in a parallel way
- An efficient pathway for the establishment of superexchange magnetic interactions between magnetic moments localized to the metal ion sites provided by proper bridging ligands
- Superexchange interactions stronger than the thermal energy in a discrete temperature range

The molecular systems that fully satisfy these conditions are 2D or 3D magnetic coordination polymers. In these systems an infinite number of paramagnetic building blocks, as for example the $[Cr(oxalate)_3]^{3-}$ complex reported in Fig. 3.2, connected through coordination bonds by means of paramagnetic metal ions of different nature, for example, Mn^{II}, can show ferromagnetic ordering (Fig. 3.7) below a critical temperature T_c (Tamaki et al., 1992).

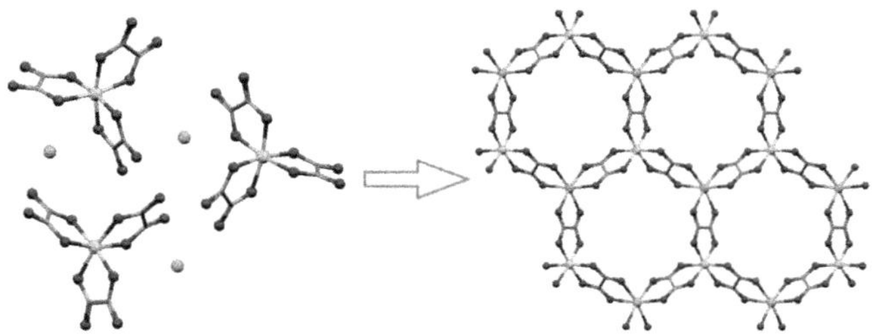

Figure 3.7 Representation of the molecular building blocks (left) responsible for the formation of a 2D molecular network through self-assembling (right) where ferromagnetic ordering can be observed due to the close contact between paramagnetic ions connected by bridging ligands (herein oxalate).

Additional Considerations

Metal ions: Although, in principle, ferromagnetic ordering can be achieved in self-assembled homopolynuclear structures, molecular systems based on metal ions of the same nature most often show antiparallel alignment of the magnetic moments (which is energetically favored), leading to antiferromagnetic, instead of ferromagnetic, ordering (cf. Section 3.1.3).

Dimensionality: One can wonder why 1D coordination polymers are not mentioned here even though they satisfy the same requirements though with a lower degree of dimensionality. This is due to the fact that magnetic ordering is a cooperative phenomenon for which the magnetic exchange interactions should extend over the three dimensions of the real space. Actually, 2D materials are placed at the limit of what can be classified as magnetic ordering since the interactions involve only two dimensions. Nonetheless, we will consider 2D materials as molecular ferromagnets since, despite theory, they exhibit exactly the same properties as 3D materials do, though with some limitations due to the less extended structure (vide infra).

This example clearly shows the potential of molecular chemistry for the preparation of magnetic molecular materials. This synthetic strategy, commonly known as the "complex as ligand" approach, applies to a wide range of paramagnetic building blocks where both the metal and the ligands can be varied and some classes of ligands can be further functionalized by changing the substituents. Furthermore, these building blocks can be combined with different partially coordinated or uncoordinated paramagnetic metal ions (both d and f) to give molecular architectures with various dimensionalities. This provides the opportunity to obtain a wide number of different systems showing tailored magnetic properties that most often can be predicted on the basis of (i) the electron configuration of the coordinated metals, (ii) the selected coordination geometry, and (iii) preliminary magnetic studies on the extent of the mediated superexchange interactions.

Following these design principles various classes of molecule-based ferromagnets have been prepared and structurally and magnetically characterized. Among them, some archetype examples are the family of 2D oxalate-based ferromagnets, already cited above (cf. Fig. 3.7), and the family of 3D ferromagnets based on the Prussian blue structural motif called Prussian blue analogs.

The family of oxalate-based ferromagnets refers to coordination polymers characterized by a 2D layered structure generated by self-assembling of preformed tris-chelated trianionic metal complexes and naked divalent metal ions in a 1:1 stoichiometric ratio. The global charge of these coordination polymers is thus −1 and is balanced by the presence of a monopositive countercation. The cation is placed between the layers and does not only have the mere function of counterbalancing the charge but also plays an important structural role in templating the overall crystal structure. This means that although the nature of the cation can be varied, not all counterions can be used to prepare these materials.

The monoanionic nature of the magnetic layers makes these systems particularly appealing because, in addition to the ferromagnetic properties, a cation bearing another functionality can be introduced. This aspect will be thoroughly discussed in Chapter 4.

It has been established that the nature of the cation does not significantly affect the resulting magnetic properties, which are strictly correlated to the 2D nature of the magnetic layers. Indeed, from the magnetic point of view, oxalate-based molecular magnets show relatively high ordering temperatures (in the 6–48 K range) (Launay and Verdaguer, 2014). This results from the restricted dimensions of the oxalate bridging ligand, which provides an efficient pathway for superexchange magnetic interactions while maintaining the paramagnetic metal ions in very close proximity. Such a simple structural moiety, unfortunately, does not offer great opportunities for chemical tunability. Indeed, only the replacement of the oxygen donor atoms with sulfur atoms (dithio- or tetrathioxalate) has been realized. However, the number of transition metal ions that can be coordinated by the oxalate is large, and by combining different pairs of metal ions an extensive modulation of magnetic properties can be achieved. Indeed, depending on the coordinated metal ions, ferromagnetic or ferrimagnetic ordering has been observed (cf. Section 2.1.12), with characteristic ordering temperatures and coercive fields (Table 3.1).

Unfortunately, both these magnetic parameters suffer from the 2D nature of these materials, as already pointed out previously (see the box titled "Additional Considerations"). The absence of long-range magnetic ordering extended in the three space dimensions does not allow communication between the paramagnetic metal centers strong enough to increase further the ordering temperature and the coercive fields. The latter are in fact usually small (less than 0.1 T, tesla); thus these materials are classified as soft ferro- and ferrimagnets.

Table 3.1 Type of magnetic ordering and ordering temperatures for selected compounds of formula $[(n\text{-Bu})_4N][M^{II}M^{III}(ox)_3]$ (Launay and Verdaguer, 2014)

Compound	**Type of ordering**	T_c **(K)**
$[(n\text{-Bu})_4N][Mn^{II}Cr^{III}(ox)_3]$	Ferromagnetic	6
$[(n\text{-Bu})_4N][Fe^{II}Cr^{III}(ox)_3]$	Ferromagnetic	12
$[(n\text{-Bu})_4N][Co^{II}Cr^{III}(ox)_3]$	Ferromagnetic	10
$[(n\text{-Bu})_4N][Ni^{II}Cr^{III}(ox)_3]$	Ferromagnetic	14
$[(n\text{-Bu})_4N][Cu^{II}Cr^{III}(ox)_3]$	Ferromagnetic	7
$[(n\text{-Bu})_4N][Mn^{II}Fe^{III}(ox)_3]$	Ferrimagnetic	25–28
$[(n\text{-Bu})_4N][Fe^{II}Fe^{III}(ox)_3]$	Ferrimagnetic	33–48
$[(n\text{-Bu})_4N][Ni^{II}Fe^{III}(ox)3]$	Ferrimagnetic	28

ox = oxalate

Another peculiarity of the crystal structure of these compounds comes from the stereochemical configuration of the tris-chelated building blocks. In principle, if a racemic mixture of the Λ and Δ enantiomers is combined with the bivalent metal ions, two structures of different topologies can be obtained: a 2D structure is attained if enantiomers of opposite chirality associate with each other; instead, if the two enantiomers self-assemble independently, a chiral 3D structure made by dodecagons results (Fig. 3.8) (Pilkington and Decurtins, 2002). The preference for one of the two topologies is dictated by the nature of the countercation.

It is worth noting that, despite the higher dimensionality, the relative positions of the metal ions with respect to each other in this 3D structure do not necessarily favor an increase of the ordering temperatures, that are, in general, higher for the compounds showing 2D structures. The main reason for this is that an inefficient pathway comes from the manifold of inequivalent directions along which the superexchange interactions can propagate (Coronado et al., 2010).

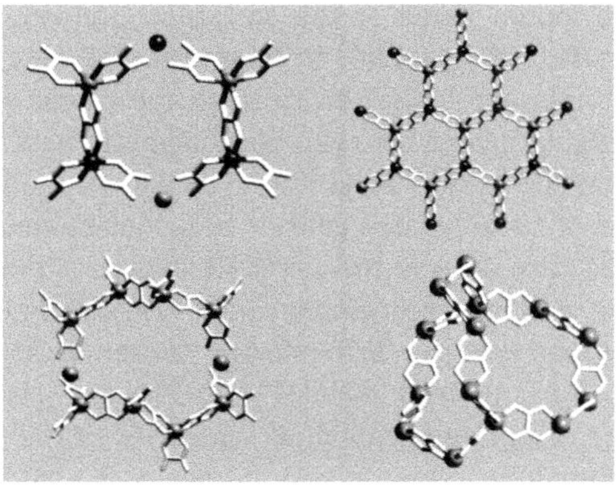

Figure 3.8 Two dimeric units formed by enantiomers of opposite chirality needed to obtain a hexagonal moiety (top left). A portion of the resulting honeycomb 2D layers (top right). Two tetrameric units formed by enantiomers having the same chirality needed to obtain a dodecagon (bottom left). A portion of the resulting 3D structure (bottom right). Adapted with permission from Pilkington and Decurtins (2002). Copyright 2002 Wiley-VCH.

These considerations are very instructive since they demonstrate that magnetic properties, and specifically the strength of the magnetic ordering, are very difficult to predict as they are affected by several electronic and structural parameters, often not easily controlled.

To conclude, the family of oxalate-based magnets represents an archetype example of molecule-based ferromagnets where the magnetic properties are modulated by the nature of the coordinated metal ions. Such systems are one of the first families of coordination polymers showing long-range magnetic ordering, both ferro- and ferrimagnetic, and have promoted the field of molecular magnetism as they have been taken as reference examples for at least two decades to prepare similar systems based on other types of organic ligands. Their great relevance is also associated with the input given to the field of multifunctional materials due to the peculiar 2D and anionic nature of the magnetic layers that allows chemists to place between the layers several types of cations and thus to prepare a multitude of molecular materials showing a combination of, or interplay between, different physical properties (see Chapter 4).

Another class of molecule-based ferromagnets worth mentioning is the family of compounds called Prussian blue analogs. Prussian blue is the common name of the dark-blue pigment of chemical formula $Fe^{III}_4[Fe^{II}(CN)_6]_3{\cdot}xH_2O$. It presents a 3D structure where Fe^{III} and Fe^{II} ions are connected through the cyano ligand, which acts as a bridging ligand between paramagnetic ions. Compounds of formula $AM^{III}[M^{II}(CN)_6]$ or $M^{III}[M'^{III}(CN)_6]$, where A is an alkali metal ion and M^{III} and M^{II} two generic hexacoordinated trivalent or divalent metal ions, present the same 3D structural motif as that of Prussian blue, and are accordingly called Prussian blue analogs. It can be mentioned that the formulae are written as $AM^{III}[M^{II}(CN)_6]$ or $M^{III}[M'^{III}(CN)_6]$, instead of $A[M^{III}M^{II}(CN)_6]$ or $[M^{III}M'^{III}(CN)_6]$, because these compounds can show vacancies of the $[M^{II/III}(CN)_6]$ units. Thus, the stoichiometric ratio between the M^{III} ion and the $[M^{II/III}(CN)_6]$ unit is not necessarily 1:1 (vide infra).

These molecule-based compounds can be prepared in a way similar to what is described above for oxalate-based compounds by following the complex-as-ligand approach. In this case the molecular building block is a hexacyanidometallate(II) anion that can further coordinate six naked M^{II} or M^{III} ions along the bond directions imposed by the hexacoordination of the metal ion and the linear molecular geometry of the ligand (Fig. 3.9). This results in a 3D structure in which the alkali metal, eventually needed to counterbalance the exceeding negative charge, is placed in the cubic cavities of this coordination polymer.

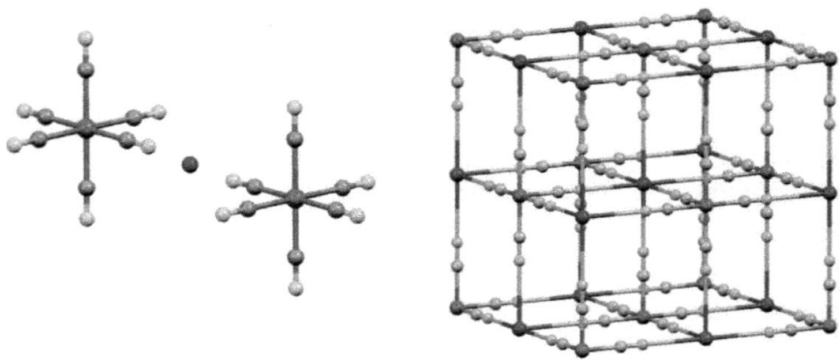

Figure 3.9 Two hexacyanidometallate anions that approach a naked metal ion along the direction imposed by the preferred octahedral geometry of the coordinated metal ion (only two out of six building blocks are reported for graphical clarity) (left). The 3D structure of a generic Prussian blue analog. obtained as a result of this molecular self-assembly (right). The alkali metal ions are omitted for graphical clarity.

The magnetic properties of these materials can be modulated by changing the nature of the coordinated metal ions and their stoichiometric ratios. Also in this case, although the ligand provides an efficient short pathway for magnetic exchange interactions, its unsophisticated nature does not allow structural modifications on the coordinating moiety. Nonetheless, ferro- or ferrimagnetic ordered materials can be obtained as a function of the nature of the coordinated metal ions (Fe^{III}, Cr^{III}, V^{III}, Mn^{II}, Fe^{II}, etc.) and in accordance with the *Goodenough–Kanamori–Anderson rules* described in Section 2.1.12. The ordering temperature can range from 5.6 K, as found for the original Prussian blue compound, to room temperature and above (376 K) by rational chemical adjustment of the stoichiometric ratio of the coordinated metals (Table 3.2).

Table 3.2 Type of magnetic ordering and ordering temperatures for selected compounds of formula $AM^{II}[M^{III}(CN)_6]$ or $M^{II}[M^{III}(CN)_6]$ (Launay and Verdaguer, 2014)

Compound	Type of ordering	T_c (K)
$Fe^{III}[Fe^{III}(CN)_6]_{3/4}\cdot 3.7H_2O$	Ferromagnetic	5.6
$Mn^{II}[Mn^{IV}(CN)_6]$	Ferrimagnetic	49
$Ni^{II}[Cr^{III}(CN)_6]_{2/3}\cdot 4H_2O$	Ferromagnetic	53
$Cu^{II}[Cr^{III}(CN)_6]_{2/3}\cdot 5\text{-}6H_2O$	Ferromagnetic	66
$Mn^{II}[Cr^{III}(CN)_6]_{2/3}\cdot 5\text{-}6H_2O$	Ferrimagnetic	66
$CsNi^{II}[Cr^{III}(CN)_6]\cdot 2\text{-}4H_2O$	Ferromagnetic	90
$CsMn^{II}[Cr^{III}(CN)_6]$	Ferrimagnetic	90
$V^{IV}O[Cr^{III}(CN)_6]_{2/3}\cdot 4.5H_2O$	Ferrimagnetic	115
$Cs_2Mn^{II}[V^{II}(CN)_6]$	Ferrimagnetic	125
$Cs_{2/3}Cr^{II}[Cr^{III}(CN)_6]_{8/9}\cdot 40/9H_2O$	Ferrimagnetic	190
$Cr^{II}[Cr^{III}(CN)_6]_{2/3}\cdot 10/3H_2O$	Ferrimagnetic	240
$V^{II}{}_{0.42}V^{III}{}_{0.58}[Cr^{III}(CN)_6]_{0.86}\cdot 2.8H_2O$	Ferrimagnetic	315
$KV^{II}[Cr^{III}(CN)_6]$	Ferrimagnetic	376

In the Prussian blue compound the low value of the ordering temperature originates from the large distances (ca. 10.4 Å) between Fe^{III} (**S** = 5/2) paramagnetic ions, the Fe^{II} ions being diamagnetic as a result of the *ls* configuration (**S** = 0). Room-temperature magnetic ordering has been achieved by combining the Cr^{III} and V^{II}/V^{III} metal ions in a nonstoichiometric ratio in the compound $V^{II}_{0.42}V^{III}_{0.58}[Cr^{III}(CN)_6]_{0.86}$·$2.8H_2O$. It is worth noting that the nonstoichiometry of this compound places it at the limit of what can be considered a molecular material.

Despite the high ordering temperatures, Prussian blue analogs have three important drawbacks. They usually show low solubility, poor crystallinity, and low values of coercive fields. These issues highlight how challenging the field of molecular materials can be. Indeed, even if the 3D crystal structure of Prussian blue analogs allows for more favorable exchange interactions (propagating in the three space directions) with respect to the oxalate-based magnets, providing high ordering temperatures, their combination with functional cations is not feasible.

Although the compound $V^{II}_{0.42}V^{III}_{0.58}[Cr^{III}(CN)_6]_{0.86}$·$2.8H_2O$ was not the first room-temperature magnet being synthetized starting from molecular precursors, compounds belonging to the family of its analogs are of particular relevance in this field since they represent a significant example where the modulation of the magnetic properties has been realized on the basis of the developed theoretical models through proper chemical design.

We have so far described the magnetic ordering properties of coordination polymers. In such compounds the unpaired electrons, which are at the origin of the magnetic moments, are localized on d-metal ions and the magnetic interactions accomplished by the presence of bridging ligands. However, magnetic ordering can also be achieved on different classes of molecular compounds, for example, those based on organic or metallorganic radicals, in which the electrons are localized on p orbitals. Among them, two of the most famous examples are the compounds of formulae $[Fe^{III}(C_5Me_5)_2]^{\cdot+}[TCNQ]^{\cdot-}$ (Candela et al., 1979) and $[Fe^{III}(C_5Me_5)_2]^{\cdot+}[TCNE]^{\cdot-}$ (Miller et al., 1986). Both compounds are obtained by combining a radical cation, the metallorganic compound $[Fe^{III}(C_5Me_5)_2]^{\cdot+}$, and a radical anion, the organic radicals $[TCNQ]^{\cdot-}$ or $[TCNE]^{\cdot-}$ (TCNQ = tetracyanoquinodimethane; TCNE = tetracyanoethylene) (Fig. 3.10).

$[Fe^{III}(C_5Me_5)_2]^{\cdot+}[TCNQ]^{\cdot-}$ is the first example of a molecular metamagnet, that is, a material that undergoes a transition between two magnetic ordered states due to the presence of an external magnetic field. $[Fe^{III}(C_5Me_5)_2]^{\cdot+}[TCNQ]^{\cdot-}$ shows an antiferromagnetic/ferromagnetic transition at a critical magnetic field of 0.15 T at 2.55 K. Thus, this system is characterized by an antiferromagnetic ground state. The $[Fe^{III}(C_5Me_5)_2]^{\cdot+}[TCNE]^{\cdot-}$ system shows instead ferromagnetic ordering with a T_c of 4.8 K and a coercive field of 0.1 T at 2 K. In these two systems, the nature of the radical anion is responsible for the different magnetic behaviors. With the TCNQ radical the stabilized ground state is the antiferromagnetic one, whereas with the TCNE the major spin density due to the smaller structure and the same spin allows for the stabilization of the ferromagnetic ground state and the observation of spontaneous ferromagnetic ordering.

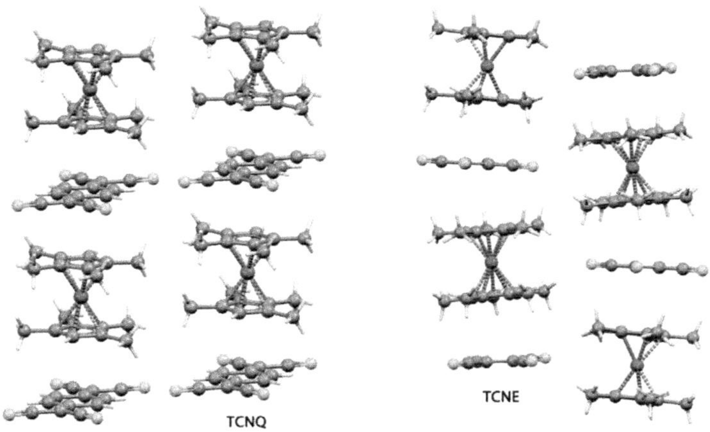

Figure 3.10 View of the crystal structure for the donor acceptor systems $[Fe^{III}(C_5Me_5)_2]^{\cdot+}[TCNQ]^{\cdot-}$ (left) and $[Fe^{III}(C_5Me_5)_2]^{\cdot+}[TCNE]^{\cdot-}$ (right).

Many other examples of molecule-based magnets are reported in the literature. However, a detailed review of such compounds is beyond the scope of this book and the reader is referred to more specialized and exhaustive texts on the topic (see suggested reads).

3.1.3 Antiferromagnetism

Antiferromagnetism is an ordering phenomenon in which magnetic moments of the same magnitude within portions of a magnetic structure are antiparallelly aligned below a certain critical temperature T_c (Fig. 3.11).

Figure 3.11 Schematic representation of an antiparallel-aligned orientation of the magnetic moments of an antiferromagnet as a result of their magnetic exchange interactions.

This phenomenon is, in origin, completely analogous to ferromagnetism in that they both originate from (super)exchange magnetic coupling interactions extended over a magnetic structure and only differ in terms of the type of alignment of the magnetic moments in the ordered state, leading to strong differences in the resulting magnetic properties. While a parallel alignment of the magnetic moments (ferromagnetic material) leads to spontaneous magnetization higher than that induced by an external magnetic field in paramagnetic materials, the antiparallel alignment of the magnetic moments (antiferromagnetic material) causes a suppression of the induced magnetization, and below T_c the magnetization of the material starts to decrease, reaching the zero value at the lowest achievable temperature.

We have already explained that the interactions between two magnetic moments aligned antiparallelly is energetically favored with respect to the parallel alignment. This explains why most of the materials based on the same type of magnetic atoms or ions behave as antiferromagnets or why simple dimeric compounds usually show antiferromagnetic instead of ferromagnetic coupling (cf. Section 2.1.13).

The thermal variation of the magnetic susceptibility above the critical temperature for an antiferromagnet can also be described by the Curie–Weiss law. However, since the magnetic exchange coefficient for antiferromagnetic coupling and ordering is negative, the Curie–Weiss law becomes:

$$\chi = \frac{C}{T + T_c} \tag{3.1}$$

For antiferromagnets T_c appears as a negative temperature below the absolute zero. Indeed, the plot of χ^{-1} against T intercepts the temperature axis in the negative region at $-T_c$. Below the corresponding absolute temperature value, the Néel temperature T_N, these materials undergo a transition from a disordered paramagnetic state to an antiferromagnetic ordered state. For temperatures lower than the Néel temperature, the susceptibility does not follow the Curie–Weiss law anymore and behaves in a more complex way as for ferromagnetism. The thermal variation of the magnetic susceptibility as a function of the temperature for a generic antiferromagnet is shown in Fig. 3.12.

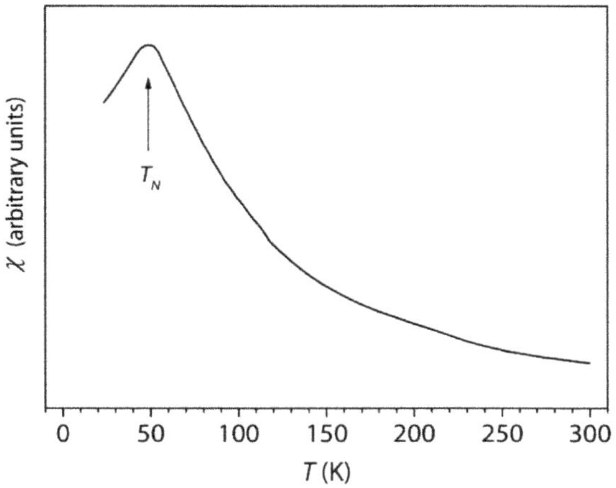

Figure 3.12 Thermal variation of the susceptibility for a paramagnetic material that undergoes a transition to antiferromagnetism at the Néel temperature T_N.

The Néel temperature of antiferromagnets is the analogous of the Curie temperature of ferromagnets. Both mark the borderline temperature above which the material is disordered and below which it is ordered. In both cases the thermal energy is equal to the coupling energy at these transition temperatures. However, it should be pointed out that these two temperatures are not equivalent. Some materials can be antiferromagnetically and ferromagnetically ordered in different temperature regions and they can have both Curie and Néel temperatures (Jiles, 2008), although such phenomenon is not very common for molecular materials and can be only rarely observed in classical inorganic compounds such as metal alloys.

Given the reduced magnetization of an antiferromagnet with respect to that induced by an external field for a paramagnet, it is easy to understand that antiferromagnetic materials cannot find as many applications as ferromagnetic ones. Nonetheless, the antiparallel-aligned ordered state associated with an ideally zero magnetization of antiferromagnets is peculiar and can find very important applications when antiferromagnetic and ferromagnetic materials are properly combined. Two phenomena arising from their interactions are (i) the exchange bias, which allows the coercivity of a ferromagnetic medium to be increased, with tremendous applications in magnetorecording, and (ii) the giant magnetoresistance employed for the preparation of magnetic field sensors.

As far as magnetic molecular materials are concerned, the structural and electronic features that a system must satisfy to behave as an antiferromagnet are the same as those it must satisfy to behave as a ferromagnetic material, the only exception being that the magnetic moments have to be of the same magnitude (i.e., homometallic systems) and must couple in an antiferromagnetic way. A portion of a magnetic structure showing antiferromagnetic ordering is reported in Fig. 3.13.

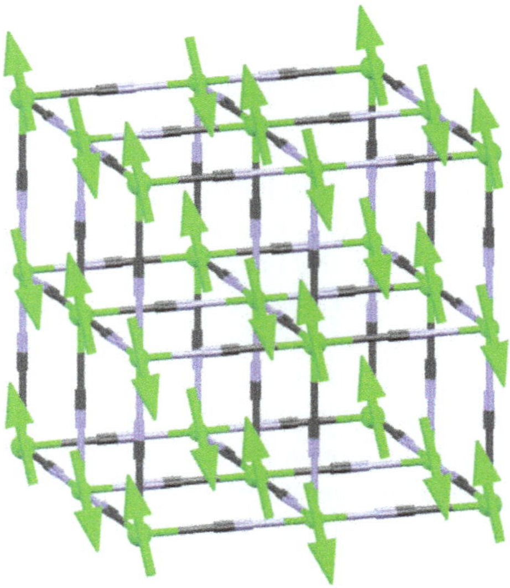

Figure 3.13 Schematic representation of a portion of a 3D molecular lattice where the magnetic moments (indicated with arrows) of each paramagnetic metal ion are antiparallelly aligned. Superexchange interactions are mediated by bridging ligands (blue and gray sticks) along the three space dimensions.

It must be stressed again that only homometallic magnetic structures can behave as pure antiferromagnets. A heterometallic system can also be subjected to antiferromagnetic interactions, but the overall magnetic behavior will be different, as will be described in Section 3.1.4.

Some antiferromagnetic materials exhibit a nonzero magnetic moment at a temperature near the absolute zero. This phenomenon is called spin-canting and arises from magnetic moments that are antiparallelly aligned with respect to each other, although not collinearly (Fig. 3.14).

Figure 3.14 Schematic representation of an antiparallel but not collinear orientation of the magnetic moments of a material showing spin-canting.

If the magnetic moments are aligned in a slightly tilted way the resulting net magnetization will be nonzero. For this reason these systems are also called weak ferromagnets. The origin of the spin-canting can be explained as a competition between the isotropic exchange interaction, which tends to align the moments in a perfect collinear way, and the antisymmetric exchange induced by spin–orbit coupling interactions, which tends to align the moments at 90° with respect to each other (Winpenny, 2012). The relative strength of these two effects determines the extent of this phenomenon.

3.1.4 Ferrimagnetism

Ferrimagnetism can be considered as a specific case of antiferromagnetism in which the magnetic moments of two sublattices A and B, although antiparallelly aligned, give rise to an overall nonzero magnetization due to the different magnitudes of the magnetic moments of A and B (Fig. 3.15).

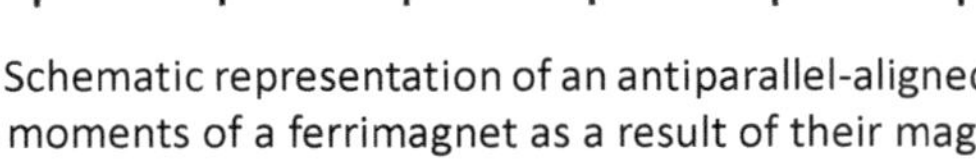

Figure 3.15 Schematic representation of an antiparallel-aligned orientation of the magnetic moments of a ferrimagnet as a result of their magnetic exchange interactions. Note the different magnitudes of the two sets of parallel-aligned magnetic moments.

Ferrimagnetic ordering was first proposed by Néel to explain the magnetic behavior of ferrites. Since ferrimagnets behave, on a macroscopic scale, as ferromagnets, for many years it was not realized that there was a distinction. In fact, they have spontaneous magnetization below the critical temperature and are organized in magnetic domains, thus exhibiting hysteresis and saturation magnetization, and only slight differences in the magnetic behaviors allow one to differentiate between ferrimagnetic and ferromagnetic ordering.

The general model used for the description of ferrimagnetic ordering considers the magnetic structure as divided into two sublattices with magnetic moments of different magnitudes. The magnetic moments are parallel aligned within the same sublattice but are antiparallelly aligned with respect to the magnetic moments of the other sublattice. The superexchange interactions responsible for the coupling of the magnetic moments give rise to an overall ferrimagnetic ordering that can be also viewed as noncompensated antiferromagnetism.

Generally, the magnetic exchange interlattice constant between two sublattices A and B, J_{AB}, is stronger than the intralattice constants, J_{AA} and J_{BB}. This leads to the following expression for the temperature dependence of the susceptibility, which is more complicated than the Curie–Weiss law but is nevertheless similar:

$$\chi = \frac{(C_A + C_B)T - 2J_{AB}C_AC_B}{T^2 - J_{AB}{}^2C_AC_B} \tag{3.2}$$

where C_A and C_B are the Curie constants for the respective sublattices (Jiles, 2008). In this case the Curie temperature T_C is given by:

$$T_C = J_{AB}\sqrt{C_AC_B} \tag{3.3}$$

From the above equations it can be seen that, unlike ferromagnets, the reciprocal ferrimagnetic susceptibility shows a considerable curvature as a function of the temperature as T_c is approached.

As a result of two ordered magnetic states that below the critical temperature give rise to macroscopically similar magnetic responses, the magnetic properties of molecular ferromagnets and ferrimagnets can be differentiated by analyzing the temperature dependence of the χT product. Consider as example two molecular compounds with a similar magnetic structure, for instance, a 3D lattice, composed of two different magnetic ions that are able to interact through ferromagnetic or antiferromagnetic interactions, to lead ferro- or ferrimagnetic ordering (black and red lines in Fig. 3.16). In the high-temperature regime, that is, above the critical temperature, both systems behave as paramagnets with randomly oriented magnetic moments. The presence of a small static external magnetic field induces the alignment of the magnetic moments with respect to the field direction, and the χT value is given by the sum of the paramagnetic contributions of the two involved metal ions. This value will be equal for the two compounds because no magnetic exchange interactions dominate at these temperatures. As the temperature decreases and the critical temperature is approached, deviations from the linear behavior expected for a noninteracting paramagnetic system are observed in both cases. In this temperature region major differences between the two systems emerge. The effectiveness of ferromagnetic interactions in parallel-aligning the magnetic moments increases continuously with temperature. When the thermal energy is equal to the exchange interaction energy the critical temperature is reached and a phase transition occurs. Below that temperature the system is ordered and the magnetic response is higher than that induced by the field due to the cooperative phenomenon. This emerges as a rapid increase of the χT value as the temperature decreases. On the contrary, since in a ferrimagnet the interactions between magnetic moments are antiferromagnetic in nature, the first effect of the temperature decrease is a reduced χT value. This can be explained by considering the magnetic moments as coupled in pairs that are uncorrelated with each other, giving rise to an antiferromagnetic interaction between pairs that globally reduces the total magnetization. When the thermal energy is of the same order of the exchange interactions, the uncorrelated couples of magnetic moments become correlated over the whole magnetic structure (ordering) and a sudden increase of the χT value is observed as the temperature decreases. In summary, as the temperature decreases both systems show an increase of the χT value with respect to that expected for the paramagnetic regime (above the critical temperature T_c), but while for a ferromagnetic material the χT increase with temperature is monotonous, for a ferrimagnetic material it is non-monotonous and is preceded by a minimum (Fig. 3.16).

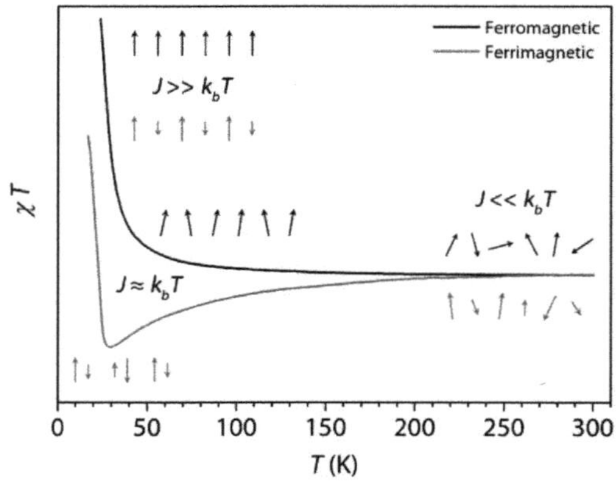

Figure 3.16 Thermal variation of the χT product for a paramagnetic material that undergoes a transition to ferromagnetism (black line) or to ferrimagnetism (red line).

Some difference can also be observed when analyzing the hysteresis loop. A ferromagnetic material shows a hysteresis loop where the saturation magnetization $\mathbf{M}_S$ is reached (more or less easily depending on the anisotropy of the system) as the strength of the external field increases. Once reached, it is maintained up to the highest applied field value (cf. Fig. 3.6) and cannot increase further because all magnetic moments are already aligned with the direction of the external magnetic field. In a ferrimagnetic material this is not equally true. First, we have to consider that the theoretical saturation magnetization will be intrinsically lower for the ferrimagnetic ordering as a result of the antiferromagnetic interactions between magnetic moments. In both cases $\mathbf{M}_S$ can be calculated according to Eq. 2.53, but with some differences. For instance, if we consider a system containing Cr^{III} and Mn^{II} metal ions, in the ferromagnetic ordering case the $\mathbf{M}_S$ value is given by the positive sum of the two ground spin states ($\mathbf{S}_T = \mathbf{S}_{Mn^{II}} + \mathbf{S}_{Cr^{III}} = 5/2 + 3/2 = 4\ \mu_B$) multiplied by the **g** factor,[3] whereas in the ferrimagnetic ordering case the $\mathbf{M}_S$ value is given by the negative sum of the two ground spin states ($\mathbf{S}_T = \mathbf{S}_{Mn^{II}} - \mathbf{S}_{Cr^{III}} = 5/2 - 3/2 = 1\ \mu_B$) multiplied by the **g** factor. The magnetic exchange interactions that force the system in an antiferromagnetic alignment, although strong enough to favor long-range ferrimagnetic ordering, are lower in strength with respect to

[3]To be considered equal to 2.0023 in a first approximation for both metal ions.

an external field of several tesla (that is usually applied to perform this type of measurements). This means that the hysteresis loop for a ferrimagnetic material will be characterized by a magnetization that, above a certain magnetic field strength, increases more than the calculated theoretical value for the ferrimagnetic case because it starts to orientate the magnetic moments in the direction of the field regardless of the antiparallel alignment imposed by the exchange interactions. We can say that the system is intrinsically ferrimagnetic but its magnetic structure is modified by the external magnetic field to an ideally ferromagnetic one. Since there is competition between the exchange interactions and the external magnetic field, this results in a continuous increase of $\mathbf{M}_S$ that, although higher than the theoretical value for the ferrimagnetic ordering, does not usually reach the theoretical value for the ferromagnetic case at the highest applied field (Fig. 3.17).

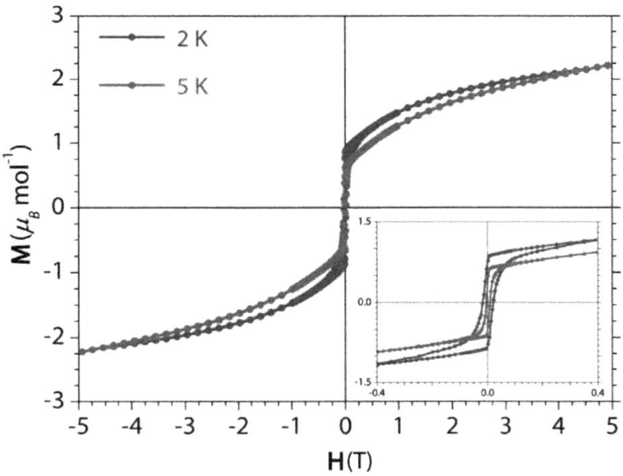

Figure 3.17 Hysteresis loops recorded at two temperatures (see legend) for a ferrimagnetic molecular material based on Mn^{II} and Cr^{III} metal ions. The inset shows a detailed view of the "opened" hysteresis loop. Note that the low $\mathbf{H}_C$ is not a consequence of the ferrimagnetic ordering but is simply related to the intrinsic soft nature of the material selected for this example.

Note that the calculated $\mathbf{M}_S$ value for the ferrimagnetic ordering (1 μ_B) is already reached at zero field as a result of the relative alignment of the magnetic moments, while the effect of the external field is to continuously increase $\mathbf{M}_S$ without reaching saturation at the highest applied field. It is instructive to compare the shape of the loop shown in Fig. 3.17 with that reported in Fig. 3.6 for a ferromagnet.

From the considerations reported so far, it emerges that molecular materials showing ferrimagnetic ordering below a certain critical temperature T_c are those systems that satisfy the same electronic and structural features as those shown by ferromagnetic materials, the only exception being that two paramagnetic metal ions with magnetic moments of different strengths are antiparallelly instead of parallelly aligned.

Given that the ferromagnetic interactions are more difficult to achieve than antiferromagnetic ones, the preparation of ferrimagnetic materials is an elegant way to obtain magnetic properties similar to that of ferromagnetic materials without contrasting the energy minimization requirement given by the antiferromagnetic interactions.

A nice example of ferrimagnetic molecular materials is represented by the family of anilate-based magnets (anilate is a generic derivative of the 2,5-dihydroxy-1,4-benzoquinone ligand) of the general formula $[Mn^{II}Cr^{III}(X_2An)_3]^-$ (X = H, Cl, Br, or I; An = anilate) (Atzori et al., 2013). These systems are molecular materials where 2D anionic coordination polymers featuring a honeycomb crystal structure, similar to that shown with oxalate-based building blocks in Fig. 3.7, alternate with layers of cations. They have been prepared following the complex-as-ligand approach illustrated in Section 3.1.2 by reacting preformed Cr^{III} paramagnetic building units with bare paramagnetic Mn^{II} ions.

While most molecular materials are realized with unfunctionalizable ligands (oxalate, cyanide, etc.) to preserve short contacts and strong interactions, these chemical entities are peculiar because the anilate moiety can be functionalized on the 3,6-position of the dihydroxybenzoquinone ring with different substituents while maintaining strong enough superexchange interactions (Fig. 3.18). The substituents are directly connected to the delocalized allyl fragment of the ring that is responsible for the superexchange magnetic interactions and have been employed as "adjusting screws" to modulate the strength of the magnetic interactions between coordinated metal ions. This ultimately results in a modulation of the magnetic properties that emerges in terms of fine-tuning of the critical temperature that scales linearly with the electronegativity of the substituents (Fig. 3.19).

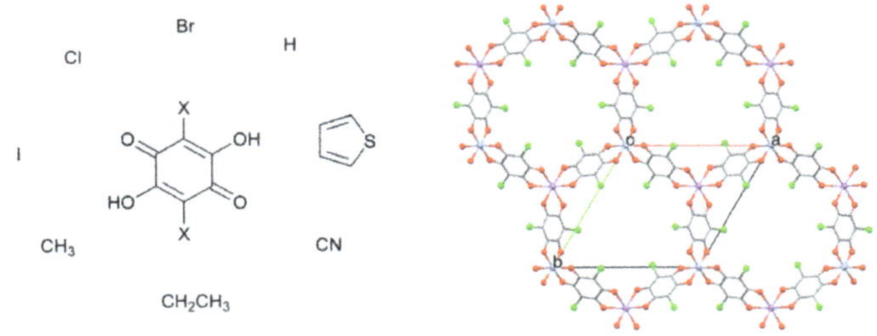

Figure 3.18 The anilate moiety (2,5-dihydroxy-1,4-benzoquinone) with various substituents employed for chemical tunability of the magnetic properties (left) and a view of the 2D honeycomb structure (right).

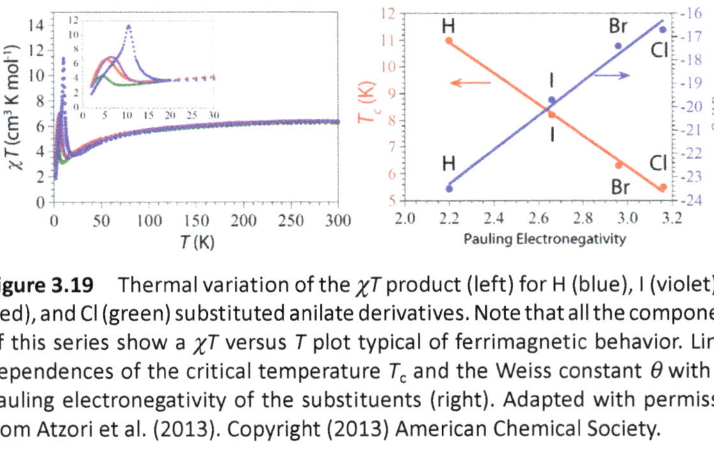

Figure 3.19 Thermal variation of the χT product (left) for H (blue), I (violet), Br (red), and Cl (green) substituted anilate derivatives. Note that all the components of this series show a χT versus T plot typical of ferrimagnetic behavior. Linear dependences of the critical temperature T_c and the Weiss constant θ with the Pauling electronegativity of the substituents (right). Adapted with permission from Atzori et al. (2013). Copyright (2013) American Chemical Society.

3.1.5 Spin Crossover

Coordination compounds of transition metal ions having a $3d^4$–$3d^7$ electronic configuration in an octahedral environment[4] can feature two possible configurations of the electrons located on the d orbitals.

[4]The tetrahedral coordination geometry can, in principle, show the same features as those of the octahedral one because both induce a separation of the d orbitals of the metal center into two sets, t_{2g} and e_g. However, the strength of the tetrahedral ligand field induces an energy difference between them, Δ_t, that is small compared to Δ_O, and consequently, most tetrahedral complexes show dominant *hs* configurations in the entire temperature range.

Depending on the strength of the ligand field, quantified by the Δ_O parameter, which represents the energy difference between the t_{2g} and e_g orbitals, *hs* or *ls* electronic configurations can be observed. The *hs* state is favored when Δ_O is lower than the energy required to pair two electrons in the same orbital (spin-pairing energy, *P*). In this case the electrons reside in both t_{2g} and e_g orbitals, and the number of unpaired electrons is maximized. On the contrary, the *ls* state is favored when Δ_O is higher than the pairing energy. In this case the electrons occupy predominantly the t_{2g} orbitals and the number of unpaired electrons is minimized (Fig. 3.20).

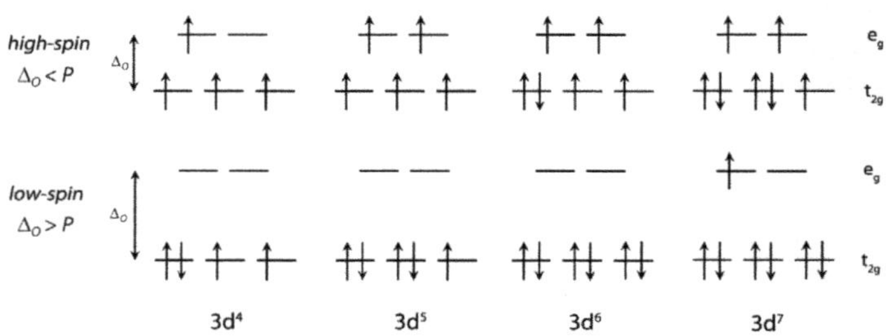

Figure 3.20 High-spin and low-spin d orbitals' occupancy for first-row transition metal ions with $3d^4$, $3d^5$, $3d^6$, and $3d^7$ electronic configurations.

Metal ions that can show *hs* or *ls* configurations are Cr^{II}, Mn^{II}, Mn^{III}, Fe^{II}, Fe^{III}, Co^{II}, and Co^{III}, all belonging to first-row transition metal ions series. On the contrary, those of the second and third rows usually show *ls* configurations due to stronger ligand field effects, which increase the separation between t_{2g} and e_g orbitals. It can be clearly concluded that the ligand field has a huge influence on the resulting magnetic moment of a certain paramagnetic metal ion by determining its spin state.

Interesting magnetic properties can emerge if Δ_O and *P* have comparable energy values so that their difference in energy is similar to the thermal energy. In this case, a competition between the *hs* and *ls* states can be observed since the electrons can assume one of the two configurations as a function of the temperature. More precisely, the *ls* configuration is the energetically favored one at low temperatures, whereas the *hs* configuration becomes predominant as the temperature is raised. In this case, an investigation of the magnetic properties of such compounds shows a strong variation in the magnetic response as a function of temperature due to the change in spin multiplicity. Those systems that exhibit *ls-hs* electronic configuration changes as a result of an external stimulus (temperature, pressure, applied magnetic or electric field, light irradiation, etc.) are commonly called spin-crossover (SCO) systems. These compounds are particularly relevant in the framework of molecular materials since they can be regarded as the smallest switches ever obtainable, as this property is a single-molecule property (cf. Section 1.4).

The fundamental electronic and structural features that a certain molecular material should have to behave as a SCO system can be summarized as follows:

- An octahedral coordination geometry of the metal ion that splits the d orbitals into two sets (t_{2g} and e_g) having different energies
- A $3d^4$–$3d^7$ electronic configuration that guarantees two possible spin states (*hs* and *ls*) as a function of the ligand field (Δ_O)
- A ligand field of appropriate strength that ensures a value of Δ_O of comparable energy with respect to the spin-pairing energy P
- A difference in energy between Δ_O and P of the same order of magnitude as that of the thermal energy so that the most thermodynamically stable state (*ls* or *hs*) can be switched with temperature

While the coordination geometry around a metal center is quite easy to chemically control through a proper selection of ligands, the modulation of the strength of the ligand field to fulfil the condition $(\Delta_O - P) \approx k_B T$ is not trivial. However, on the basis of a statistic number of experimental results, it is possible to phenomenologically identify the type of ligands that could satisfy this condition for a certain metal. For example, many SCO compounds have been obtained with Fe^{II} or Fe^{III} as metal ions where the coordinated ligands provide two O and four N donor atoms. Depending on the nature of the ligand (mostly its electron-donating or electron-withdrawing ability) and, in case, the nature of the substituents, SCO can occur over a wide temperature range. Other types of ligand fields can also be appropriate to observe this phenomenon, but a detailed review of such features is beyond the scope of this book and the reader is referred to more specialized texts, such as the textbook *Spin-Crossover Materials: Properties and Applications* (Halcrow, 2013).

Looking in more detail at the SCO phenomenon, it can be inferred that it can be regarded as an intra-atomic charge transfer (CT) between the t_{2g} and e_g orbitals. Since the t_{2g} orbitals usually have a bonding/nonbonding character and the e_g orbitals typically are antibonding in nature, population or depopulation of the latter results in important optical and structural modifications. In fact, while in the *ls* state most of the electrons occupy orbitals having a bonding/nonbonding character, part of them are promoted to orbitals of antibonding character in the *hs* state. This results in metal–ligand bond lengths that are, on average, 0.2 Å longer in the *hs* state (lower bonding character) than in the *ls* state (higher bonding character). Conversely, a similar change in the molecular arrangement would allow for extending the influence of an external stimulus other than the temperature in order to switch the system from one state to the other. For instance, a pressure applied on a system that is in the more voluminous *hs* state can favor a switch to the less voluminous *ls* state (Fig. 3.21).

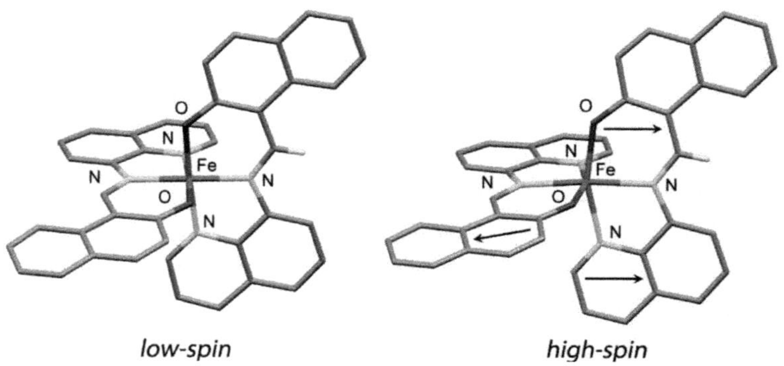

Figure 3.21 Comparison between the molecular structures of an archetype example of a spin-crossover system (Kuroda-Sowa et al., 2008) in the *ls* and *hs* states. Note the less distorted octahedral coordination geometry around the Fe^{II} center in the *ls* state with respect to that of the *hs* state. Longer bond lengths result in a higher molecular volume. This can be clearly appreciated here by noticing the longer distance between the napthaldehyde and quinoline fragments that are closer to the observer (compare the left and right structures).

From a thermodynamic point of view, the transition between the *ls* and *hs* states is accompanied by an augmented enthalpy of the system. This energetically unfavorable circumstance is compensated for by the increased entropy of the system associated with a higher spin multiplicity and by the increased density of vibrational levels associated with the *hs* state with respect to those of the *ls* state (Fig. 3.22). For this reason the SCO phenomenon is also defined as an entropy-driven process (Halcrow, 2013).

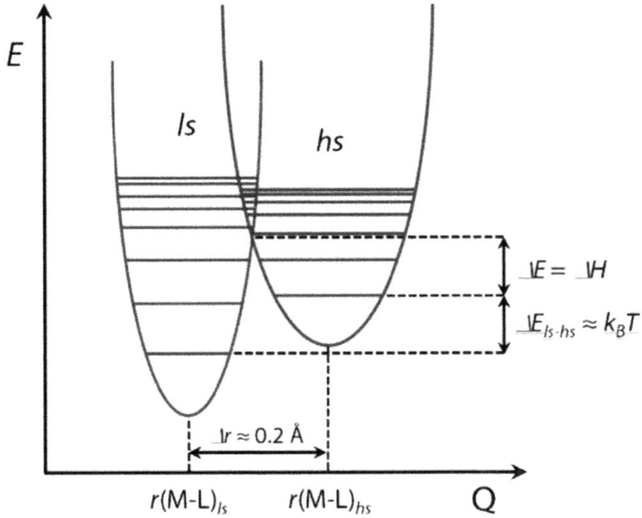

Figure 3.22 Schematic picture of the potential energy (E) of the two spin states as a function of the nuclear coordinate (Q) chosen as the M–L bond distance r. $\Delta E_{ls\text{-}hs}$ represents the energy required to switch the system from the *ls* to the *hs* state (comparable to the thermal energy k_BT). ΔE represents the additional energy required to pass from the lowest vibrational level of the *ls* state to the lowest vibrational level of the *hs* state and corresponds to a positive value of the enthalpy ΔH. This positive enthalpy is balanced by the higher entropy ΔS of the *hs* state. Energy curves were drawn as harmonic models for simplicity sake.

From a magnetic point of view, the most interesting and investigated SCO systems are those based on the $3d^6$ Fe^{II} metal ion in view of the fact that the crossover between the two configurations goes from a completely diamagnetic state (**S** = 0) to a paramagnetic state with a magnetic moment of relevant strength (**S** = 2). This type of system can be regarded as the most suitable for applications as molecular switches since the presence or absence of a magnetic moment can be, in principle, easily detected by a hypothetically interfaced detector with respect to two paramagnetic states of different strengths (e.g., the $3d^7$ configuration is associated with a *ls* state with **S** = 1/2, 0.375 $cm^3\ K\ mol^{-1}$, and a *hs* state with **S** = 3/2, 1.875 $cm^3\ K\ mol^{-1}$).

Once again the χT versus T plot is very useful to evaluate the different magnetic responses of this type of magnetic material as a function of temperature and allows the determination of the SCO transition temperature, indicated as $T_{½}$. In this view, SCO materials can be considered as a subclass of paramagnetic materials (vide supra), as their magnetic properties are those of single and noninteracting metal centers with the extra property of changing the spin state upon an external stimulus. Consequently, the χT versus T plot for a SCO compound is expected to (i) be almost constant at low temperature when associated with the *ls* state, (ii) start increasing, more or less abruptly, depending on the nature of the investigated sample, its crystallinity, and some structural features (vide infra) as the temperature increases to values close to the transition temperature $T_{½}$, and (iii) reach the value of the *hs* state, which is maintained constant up to the highest temperature (Fig. 3.23).

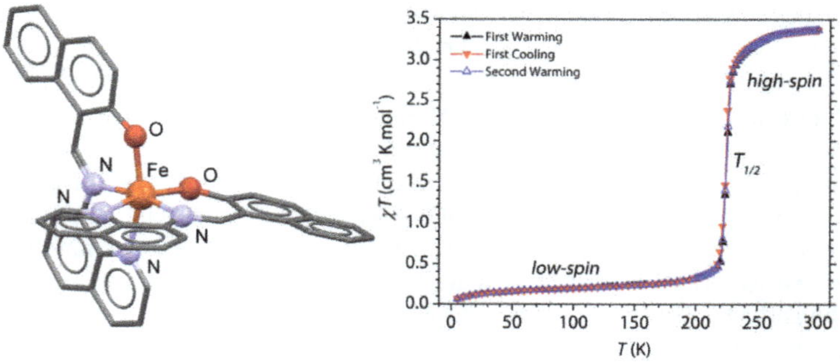

Figure 3.23 Molecular structure of an archetype example of spin-crossover complex based on the Fe^{II} ion (left) and the relative χT versus T plot (right) (Kuroda-Sowa et al., 2008). The χT value is almost zero at low temperature (0–200 K), where only a small residual paramagnetic signal is observed. Above 200 K the χT value abruptly increases, reaching the *hs* value of ca. 3.4 $cm^3\ K\ mol^{-1}$, in good agreement with the expected value for an Fe^{II} ion with **S** = 2 and **g** = 2.1.

The transition temperature $T_{½}$ is defined as the temperature at which a half of the spin population is converted. This parameter can be easily determined by calculating the first derivative of the experimental data, as the inflection point of the sigmoidal trend would correspond to a maximum. Since these materials have been proposed and presented herein as molecular switches, one fundamental requisite for them is the ability to retain the switching properties. In other words, the *hs–ls* transition has to be reversible. This can be proved by performing magnetic measurements in both warming and cooling modes and repeating a third measurement to ensure that the thermal treatment did not irreversibly alter the chemical structure of the material and consequently its magnetic properties (Fig. 3.23).

The example reported above clearly shows a very good reversibility in both warming and cooling modes. However, this behavior is not the only one that can be observed. Warming and cooling measurements can display magnetic trends not perfectly superposed, that is, associated with two $T_{½}$. In this case, it is a common practice to indicate as $T_{½}\uparrow$ the transition temperature for the warming process and $T_{½}\downarrow$ that for the cooling process. Some systems are characterized by significant differences in the $T_{½}$ value of the warming and cooling process, so the χT versus T plot shows thermal hysteresis (Fig. 3.24).

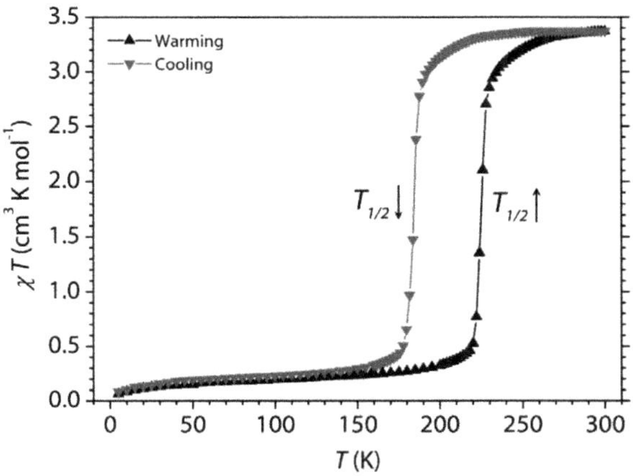

Figure 3.24 χT versus T plot for a generic spin-crossover system based on the Fe^{II} metal ion that shows hysteresis and bistability.

This introduces another fundamental concept that characterizes this class of materials: bistability. Whenever a physical property is associated with a hysteresis cycle, as already shown for the magnetization of ferro- and ferrimagnets, the system can be arbitrarily placed in one of the two states by acting on the external stimulus (magnetic field, temperature, pressure, etc.). For instance, in the example reported in Fig. 3.24 it is possible to place the system in the *hs* state at 200 K as it is cooled down from higher temperatures and in the *ls* state at the same temperature if warmed up from lower temperatures. In the temperature range defined by the width of the hysteresis loop the system is thermodynamically stable in both states (bistable), and these states can be reached with an external thermal control. It is interesting to highlight that bistability can be viewed as an opportunity to store information since a bistable material can be placed at the same temperature into two very different spin states that can be arbitrarily associated, for example, with the 0 and 1 states of a bit. It is also important to note that such bistability derives from a single-molecule property, contrary to what is observed in the magnetization of ferro- and ferrimagnets, where the contribution of the overall magnetic structure is required. This opens new perspectives into the miniaturization of information storage since, in principle, if we were able to act on each single molecule, we would manage to store a bit of information in each Fe^{II} metal complex.

The hysteretic behavior as well as the abruptness of the SCO transition can be related in some cases to cooperative effects in the solid state. We have already shown that the SCO phenomenon is accompanied by a structural modification of the coordination shell and the overall molecular volume. Since most of these systems are studied in the bulk phase, both as crystals and microcrystalline powders, where a high number of molecules interact in a crystal structure, the entity of the supramolecular interactions between molecules plays a role in determining the magnetic response. This is related to the fact that although the phenomenon is molecular in origin, the change in volume can be propagated through the crystal structure via supramolecular interactions as a sort of domino effect. More specifically, if the system is in the *ls* state and the temperature is gradually increased, initially there will be an amount of energy only sufficient to switch a low number of molecules in the *hs* state. Then, as the temperature slowly goes up, the conversion increases accordingly. If the material shows good crystallinity, the effect induced by the temperature will be flanked by the propagation effect of the change in volume, which acts as an additional force that helps the molecules in the *ls* state to adopt the *hs* state and vice versa in the cooling process. In this case, it is expected that the SCO transition will be abrupt, while in the cases of low crystallinity or crystal structures where the metal complexes are well isolated from each other and the relevance of supramolecular interactions between them is low, the SCO will be only driven by the temperature and the conversion will be observed over a higher temperature range (Fig. 3.25).

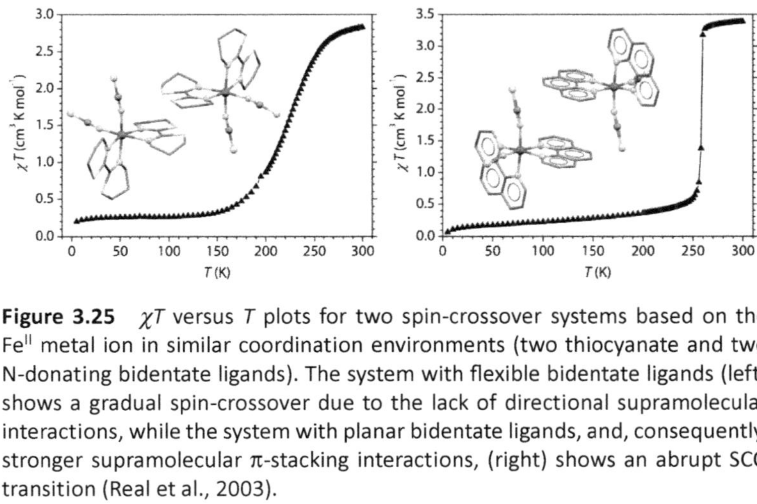

Figure 3.25 χT versus T plots for two spin-crossover systems based on the Fe^{II} metal ion in similar coordination environments (two thiocyanate and two N-donating bidentate ligands). The system with flexible bidentate ligands (left) shows a gradual spin-crossover due to the lack of directional supramolecular interactions, while the system with planar bidentate ligands, and, consequently, stronger supramolecular π-stacking interactions, (right) shows an abrupt SCO transition (Real et al., 2003).

In general, we can say that for a system presenting non-negligible supramolecular interactions in the solid state, the higher the crystallinity of the investigated sample, the more abrupt the conversion. In the case of cationic or anionic compounds the nature of the counterion can also induce modifications in the magnetic response. Each different counterion occupies a different portion of the crystal structure, and their relative positions and steric hindrance may or may not favor interactions between the magnetic molecules. This is the reason why the same metal complexes associated with different counterions can show different SCO behaviors.

Supramolecular interactions can also hamper the SCO process. Taking again into account the example reported above about the switch from the *ls* to the *hs* state, strong supramolecular interactions, instead of favoring a quick conversion, can hinder molecular rearrangement, providing an additional force that has to be overcome by the molecules before they switch to the *hs* state. This results in a higher $T_{½}$ than that expected in the absence of such interactions. The same holds when passing from the *hs* to the *ls* state, resulting in a lower $T_{½}$ than that expected in the absence of such interactions. The overall result is that the system will be characterized by two $T_{½}$ ($T_{½}\uparrow$ and $T_{½}\downarrow$) with the appearance of hysteresis related to the thermal dependence of magnetic susceptibility (vide supra) (Fig. 3.24).

So far we have analyzed in detail the behavior of SCO systems upon temperature change. However, other external stimuli can be employed to induce a *ls*–*hs* transition (vide supra). Among them, light irradiation is one of the most interesting and investigated ones since it requires relatively simple and readily available instrumentation to be realized and offers the opportunity to combine the mobility of photons with the almost static properties of the spin in a particularly useful manifestation of light–matter interaction (cf. Section 2.3.1) for the realization of a molecular device.

We have already shown in Fig. 3.22 the ground states associated with the *ls* and *hs* configurations as a function of the energy for a generic SCO system, and now we will discuss in more detail the electronic transitions involving excited states at higher energies that allow a photoconversion of the system.

Taking into account the case of a Fe^{II} d^6 metal ion, its electronic configuration is associated with the term symbol $^5D_0$ (or simply ^5D), with $^1A_{1g}$ representation for the lowest energy level of the *ls* state and $^5T_{2g}$ representation for the lowest energy level of the *hs* state (cf. Section 2.3.7.1). Excited states of allowed spin multiplicity can be populated upon photoexcitation with a laser of appropriate wavelength, usually falling in the visible region. In the case of Fe^{II} the photoexcitation induces the $^1T_{1g} \leftarrow {}^1A_{1g}$ allowed transition. Relaxation to the *ls* ground state occurs through double internal conversion; however, there is also the possibility that the $^5T_{2g}$ *hs* metastable state becomes populated via intersystem crossing (ISC). Since the $^5T_{2g} \rightarrow {}^1A_{1g}$ transition is spin forbidden, at low temperature the lifetime of the metastable *hs* state is quite long so the system is "trapped" in this state and survives long enough to observe a *ls-hs* conversion (Fig. 3.26).

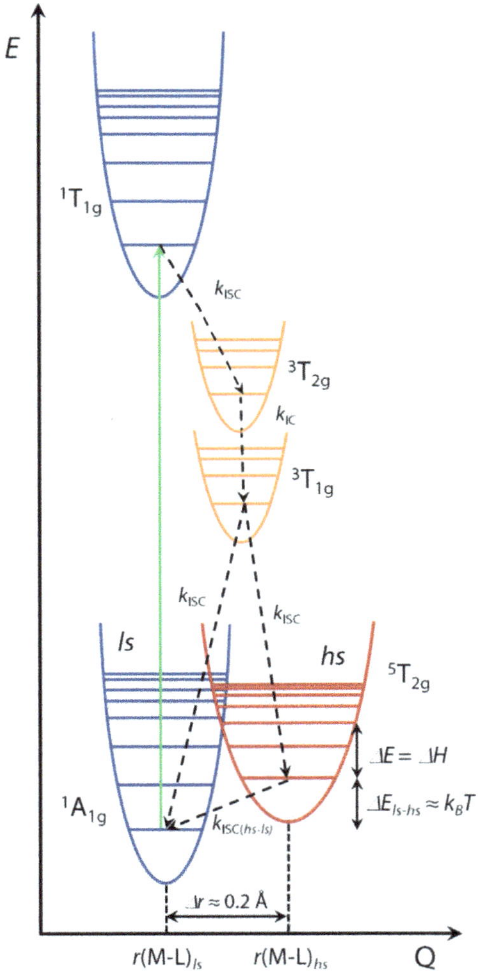

Figure 3.26 Schematic picture of the photocycle for the photoexcitation and photoconversion processes of a generic d^6 system at low temperature. The green arrow indicates the photoexcitation through light irradiation from the ground to the excited *ls* state. Then a mechanism of intersystem crossing and internal conversion (dashed arrows) allows for photoconversion to the metastable *hs* state, which slowly relaxes to the most stable *ls* ground state: $k_{ISC(hs\text{-}ls)}$. Note that the energy curves are depicted as harmonic and spatially separated for graphical clarity.

The phenomenon described above, called light-induced excited-state spin-trapping (LIESST) effect, is more frequently observed for those systems that are characterized by low $T_{½}$ temperatures since the stabilization of the metastable *hs* state is expected to be energetically favored, that is, there is a low energy difference between the *hs* metastable state and the *ls* ground state.

Experimentally, once the system has been converted to the *hs* state, it is a common practice to investigate the evolution of the populated metastable state as a function of the temperature to quantify the ability of the system to retain this state. A typical experiment consists in:

- Sample photoexcitation at low temperature (usually at 5 or 10 K) with a laser of appropriate wavelength for a finite time[5]
- Measurement of the magnetic response of the system after irradiation by continuously increasing the temperature at a fixed warming rate (usually 0.5 K/min.)

What is commonly observed is that the magnetic signal associated with the *hs* metastable state shows at first an increase, then it reaches a maximum or an almost constant value over a relatively wide range of temperatures, and then it starts to drop to the χT value of the pristine system (Fig. 3.27). This trend can be explained by taking into consideration that thermal energy can favor slow processes such as ISC, both between excited energy levels and in regard to relaxation to the ground *ls* state from the higher *hs* state.

The temperature at which the χT versus T plot presents the maximum variation after photoconversion is called T_{LIESST} and, analogously to the determination of the $T_{½}$, is obtained through the first derivative of the experimental data (the temperature of the minimum is the T_{LIESST}). In general we can say that the LIESST effect is observed at low temperatures, usually in the 2–100 K range, since at higher temperature values the thermal energy favors deactivation of higher *ls* states through competitive nonradiative channels.

A partial photoconversion of the system is also commonly observed. This is related to several reasons, such as (i) the absorption properties of the material, which can limit the efficiency

[5]The irradiation time is a function of the ability of the system to be converted to the metastable state. The conversion is monitored by looking at the magnetic response of the system before and during the irradiation process.

of the conversion, (ii) the nominal power of the laser, (iii) the power attenuation of the laser through the optical fiber needed to reach the sample,[6] and (iv) the shape and thickness of the sample. The latter is particularly relevant because while a standard magnetic measurement furnishes a concerted response related to the entire mass of the sample, photophysical processes are associated with single molecules (one photon excites one molecule) and the photoconversion can easily take place on the amount of sample lying on the surface but not on that located in the inner bulk due to light penetration limitation. This justifies lower χT values for the photoinduced *hs* (metastable) state with respect to that of the *hs* (ground) state induced by the temperature.

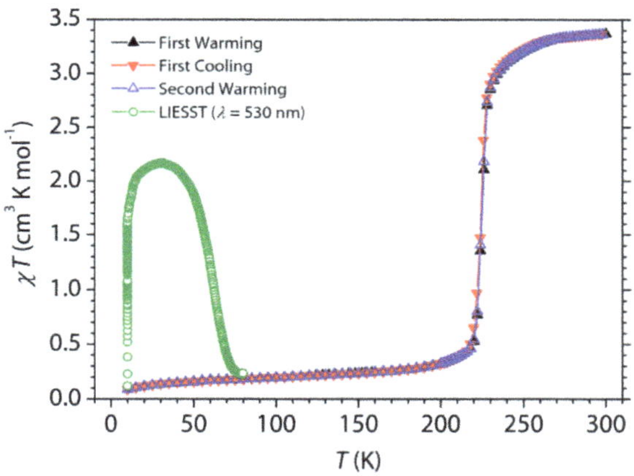

Figure 3.27 χT versus T plot for the spin-crossover complex shown in Fig. 3.23 together with photomagnetic measurements associated with the LIESST effect. At 10 K (photoexcitation temperature) the χT value increases, and then it reaches a maximum and starts to decrease as the temperature increases. The T_{LIESST} in the present case is ca. 60 K.

3.1.6 Valence Tautomerism

In chemistry, tautomerism is the structural rearrangement of the chemical bonds of a given chemical species as a consequence of an

---

[6]In a standard photomagnetic measurement, the sample is placed inside a magnetometer and the distance between the light-emitting diode and the sample cannot be shortened below the limit of some meters for practical reasons.

intermolecular chemical reaction (tautomerization). This reaction consists in the intramolecular migration of a hydrogen atom from a functional group to another, as for example in the conversion between the keto- and enol- forms of ketones or enolates. Similarly, in some chemical systems, valence electrons (usually one) can migrate from a portion of the molecule (e.g., an organic ligand) to another (e.g., a metal center). This process, where the topology of the molecular bonds remains unchanged before and after the conversion, is commonly referred to as valence tautomerism (VT) or redox isomerism, since the electron transfer implies a change in the redox states of the two components of the molecular system. Such a phenomenon is particularly interesting in the field of molecular magnetism because the electron transfer can be responsible for the conversion between two spin states upon application of an external stimulus (temperature change, light irradiation, etc.), as illustrated above for the case of SCO systems, leading to switchable and/or bistable molecular materials.

VT has been found in materials belonging to different classes (magnetic nanoparticles, coordination polymers, molecular complexes, etc.), but only the case of valence tautomers belonging to the class of molecular materials will be presented herein, with more emphasis on the simplest case of mononuclear metal complexes where a metal center and an organic ligand are involved in the valence tautomeric interconversion.

Materials showing VT are not easy to obtain since the conversion from one form to the other has several structural and electronic requisites that are quite difficult to fulfill in a single molecular entity. However, some guidelines for their preparation can be summarized as follows:

- The two components of the molecular systems involved in the electron exchange must be redox active, that is, both the metal and the organic ligand must have accessible and stable redox states.
- The redox potentials of the two redox couples involved in the valence interconversion process must have close values.
- The electronic exchange between the two components should be intermediate and reversible; in other words, they should belong to Class II mixed-valence species in the Robin and Day

classification (see the box titled "Robin and Day Classification of Mixed-Valence Species").

- The frontier orbitals of the two components must lie close in energy to allow for a reversible conversion between the two thermodynamically stable states (cf. Fig. 3.30).
- The two electronic states must be characterized by differently populated antibonding molecular orbitals (MOs).

An archetype molecular example that fulfills all these requirements is a $Co^{II/III}$-dioxolene metal complex, as the one reported in Fig. 3.28.

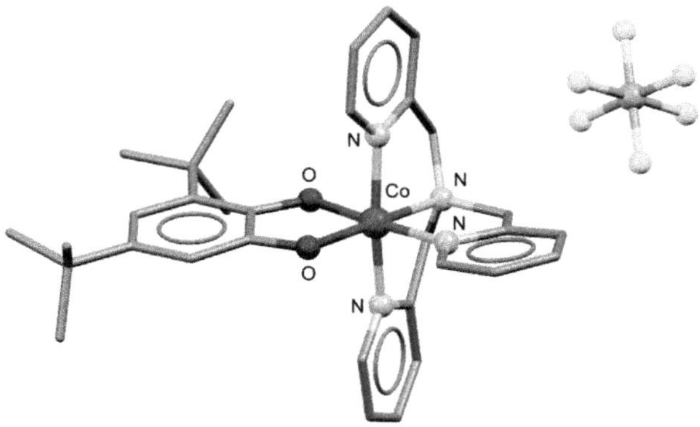

Figure 3.28 Molecular structure of an archetype example of a valence tautomer based on the redox active $Co^{II/III}$ metal ion and bis(3,5-di-tert-butyl)-dioxolene ligand (Beni et al., 2008).

Cobalt has two accessible and stable oxidation states (Co^{II} and Co^{III}) differing for one single electron, while the orthoquinone ligand has three accessible and chemically stable electronic states differing for one electron. These are the neutral and diamagnetic (**S** = 0) quinone form, the monoanionic and paramagnetic (**S** = 1/2) semiquinone (SQ) form, and the dianionic and diamagnetic catecholate form (Fig. 3.29).

Although three stable forms can be accessible, the two charged forms are the most valuable in the preparation of molecular complexes showing VT because of their higher tendency to bind metal ions with respect to the neutral one.

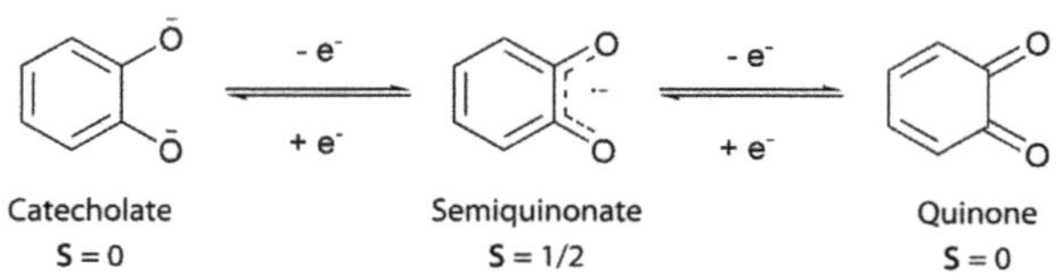

Figure 3.29 Redox equilibria between the three stable forms of a generic orthoquinone dioxolene ligand.

Robin and Day Classification of Mixed-Valence Species

Mixed-valence species are chemical compounds in which two molecular sites can reversibly exchange one electron, and consequently switch oxidation states, giving rise to two different chemical species. The degree of mixing between the two electronic states can vary between two limit situations: the case in which each site has its well-defined valence and the case in which the valence is completely delocalized over the molecular or crystal structure. In 1968, Robin and Day proposed a classification of the degree of mixing of the mixed-valence species that is very commonly used (Robin and Day, 1968).

Class I: Each site has its well-defined valence, and the two possible chemical species cannot be easily converted from one to the other. This means that the properties of the resulting material are given by the sum of the contributions of the two distinct species.

Class II: Each site has its own valence, but a small amount of energy (thermal energy, light irradiation, etc.) can be enough to allow for electron transfer from one site to the other, with concomitant change of the valence states. The properties of such materials are affected by the possibility that the electron undergoes electron transfer and peculiar behavior related to this may result.

Class III: Complete electron delocalization hampers the discrimination between molecular sites that are thus understood to have the same valence. The oxidation states of each site are then taken as fractional as a result of the delocalized mixed valence. The observed properties are completely different from those of the two limit species taken separately.

In this example, the Co^{III} ion shows a slightly distorted octahedral coordination geometry, with the bis(tert-butyl)-catecholate as the redox-active ligand. The coordination geometry is completed by a tetradentate neutral ligand, and the total +1 charge of the cationic complex is balanced by a PF_6^- anion. In this state, both the ligand and the metal ion have a paired electron configuration, since the d^6 electronic configuration of the Co^{III} metal ion in an octahedral environment favors the *ls* state, while the ligand is in the dianionic form (Cat). This state is thermodynamically favored at low temperature.

It must be noted that these types of complexes can be prepared with more than one redox-active ligand like the bis(3,5-di-tert-butyl)-catecholate herein presented. However, since the metal ion is only capable of switching from oxidation states that differ by one electron, only one ligand will be involved in the valence tautomeric process. This archetype example has then been chosen to point out that at least one redox-active ligand must be present, although more than one can coordinate the metal.

If the system in the Co^{III}-Cat form is exposed to an external stimulus of appropriate energy (e.g., the temperature is increased or the sample is irradiated by light of appropriate wavelength), a valence tautomeric interconversion can occur with an electron transfer from the ligand to the metal and concomitant change of the oxidation and spin states of the two components. In this case, the ligand loses one electron when passing to the SQ form whereas the metal ion gains one electron by transitioning to the Co^{II} oxidation state. The system in the Co^{II}-SQ form will be characterized by a completely different spin state than that of the Co^{III}-Cat form since the d^7 electronic configuration of the Co^{II} metal ion will now favor the *hs* state (three unpaired electrons, vide supra), whereas the ligand goes from the **S** = 0 spin state of the dianion to the **S** = ½ spin state of the SQ. The final result is that the system goes from a global diamagnetic ground state (**S** = 0) at low temperature to a global paramagnetic ground state at high temperature (**S** = 1 or 2, depending on the nature of the magnetic coupling between the unpaired electron located in the orbital of the organic ligand and one of the d electrons), similar to what is observed in SCO systems, but as a result of a different driving force (Fig. 3.30).

The case of Co-dioxolene systems is somehow special since the valence tautomeric interconversion is not simply accompanied by a one-electron transfer from the ligand to the metal and vice versa, but this electron transfer also induces a significant variation in the spin state of the metal ion, which goes from a *ls* to a *hs* state, enhancing the difference in the magnetic response and thus leading to a switchable and/or bistable material similar to SCO compounds. However, it must be highlighted that in this instance the metal ion also changes its oxidation state, contrary to what occurs in SCO systems, where the oxidation state of the metal ion is retained. Thus VT cannot be regarded as a simple SCO behavior.

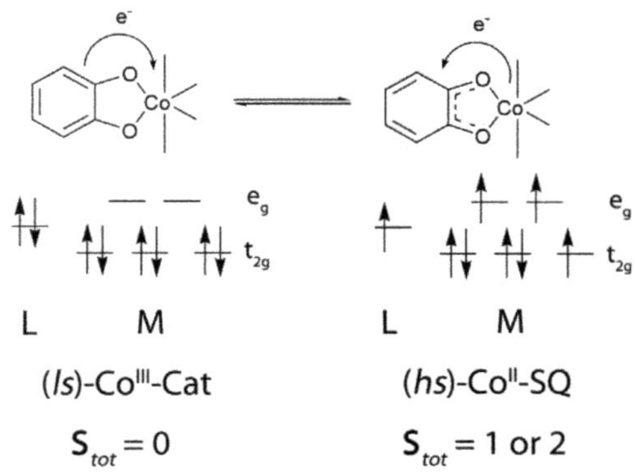

Figure 3.30 Valence tautomeric interconversion between the (*ls*)-Co^{III}-Cat and (*hs*)-Co^{II}-SQ forms of a generic Co-dioxolene system. L and M indicate the frontier orbitals of the ligand and the metal, respectively. The (*hs*)-Co^{II}-SQ form shows $\mathbf{S}_{tot}$ = 1 if the interaction between the unpaired electron on the ligand and one of those of the metal is antiferromagnetic, and $\mathbf{S}_{tot}$ = 2 if this interaction is ferromagnetic.

It is now easy to imagine how the magnetic properties of these materials change with temperature. A typical χT versus T plot of such a system will be characterized by an almost zero value of the χT product at low temperature (**S** = 0), followed by an increase whose slope depends on the nature of the system, its crystallinity, and some structural features (vide supra), to reach, upon further heating, the expected high temperature value of the χT product (Fig. 3.31).

It must be noted that since the Co^{II} ion in an octahedral coordination geometry peculiarly has a non-negligible orbital contribution with respect to the other transition metal ions,[7] the spin-only magnetic moment formula (Eq. 2.41) does not work well

[7]A d^7 ion in an octahedral coordination geometry should have an unpaired electron in degenerate t_{2g} orbitals. Such degeneration induced by the crystal field is not energetically favored, and it is broken by the spin–orbit coupling, which further splits the t_{2g} orbital manifold into two subsets. This introduces a significant deviation of the **g** factor with respect to the free electron value (ca. 2.00), with a strong deviation in the Curie behavior as well (Gatteschi et al., 2006).

to estimate the χT value at high temperature (1.875 cm^3 K mol^{-1}). The value of the χT product should be ca. 3.0–3.4 cm^3 K mol^{-1}, as expected in the case of isolated Co^{II} ions with an unquenched orbital contribution, plus or minus 0.375 cm^3 K mol^{-1} depending on the sign of the exchange magnetic interaction with the extra electron on the organic radical.

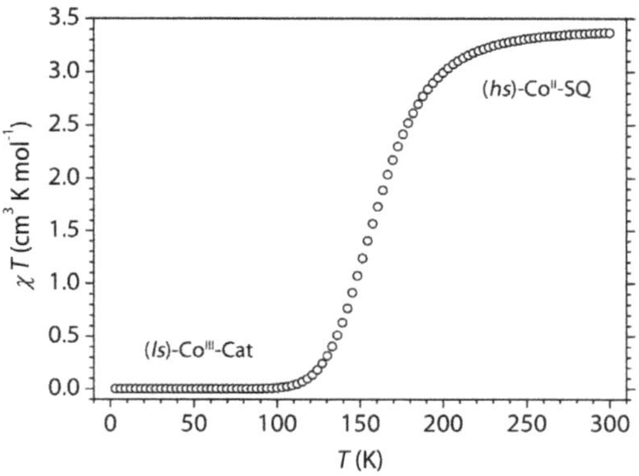

Figure 3.31 Simulated χT versus T plot for a hypothetic Co-dioxolene system exhibiting valence tautomeric interconversion. At low temperature, the (*ls*)-Co^{III}-Cat is the thermodynamically stable form, whereas as the temperature increases, a gradual conversion toward (*hs*)-Co^{II}-SQ occurs. Once the temperature assumes a sufficiently high value, all the (*ls*)-Co^{III}-Cat species can be considered converted, as evidenced by the plateau of the χT product that is in the range expected for an *hs* Co^{II} ion plus the contribution of one extra electron of the organic ligand.

The role played by the tetradentate ligand has been so far ignored to focus on the valence interconversion process. However, this ligand, which in the field of VT is commonly called ancillary to point out that it is not involved in the electron transfer process, has at least two main functions. Experimental evidence indicates that for the complex to display VT the coordination sphere of the Co ion must be octahedral and feature at least two N donating atoms, which can be provided by two monodentate or a bidentate ligand (pyridine; 2,2'-bipyridine and its derivatives; or 1,10-phenantroline and its derivatives, etc.), in addition to two O donor atoms of the dioxolene redox-active ligand. The rest of the coordination sphere can be filled by other two donor atoms that can be either N atoms (as in the case of the tetradentate ligand reported in Fig. 3.28 or an additional bidentate ligand) or O atoms (as, for example, another dioxolene that does not participate in the valence interconversion process). This combination seems fundamental to observing valence tautomeric interconversion in such systems because this ligand field provides a redox potential to the Co^{II}/Co^{III} redox couple that matches those of the Cat/SQ redox couple of the ligand, one of the fundamental requisites to observing this phenomenon. Slight modifications of the periphery of the ancillary ligand with various types of substituents allows fine-tuning of the redox potential of the metal ion, determining a modulation of the temperature at which the valence interconversion process takes place ($T_{1/2}$), as elegantly demonstrated by the series of compounds of general formula [Co(3,5-di-tert-butyl-dioxolene)(Me_ntpa)]PF_6 (tpa = tris(2-pyridylmethyl)amine; n = 0, 1, 2, and 3) (Fig. 3.32) by Dei and coworkers (Beni et al., 2008).

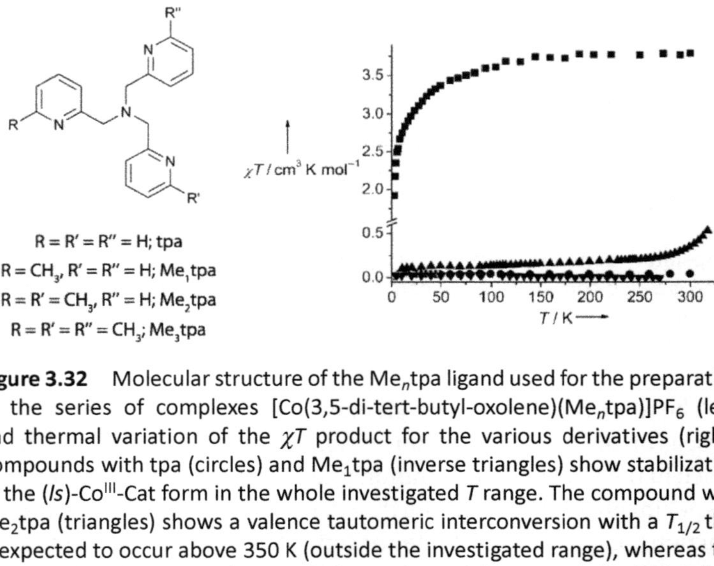

Figure 3.32 Molecular structure of the Me_ntpa ligand used for the preparation of the series of complexes [Co(3,5-di-tert-butyl-oxolene)(Me_ntpa)]PF_6 (left) and thermal variation of the χT product for the various derivatives (right). Compounds with tpa (circles) and Me_1tpa (inverse triangles) show stabilization of the (*ls*)-Co^{III}-Cat form in the whole investigated T range. The compound with Me_2tpa (triangles) shows a valence tautomeric interconversion with a $T_{1/2}$ that is expected to occur above 350 K (outside the investigated range), whereas the compound with Me_3tpa (squares) shows the stabilization of the (*hs*)-Co^{II}-SQ form in the whole investigated T range. Adapted with permission from Beni et al. (2007). Copyright 2007 Wiley-VCH.

Furthermore, such a ligand field ensures a Δ_O suitable for a switch from the *ls* electronic configuration of Co^{III} to the *hs* state of Co^{II} once an extra electron occupies the antibonding e_g orbitals. In the presence of a stronger ligand field, the *ls*-*hs* conversion would not be possible and the significant variation of the ground spin state would be then lost.

As in the case of SCO complexes, valence tautomers can be triggered through light irradiation of a proper wavelength (it must fall in the absorption range of the ligand-to-metal charge transfer, LMCT, band; cf. Section 3.3.1) to promote the electron transfer from the ligand to the metal also at temperatures at which the (*hs*)-Co^{II}-SQ state is not the most thermodynamically stable. Analogously to SCO complexes, this leads to a photoinduced metastable state that is formally the same as that observed in the high-temperature regime, with the only exception that at these temperatures this state would be short-lived. What can be experimentally observed in such compounds is a phenomenon of LIESST, with a characteristic time decay and a given temperature range where the metastable species can have a lifetime long enough to be observed. Such species being metastable in nature, it is quite a common finding that the photoinduced interconversion would only concern a restricted number of species, that is, the photoconversion only partially occurs.

Finally, it must be remarked that the valence tautomeric interconversion leads to strong variations also in the optical absorption properties of these materials, as it can be expected as a result of a different electronic distribution over the frontier MOs of the two electronic states. However, as these effects do not affect the magnetic properties of the compounds, they will not be discussed further herein.

3.1.7 Slow Magnetic Relaxation

The physical properties of the magnetic materials described so far can be categorized into two main classes: (i) materials in which a multitude of magnetic ions interact with each other through superexchange interactions to furnish a cooperative magnetic property (ferromagnetism, antiferromagnetism, or ferrimagnetism) and (ii) materials in which almost isolated magnetic ions, being surrounded by diamagnetic ligands, give a magnetic response related to an atomic magnetic moment (paramagnetism, SCO, or VT). Intriguing magnetic properties can be observed in those molecular materials that, in a certain sense, present a combination of the features of both the above-mentioned classes: an assembly, although limited to a discrete number, of magnetic ions that strongly interact through superexchange interactions to give a global high magnetic moment and a chemical structure in which the magnetic ions of each discrete molecule are surrounded by diamagnetic ligands that isolate each entity from the others, hampering supramolecular magnetic exchange interactions. Since the magnetic properties of such molecular materials are related to a, however sophisticated, single-molecular entity and resemble those of permanent magnets (vide infra), these systems are commonly called single-molecule magnets (SMMs). The origin of the magnetic behavior of SMMs is, however, very different with respect to that of ordered magnetic materials and is related to the slow relaxation of the magnetization. The slow relaxation of the magnetization is a fundamental feature that characterizes the magnetic molecular materials described in this section. For each of them, the origin of the magnetic behavior will be explained and the key chemical and structural features that are required to observe such properties highlighted.

3.1.7.1 Single-molecule magnets

The magnetic properties of matter can be very sensitive to the dimensions of the "magnetic object." The most evident manifestation of this assumption is the drastic change of magnetic properties that classical inorganic materials (e.g., ferrites) show when the material dimensions are reduced to the nanoscale. Magnetic nanoparticles are characterized by a peculiar magnetic behavior called superparamagnetism, which bulk materials cannot display. When the dimensions of a magnetic material are reduced to dimensions comparable to those of the magnetic domains, it is energetically unfavorable for the material to assume the typical multidomain magnetic structure, and a monodomain structure in which the magnetic ions constituting the nanoparticle behave as a "giant" and unique spin is preferred. Each nanoparticle thus behaves as a super paramagnetic atom or ion, with a magnetic moment that, in a first approximation, is given by the sum of the magnetic moments of hundreds or thousands of magnetic ions (depending on the nanoparticle dimensions). For this reason the property is referred to as superparamagnetism.[8]

---

[8]Note that this behavior is displayed by magnetic materials that show cooperative magnetic behavior in a bulk phase, such as ferromagnetism and ferrimagnetism. In this case, the cooperative interaction is extended within the magnetic domain that constitutes each nanoparticle. Thus, the superparamagnetic behavior can only emerge at temperatures lower than the ordering temperature since magnetic interactions between atomic or ionic magnetic moments are needed to obtain a "giant" spin.

For a spherical magnetic nanoparticle it can be reasonably assumed that its intrinsic magnetocrystalline anisotropy generates a unique axis along which the overall magnetic moment tends to be aligned. This axis is called the "easy axis" of magnetization, and its direction depends on the magnetic anisotropy, *K*, of the system. In this particular case, it is called uniaxial anisotropy.

It must be stressed here that the presence of a certain degree of anisotropy, whether related to the intrinsic electronic properties of the material or its geometrical features, is fundamental to observe an easy axis for the magnetization, as will be better evidenced further in this section for anisotropic magnetic molecules. As a consequence of the presence of an easy axis, the giant magnetic moment of a magnetic nanoparticle has two energetically favored orientations, the magnetic moment along the easy axis in a spin-up configuration and that in a spin-down configuration (Fig. 3.33).

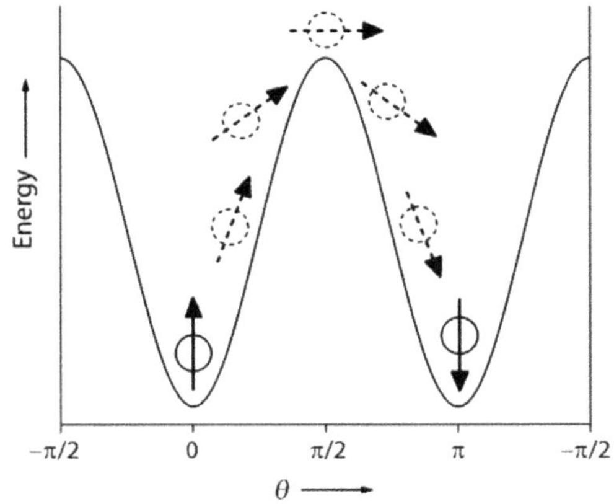

Figure 3.33 Energy profile of the magnetization for a magnetic nanoparticle as a function of the angle with respect to the easy-axis direction (θ). The energy minima correspond to the magnetization parallelly aligned ($\theta = 0$) or antiparallelly aligned ($\theta = 0$) with respect to the easy axis, while the energy maximum corresponds to the reciprocal perpendicular orientations ($\theta = \pm\pi/2$). The reversal of the magnetization over the energy barrier (according to the Stoner–Wohlfarth model $E_A = KV\sin^2\theta$, where K is the anisotropy constant and V is the volume of the nanoparticle) is also schematized.

These two states thus define two energy minima separated by an energy barrier that has to be overcome to reverse the magnetization from one state to the other. This energy barrier is a function of the magnetic anisotropy constant and the volume of the particle (Stoner–Wohlfarth model; see Fig. 3.33). Since the volume of a nanoparticle is very small, the amount of energy required to reverse the magnetization is comparable to the thermal energy k_BT. If the thermal energy is higher than the energy barrier, the magnetization will be easily reversed from one state to the other and the total magnetization is averaged to zero. If the thermal energy is lower than the energy barrier, the system will be unable to spontaneously reverse the magnetization from one state to the other and the magnetization assumes a nonzero value. Thus, the system is said to be unblocked at high temperatures and blocked at low temperatures. The time required for the magnetization to reverse to equivalent minima is called relaxation time τ[9] and has an exponential dependence on temperature (Néel–Brown theory; see Fig. 3.33), whereas the temperature that defines the boundary between the blocked and unblocked behavior is called blocking temperature T_B and ultimately defines the limit below which a nanoparticle shows superparamagnetism.

A similar superparamagnetic-like magnetic behavior can be observed in molecular systems containing a high number of magnetic ions (typically ≥ 4), that is, polynuclear metal complexes surrounded by some capping ligands. An archetype example of this class of molecular materials is the manganese(III/IV) polynuclear metal complex $[Mn_{12}O_{12}(CH_3COO)_{16}(H_2O)_4]$, usually called Mn_{12}, reported in Fig. 3.34, which represents the most famous example (Sessoli et al., 1993).

It is worth noting that the dimensions of these "magnetic objects" are in the low nanometer range (ca. 1–3 nm) and place them at the frontier between typical mononuclear complexes (<1 nm) and magnetic nanoparticles (ca. 3–100 nm). Indeed, in the scientific literature they are often defined as "molecular nanomagnets" to their dimensions on the nanometer scale.

Although such molecular materials and magnetic nanoparticles are similar from a certain point of view, some striking differences between them can be pointed out, as summarized below in reference to the archetype Mn_{12} compound:

---

[9]This parameter is not to be confused with the luminescence lifetime described in Section 2.3.4.3, which is indicated with the same symbol τ.

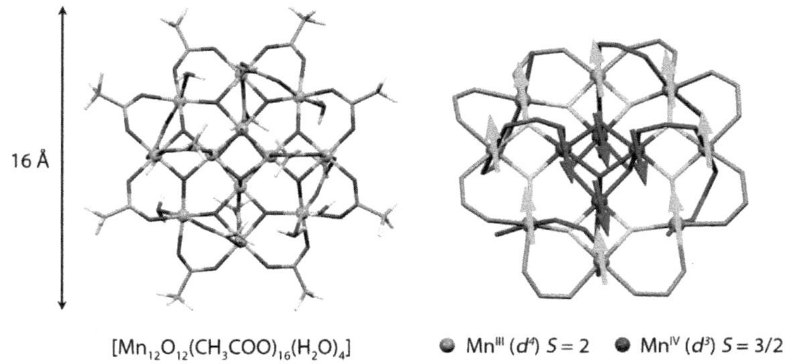

Figure 3.34 Molecular structure of the $[Mn_{12}O_{12}(CH_3COO)_{16}(H_2O)_4]$ polynuclear metal complex, an archetype example of a single-molecule magnet, with the dimension of the magnetic object highlighted (left). A detailed view of the relative orientation of the magnetic moments of the Mn^{III} and Mn^{IV} ions with respect to each other (right).

- The discrete number of magnetic ions, which communicate through specific magnetic exchange-coupling interactions, gives rise to a well-defined magnetic ground state **S**. In the specific case of Mn_{12}, a global **S** value of 10 comes from the uncompensated antiferromagnetic interactions between the eight peripheral Mn^{III} ions (**S** = 2) and the four Mn^{IV} ions (**S** = 3/2) located in the core ($\mathbf{S}_T = (8 \times 2) - (4 \times 3/2) = 10$) (Fig. 3.34).
- This large magnetic ground spin state is characterized by a relevant magnetic anisotropy that mainly results from that of the eight Mn^{III} ions, which display Jahn–Teller elongation. These crystal field distortions, together with the spin–orbit interaction, induce an axial zero-field splitting (ZFS) (cf. Section 2.1.9) for the Mn^{III} ions[10] and a preferential axis along which the magnetization tends to be aligned (easy axis of magnetization).
- The presence of an axial ZFS leads to a splitting of the **S** = 10 state into 21 levels (2**S** + 1), each characterized by a spin projection quantum number, m_S, ($+\mathbf{S} \geq m_S \geq -\mathbf{S}$), with an energy $E(m_S) = Dm_S^2$. Since D has a negative sign, the lowest-energy states are those having the maximum absolute value of m_S (m_S

[10]The zero-field splitting parameter D has a value of -0.50 cm^{-1}.

= ±10), while the highest-energy state has $m_S = 0$ (cf. Section 2.1.9). The values of the m_S energy states as a function of the angle θ between the directions of the magnetization and the easy axis of magnetization can thus be represented as shown in Fig. 3.35.

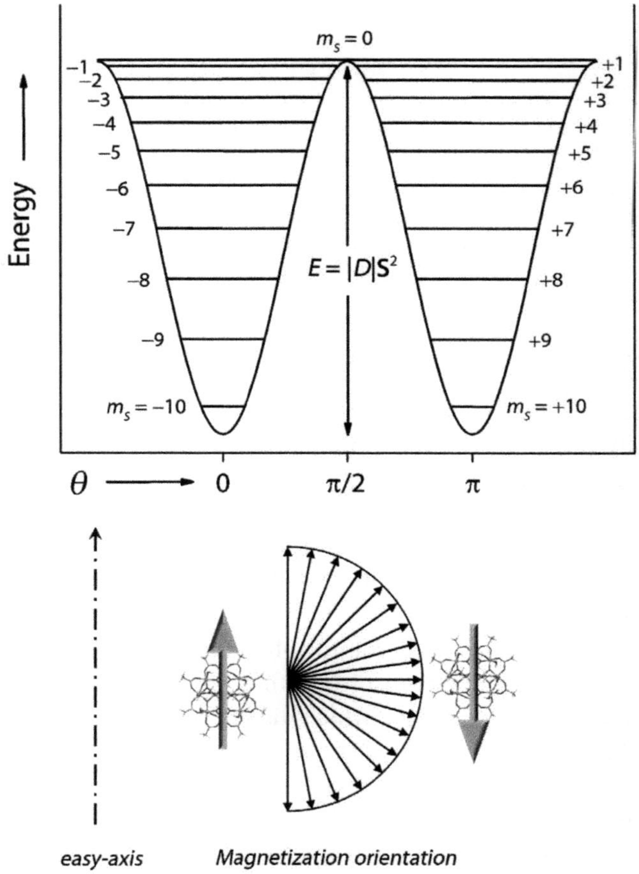

Figure 3.35 Energy profile of the magnetization for a single-molecule magnet (Mn_{12} in this specific case) as a function of the angle with respect to the easy-axis direction (θ). The energy minima correspond to the magnetization parallelly aligned ($\theta = 0$) or antiparallelly aligned ($\theta = 0$) with respect to the easy axis, while the energy maximum corresponds to the reciprocal perpendicular orientations ($\theta = \pm\pi/2$). The reversal of the magnetization over the energy barrier is also schematized at the bottom with reference to the quantization of the spin projection quantum number.

The potential energy profile as a function of the angle θ between the magnetization of a molecular nanomagnet and the direction of the easy axis closely resembles that shown by a magnetic nanoparticle. However, while in the nanoparticle case the energy barrier is a function of the intrinsic nature of the material through the magnetoanisotropy constant K and the size of the nanoparticle through its volume V, here the energy barrier is purely correlated to the molecular magnetic properties: (i) the value of the global ground spin state **S** (that is determined by the nature of the involved metal ions and their reciprocal interactions) and (ii) the magnitude of the magnetoanisotropy of the system through the absolute value of the zero-field-splitting parameter D according to the relations:

$$E = |D|\mathbf{S}^2 \text{ and } E = |D|(\mathbf{S}^2 - 1/4) \tag{3.4}$$

which refer to the case of integer and half-integer ground spin states, respectively. A high ground spin state **S** and a high anisotropy (D) provide a relevant value of the energy barrier E.

The extent of the magnetic anisotropy in a molecule is basically related to its geometrical anisotropy, particularly to the elongation along a specific direction, and determines the entity of the energy barrier separating the two possible energetically stable configurations. As a result, the energy barrier can be modulated by playing with the nature and number of coordinated metal ions, their relative disposition in the molecular structure, and the type of ligands, which can introduce different degrees of magnetoanisotropy in the system. Another relevant aspect that has to be pointed out, even if for coordination chemists it may sound rather obvious, is that all these molecular magnetic objects have the same size, contrary to what is most often observed for magnetic nanoparticles, where a certain dimensions' distribution, however controllable, is always present. This has important implications since the energy barrier is a function of the volume of the magnetic object, implying that nanoparticles within a range of different dimensions will show a distribution of values for the energy barrier. This drawback is, of course, not present in molecular compounds in which the stoichiometry is well defined, and the energy barrier is equal for all the molecules constituting the material.

Now that the origin of the energy barrier has been defined through specific electronic and structural parameters, we can further deepen the discussion by explaining how the energy barrier is related to the slow relaxation of the magnetization, which appears more as a temporal, rather than energetic, aspect, once again by taking Mn_{12} as a reference.

Similar to what is explained above for superparamagnetic behavior, in order to reorient the spin or to reverse the magnetization of a Mn_{12} molecule from the two minimum energy values $m_S = \pm 10$, a certain energy defined by Eq. 3.4 has to be furnished to the system. For this specific case, where $\mathbf{S} = 10$ and $|D| = 0.50\ \text{cm}^{-1}$ = ca. 0.72 K,[11] the amount of energy is ca. 72 K. As this process of spin reorientation is thermally activated, the time for the reorientation (relaxation time τ) of the magnetization depends on the energy barrier and the thermal energy through an exponential Arrhenius-type law:

$$\tau = \tau_0\, e^{\left(\frac{\Delta E}{k_B T}\right)} \tag{3.5}$$

This means that the fit of a plot of $\ln(\tau)$ versus $1/T$ gives the value of the energy barrier, which is commonly indicated in the scientific literature as U_{eff}. Thus, if the thermal energy is lower than that of the energy barrier, in principle, there is not enough energy available for the magnetization to be reversed and the system is said to be blocked, whereas if the thermal energy is higher than the energy barrier, the system is unblocked and fast relaxation occurs. Fast relaxation means that, in the absence of an energy barrier, the time for the magnetization to relax back to the ground state once the system is magnetized by an external magnetic field is of the order of submicroseconds. This timescale is comparable with that of the same process that actually occurs, though too fast to be detectable, in the majority of nonanisotropic magnetic systems. When the energy barrier and the thermal energy are comparable, there will be a characteristic relaxation time for each temperature value. Indeed, by studying the relaxation time as a function of the temperature with dynamic magnetic measurements (see the box titled "Physical Phenomena and Investigation Techniques Timescales"), such as alternate current (AC) susceptometry, it is possible to follow the thermal variation of the relaxation time and to extract the effective

---

[11] $1\ \text{cm}^{-1} = 0.695$ K.

energy barrier that the system is subjected to (vide supra). The amount of energy necessary to reverse the magnetization for a SMM is, in general, low with respect to the magnetic nanoparticles, so these properties are commonly observed at low temperatures (<10 K). For these reasons, one of the main challenges in this research field is to increase as much as possible the blocking temperature of these molecular materials.

Physical Phenomena and Investigation Techniques Timescales

When dealing with the investigation of the dynamic physical properties of matter, as, for example, the relaxation of the magnetization in magnetic nanoparticles or SMMs, it is important to point out that the observation of a certain behavior is dependent on the measurement timescale of the employed technique. Given the measurement time τ_m and the timescale τ of the physical phenomenon, if $\tau_m > \tau$, the technique furnishes information on the state of the system that is already evolved. If $\tau_m < \tau$, the technique provides information on the state of the system during its evolution.

Let's take as an example the case of the slow relaxation of the magnetization for SMMs. This phenomenon occurs within a timescale that is of the order of milliseconds. A technique that only allows a delay of a second between successive measurements will not be able to provide information about the evolution of the system. That will therefore appear "static" to the observer. Instead, if the same phenomenon is investigated through a technique with a temporal resolution in the millisecond timescale, the dynamics of the system's evolution from the initial to the final state can be disclosed.

These considerations point out that misleading information about the physical properties of a certain system can be provided if the experimental technique is not properly chosen depending on the timescale of the physical phenomena under investigation.

Typical timescales for experimental techniques commonly employed in the field of magnetic materials are (Launay and Verdaguer, 2014):

- Direct current (DC) magnetometry: 10^{1}–10^{2} s
- Alternate current (AC) susceptometry: 10^{-5}–10^{0} s
- Nuclear magnetic resonance (NMR): 10^{-8}–10^{-3} s
- Electron paramagnetic resonance (EPR): 10^{-11}–10^{-7} s
- Mossbauër: 10^{-10}–10^{-7} s
- Muon spin resonance: 10^{-6}–10^{-3} s

Due to the limited number of magnetic ions, SMMs, unlike nanoparticles, show well-defined allowed energy levels for the magnetization (Fig. 3.35).[12] This means that quantum size effects, which affect the reversal of the magnetization process, can emerge in these molecular materials.

It is important to highlight that SMMs show the magnetic behavior described so far also at zero field, as the two minima are generated by ZFS. If the material is subjected to the action of an external magnetic field, one of the minima will be energetically stabilized (the m_S state antiparallel-oriented with respect to the field vector **H**), while the other level will be energetically destabilized (the m_S state parallel-oriented with respect to the field vector **H**), as depicted in Fig. 3.36.

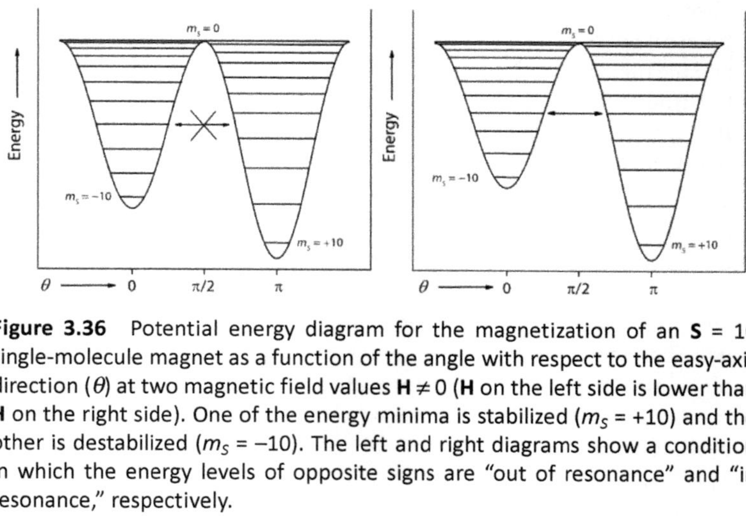

Figure 3.36 Potential energy diagram for the magnetization of an **S** = 10 single-molecule magnet as a function of the angle with respect to the easy-axis direction (θ) at two magnetic field values **H** $\neq$ 0 (**H** on the left side is lower than **H** on the right side). One of the energy minima is stabilized (m_S = +10) and the other is destabilized (m_S = −10). The left and right diagrams show a condition in which the energy levels of opposite signs are "out of resonance" and "in resonance," respectively.

As the strength of the field increases, the energy of the m_S levels changes proportionally. The population on the m_S states is affected by this perturbation according to a Boltzmann distribution. At low temperature only the ± 10 m_S states are populated, and they are equally populated at zero field. The application of an external field determines a change in the occupancy of the m_S states and only the

[12]This can be seen as analogous to the difference between the well-defined MO energy levels for a molecule and the continuum of energy levels in the band structure of the metal shown in Section 2.2.3. In the case of magnetic nanoparticles, the high number of atoms generates a continuum of energy levels so all possible orientations are allowed.

energetically stabilized m_S state will be populated. When the field is removed, an equilibrium Boltzmann distribution must be re-established; thus the spin population transferred to the +10 m_S state must return to the −10 m_S. In other words, the orientation of the spins has to flip for the magnetization to relax back to zero. This occurs through energy transfer (ET) from the environment (the lattice) to the spin via what is called spin-lattice relaxation, with a characteristic spin-lattice relaxation time τ. The energy is absorbed through the vibrational modes of the lattice (phonons) via spin-phonon interactions according to the selection rules $\Delta m_S = \pm 1$ or ± 2 depending on the type of relaxation mechanism (see the box titled "Spin-Lattice Relaxation Mechanisms") undergone by the system. Thus, in order to reverse the magnetization from $m_S = +10$ to $m_S = -10$, the molecule should consequently absorb phonons until the $m_S = 0$ state is reached and then it can relax to the $m_S = -10$ state through phonon emission. The reverse situation of stabilization/destabilization of the m_S states will be observed if the field has the opposite sign.

Spin-Lattice Relaxation Mechanisms

Different types of spin-lattice relaxation mechanisms based on spin-phonon absorption can occur. They differ from the energy and number of spin states involved, and each one has its characteristic temperature dependence.

Direct mechanism: In the direct process, the spin system relaxes by transferring its energy to the lattice through a phonon that has the same energy as the Δm_S transition. The selection rule is $\Delta m_S = \pm 1$ or ± 2. This process operates at low temperature as it involves low-energy phonons. The temperature dependence of the relaxation time is proportional to $1/T$.

Raman mechanism: In the Raman process, the spin system relaxes by transferring its energy to the lattice through a phonon that has the same energy as the Δm_S transition. However, differently from the direct process, this energy resonance is achieved when the lattice absorbs a high-energy phonon, up to a virtual excited state, and partial relaxation occurs through phonon emission toward a low-energy vibrational excited state whose energy matches that of the excited state of the system. This two-phonon process occurs at higher temperatures with respect to the direct process since it involves phonons of higher energy. The temperature dependence of the relaxation time is proportional to $1/T^n$, with $5 < n < 9$ depending on the nature of the system.

Orbach mechanism: In the Orbach process, the spin system relaxes through a two-phonon mechanism, as in the Raman process, but involves real, not virtual, intermediate excited states of both the lattice and the spin system. The temperature dependence of the relaxation time has the form of an Arrhenius-type law and allows one to associate the relaxation time with a characteristic energy barrier (Eq. 3.5).

A schematic representation of the energy diagram of the lattice and the spin system and the different relaxation mechanisms is depicted in the figure below (Liddle and van Slageren, 2015).

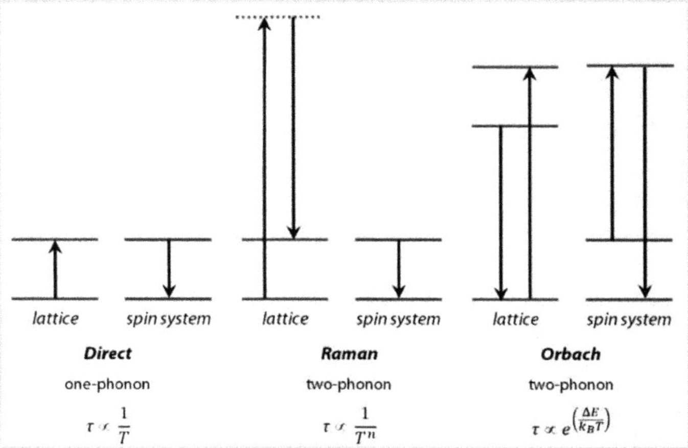

An important relaxation mechanism not involving phonon absorption also exists:

Quantum tunneling: In the quantum tunneling mechanism of relaxation, the magnetization relaxes under the energy barrier through the superposition of $\pm m_S$ spin states (vide infra).

The coexistence of different relaxation mechanisms can be often observed when describing the slow relaxation of the magnetization of SMMs. In this case, the temperature dependence of the relaxation time is given by the sum of the involved contributions, each one with its own relative coefficient.

Thus, as already mentioned previously, if there is no sufficient thermal energy to allow for the magnetization to relax, the magnetization cannot be quickly reversed by the application of an external field of opposite strength, the way it occurs for simple paramagnets, so the system is blocked and the field dependence of the magnetization shows a hysteretic behavior (remanent magnetization and coercive field) analogous to that shown by cooperative magnetic materials as ferro- and ferrimagnets. This introduces bistability at the single-molecule level , meaning that an SMM can be used, in principle, as a nanosized single-molecule memory for information storage applications.

The hysteresis cycle of SMMs can show additional features related to the quantization of the energy levels. Bearing in mind what is explained above about the stabilization/destabilization of the energy levels, there will be specific magnetic field values for which the energy of a $-m_S$ state will correspond to that of a $+m_S$ state (see Fig. 3.36, right). This means that the levels are "in resonance" and, for such field values, the magnetization can relax "under the barrier" through quantum tunneling. In such a condition, there is no reason for the magnetization to prefer one of the two orientations, so the opposite orientation also becomes allowed (admixing of states). The consequence is that the overall magnetization is averaged, and it suddenly drops or increases as the field changes (Fig. 3.37) due to an immediate reversal through the quantum tunneling "shortcut" below the potential energy wall.

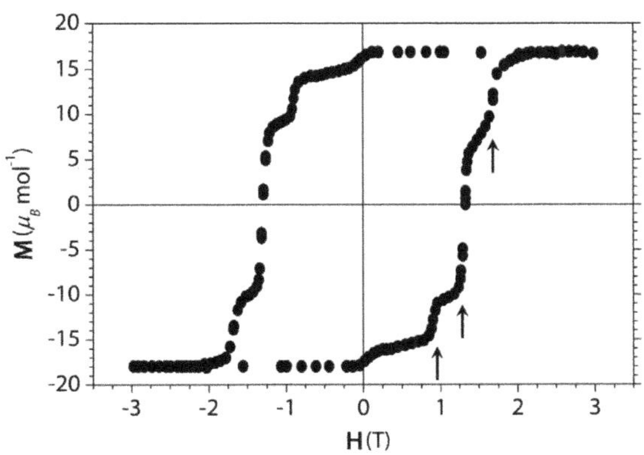

Figure 3.37 **M** versus **H** hysteresis loop for the compound Mn_{12} recorded at T = 2.1 K. The vertical "steps" (some of them indicated with arrows) correspond to critical values of the field where resonant magnetization tunneling is allowed and observed. Adapted from Christou et al. (2000), reproduced with permission.

For all the other field values where the system is "out of resonance," quantum tunneling is forbidden and only slow relaxation mediated by thermal effects (above the energy barrier) determines the dependence of **M** on **H**.

To summarize what is explained until now, SMMs are molecular materials characterized by a relevant easy-axis anisotropy (D) that associates with the ZFS of the high ground spin state (**S**) and the energy barrier (E). This energy barrier, which is proportional to both D and **S** (Eq. 3.4), defines the energy required to reverse the magnetization of the system between two minima (the most energetically stabilized spin states $\pm m_S$) in both zero and nonzero fields. The process to reverse the magnetization is thermally activated, which means that below a certain temperature there is not enough energy to reverse the magnetization. This also implies that the magnetization can relax only slowly and once the material is magnetized this condition is retained for a long time if the temperature is kept below the blocking temperature. Such slow relaxation leads to the observation of an opened hysteresis cycle, similar to that observed for ferro- or ferrimagnetic materials, but with an origin that is completely different, it being a single-molecule property. The presence of such hysteretic behavior at the single-molecule level makes these molecular materials of great interest as nanosized magnetic objects for quantum information storage. An additional feature related to the quantum nature of these magnetic properties is the observation of clear steps in the hysteresis cycle where the slow relaxation is speeded up due to a competitive relaxation phenomenon with respect to the thermally activated one, that is, the quantum tunneling of the magnetization.

With these concepts in mind it is now easy to summarize the main electronic and structural features that a system should have to show slow magnetic relaxation and behave as a SMM:

- The molecular structure of a polynuclear metal complex where several high-spin magnetic ions (typically ≥ 4) communicate through specific magnetic exchange-coupling interactions to give rise to a well-defined and high magnetic ground state **S**
- The presence of some capping ligands at the complex periphery to isolate each molecule from the others and exclude the presence of intramolecular magnetic exchange interactions

- A metal ion's coordination environment that can result in a local magnetic anisotropy (spin–orbit coupling and ZFS)
- A molecular structure in which the metal ions are arranged in an elongated or compressed shape to favor a uniaxial orientation of the magnetization

The rational preparation of a polynuclear metal complex with a large ground spin state is not a simple task for synthetic chemists, especially since it is difficult to control the number and position of the coordinated metal ions in the overall molecular structure. The complex-as-ligand approach described before for the preparation of long-range ordered molecule-based magnets applies also in the field of SMM. Indeed, by linking relatively small building blocks containing two to four coordinated metal ions with purely inorganic ligands (OH^-, RO^-, $RCOO^-$, R_3SiO^-, where R is a generic alkyl group), polynuclear metal complexes with a higher nuclearity can be obtained (Christou et al., 2000). Alternatively, polydentate ligands able to link several metal ions have been developed to prepare such polynuclear metal complexes. It must be highlighted, however, that the type of superexchange interactions also has to be controlled. Polynuclear metal complexes in which all the metal ions are coupled in an antiferromagnetic way to give a global small **S** ground spin state will not behave as good SMMs. Also, molecules of high symmetry where the ZFS is negligible ($D \approx 0$) will not behave as good SMMs however large is the **S** ground spin state achieved with metal ions.

Following these design principles, a multitude of SMMs with increasingly larger spins have been prepared and their magnetization dynamics characterized. Among them we can cite as some of the very first examples the compound $[Fe_8O_2(OH)_{12}(tacn)_6]Br_8$ (tacn = triazacyclononane) (Wieghardt et al., 1984), or Fe_8, which has a larger anisotropy with respect to the Mn_{12} system due to the higher asymmetry of the coordination sites, and the compound $[Fe_4(OCH_3)_6(dpm)_6]$ (dpm = anion of dipivaloylmethane) (Barra et al., 1999), or Fe_4, which has been the subject of intense investigations for the ability it has shown to be anchored on solid surfaces without losing its magnetic properties (see Chapter 5) (Fig. 3.38).

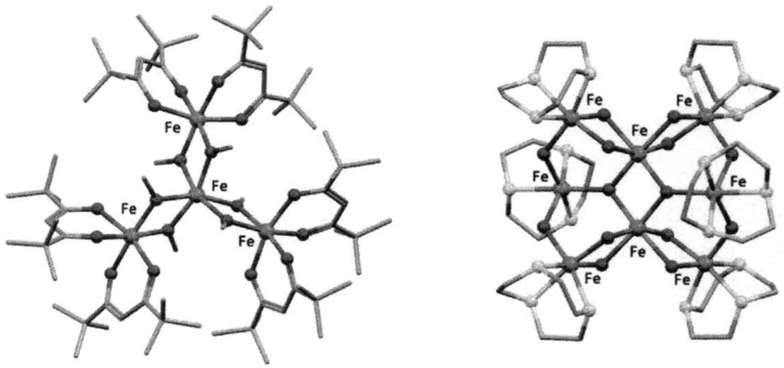

Figure 3.38 Molecular structures of the $[Fe_4(OCH_3)_6(dpm)_6]$ (left) and $[Fe_8O_2(OH)_{12}(tacn)_6]^{8+}$ (right) polynuclear metal complexes, two archetype examples of single-molecule magnets.

One of the main challenges in this field is thus to increase as much as possible the blocking temperature of the slow relaxation of the magnetization by increasing the energy barrier, U_{eff}. This will allow exploitation of these materials at a useful temperature for their practical application. Nowadays the record is held by a mononuclear Dy-based SMM (see Section 3.1.7.2) that shows hysteresis up to 60 K (Goodwin et al., 2017; Guo et al., 2017).

As briefly mentioned earlier, another very challenging direction in this field is the deposition of such molecules on surfaces and the investigation of the magnetic properties once they have been deposited. This can be realized by functionalizing the ligand periphery with suitable anchoring functional groups. This topic will be one of the themes of Chapter 5.

3.1.7.2 Single-ion magnets

The discovery of the first SMMs prompted the chemists' community to make efforts to increase the number of magnetic ions within a molecular unit. This was aimed to increase the energy barrier between the two energetically equivalent magnetic ground states (cf. Eq. 3.4) in the effort to raise the temperatures of the slow relaxation process to operating conditions useful for practical applications in technological devices (vide supra). Unfortunately, although nice chemical structures comprising more than 80 magnetic ions have been prepared and characterized so far, the energy barriers, and thus the blocking temperatures, were not as high as predicted by the theory. These results induced scientists to pursue a different strategy. The quest for the highest ground spin state **S** was partially abandoned in favor of the pursuit of larger axial anisotropy values (*D* parameter). By taking advantage of the higher magnetic anisotropy that lanthanides intrinsically show with respect to the less anisotropic d elements, another class of magnetic molecules that display very similar properties with respect to SMMs has been developed. Accordingly, it has been reported that mononuclear complexes of lanthanide ions show slow relaxation of magnetization in both zero and nonzero magnetic field. In these systems the high ground spin state is provided by a unique metal ion instead of a multitude, as seen, for example, in the Mn_{12} compound. Consequently these materials have been called single-ion magnets (SIMs), and they can be considered the second generation of SMMs.

The archetype example of this class of compounds is the $[Tb^{III}(Pc)_2]^-$ metal complex (Pc = phthalocyaninate ligand, the dianion of the phthalocyanine). $[Tb^{III}(Pc)_2]^-$ is a sandwich compound in which the Tb^{III} metal ion is coordinated by eight nitrogen atoms of two planar Pc ligands disposed above and below the metal, which features a square antiprismatic coordination geometry (Fig. 3.39).

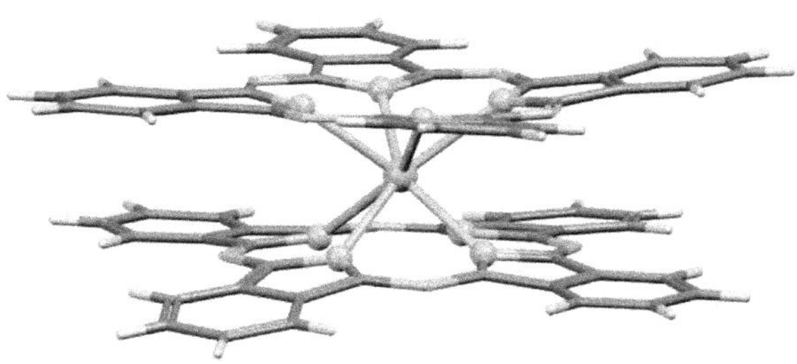

Figure 3.39 View of the molecular structure of the $[Tb^{III}(Pc)_2]^-$ metal complex.

The Tb^{III} ion has a $4f^8$ electronic configuration in which the eight electrons occupy the seven degenerate 4f orbitals. Thus, it is characterized by a set of **S**, **L**, and **J** quantum numbers of 3, 3, and 6, respectively, which give rise to a ground state term symbol $^7F_6$ (see Section 2.1.6.1). As mentioned in Section 2.3.7, for lanthanide ions the spin–orbit coupling has more influence than the ligand/crystal field in determining the ground spin state, and the $^7F_6$ state is further split by the ligand/crystal field into 2**J** + 1 m_J sublevels. In this particular case, the lowest m_J levels are those characterized by $m_J = \pm 6$. Thus, Tb^{III} in $[Tb^{III}(Pc)_2]^-$ has a negative axial ZFS parameter *D*. The direction of the anisotropy axis is perpendicular to the planes of the ligands. Both the negative *D* parameter and the high spin state **S** have been previously highlighted as fundamental prerequisites for the preparation of molecules that show slow relaxation of magnetization. The $[Tb^{III}(Pc)_2]^-$ system is thus characterized by a relative high spin state **S** = 3 and, more importantly, a strong axial spin–orbit coupling, which induces a large separation in energy (more than 400 cm^{-1}) between the lowest occupied ($m_J = \pm 6$) and the "first excited" energy sublevels ($m_J = \pm 5$) (Fig. 3.40) (Launay and Verdaguer, 2014).

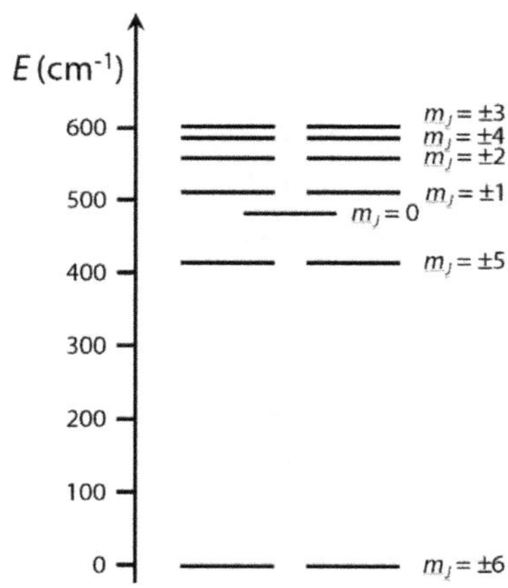

Figure 3.40 Energy diagram of the $^7F_6$ ground multiplet sublevels for the $[Tb^{III}(Pc)_2]^-$ metal complex.

This means that, analogously to SMMs, there is an energy barrier that has to be overcome to reverse the magnetization also in SIMs. Notably, in SIMs the requirements needed to observe slow relaxation of magnetization are fulfilled by means of a single metal ion with a large axial anisotropy. Although the energy barrier (ca. 230 cm^{-1}) is not as high as expected on the basis of theoretical predictions, slow relaxation of the magnetization is observed up to ca. 40 K, a temperature value that, compared to that shown by the Mn_{12} compound (ca. 6 K), represents a very good step forward toward the observation of such phenomena at ambient temperatures.

By following these criteria, a large number of SIMs have been realized by selecting suitable lanthanide ions with the highest intrinsic anisotropic properties, which are not constant across the series of these elements. Most SIMs have been prepared by taking advantage of the strong axial anisotropy shown by the Dy^{III} ion, as well as by combining different lanthanides into a single-molecular entity (polynuclear compounds). One of the main challenges in this field arises from the difficulties in controlling the chemistry of lanthanide ions, which do not follow the rules common to transition metal ions (cf. Section 3.3.4), and consequently meeting the geometrical requirements needed to achieve SIMs' properties. A very well thought design of the coordinating ligands is therefore crucial to template the synthesis of a complex with the desired symmetry, and this may be a rather tricky task.

Recently, SIMs based on low-coordinated Co^{II} mononuclear compounds have been also reported (Rechkemmer et al., 2016). The origin of the slow relaxation of the magnetization in these SIMs based on a transition metal ion is related to the relatively high axial ZFS parameter due to the non-negligible spin–orbit coupling of the anisotropic Co^{II} ion (c.f. Section 3.1.6 and footnote 7).

3.1.7.3 Single-chain magnets

In the previous sections we have described the slow relaxation of the magnetization in discrete molecular units composed of a multitude of interacting metal ions (SMMs) or a single ion with a high axial magnetic anisotropy (SIMs). Both systems can be categorized as 0-D "molecular objects" as their three dimensions in space assume comparable values, that is, their shape can be approximated to that of a sphere. Such molecular dimensionality is fundamental to observing this peculiar magnetic behavior since it can be more easily observed if the magnetic interactions between different molecules are negligible (vide supra). Accordingly, systems with 2D or 3D dimensionality are, in principle, not the best suited to show this phenomenon because extended magnetic interactions between metal centers will not allow large and well-defined ground spin states. Molecular chains, that is, systems in which one of the three space dimensions is larger than the other two, are 1D systems that are placed at the boundary between the molecular arrangements and the dimensionalities cited above. In such systems, magnetic interactions between metal centers are able to provide a large ground spin state due to short-range magnetic interactions and the presence of diamagnetic ligands on the chain periphery can limit the interactions between them. In the presence of such conditions, analogously to SMMs, whenever they are composed of metal ions showing an easy-axis magnetic anisotropy, such molecular materials can show slow relaxation of the magnetization with long spin-lattice relaxation times and opened hysteresis cycles. Given their dimensionality and the magnetic behavior analogous to SMMs, these systems are known as single-chain magnets (SCMs).

Similar to what is reported for SMMs and SIMs the main electronic and structural features that a system should have to show slow magnetic relaxation and to behave as a SCM are the following:

- The molecular structure of a 1D coordination polymer where several magnetic ions communicate through strong and specific short-range magnetic exchange-coupling interactions to give rise to a well-defined and high magnetic ground state **S**[13]
- The presence of some capping ligands at the complex periphery to isolate each molecular chain from the others in order to limit intramolecular magnetic interactions between them
- A strong easy-axis magnetic anisotropy of the metal ions constituting the magnetic chain

The first reported example in which such properties were found is the $[Co^{II}(hfac)_2(NITPhOMe)]$ ($hfac^-$ = hexafluoroacetylacetonate; NITPhOMe = methoxyphenyl-nitronylnitroxide) system (Caneschi et al., 2001), which represents an archetype example (Fig. 3.41).

[13]The number of interacting magnetic ions is limited by the length of those portions of the magnetic structure that are called magnetic domains. They are generated by the spontaneous reversal of the spin orientation or can be created by introducing defects on the magnetic chain. The latter can be generated by adding a certain amount of diamagnetic ions that substitute the paramagnetic ones without altering the overall crystal structure.

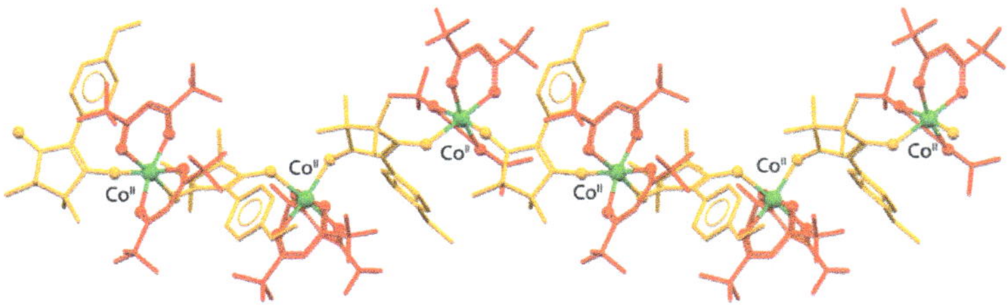

Figure 3.41 Portion of the molecular structure of the [$Co^{II}(hfac)_2$(NITPhOMe)] 1D coordination polymer, an archetype example of SCMs. The $hfac^-$ and methoxyphenyl-nitronylnitroxide ligands are reported in red and yellow, respectively. Only one chain of the crystal structure is reported for graphical clarity.

This system has been obtained by combining preformed [$Co^{II}(hfac)_2(H_2O)_2$] complexes with the neutral organic radical NITPhOMe, which acts as a bridging ligand. The presence of an unpaired delocalized electron on the bridging ligand ensures strong magnetic exchange interactions between the localized magnetic moments of the metal ions, while the $hfac^-$ ligands and the bulky substituent on the nitronylnitroxide limit magnetic interactions between chains.

In this magnetic material, the Co^{II} ion behaves, at low temperature, as an effective **S** = ½ spin center with high anisotropic **g** values while the **S** = ½ spin centers of the organic radicals have typical free electron **g** values (ca. 2). This results in noncompensated antiferromagnetic interactions typical of ferrimagnetic short-range ordering. Here, the short-range interactions are established, thanks to the 1D nature of the molecular structure. At low temperature, the system shows slow relaxation of the magnetization with relaxation times that have an Arrhenius-type dependence on temperature, similar to that shown by SMMs and SIMs (cf. Eq. 3.5). This behavior was predicted by Glauber in the 1960s for a 1D chain showing ferromagnetic ordering. He defined the energy barrier to reverse the magnetization as $E = 4J\mathbf{S}^2$, where J is the exchange coupling constant and **S** the ground spin state:

$$\tau = \tau_0 \, e^{\left(\frac{4JS^2}{k_B T}\right)} \tag{3.6}$$

where $4J$ is the energy cost to reverse one spin of the chain from the up configuration to the down configuration, by taking into account that each spin inside the chain interacts with an energy $2J$ with its two nearest neighbors' spin (Fig. 3.42).

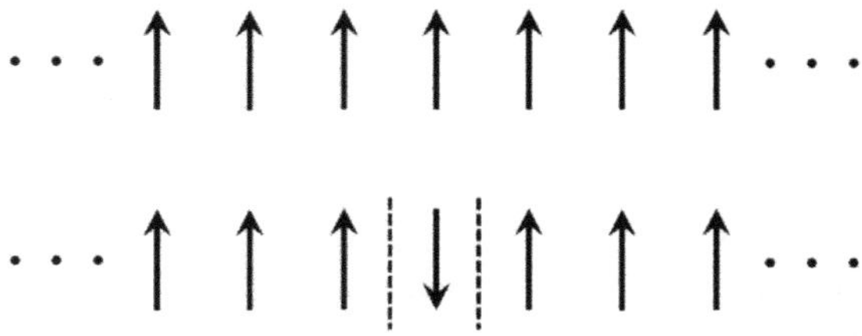

Figure 3.42 Schematic representation of an infinite 1D chain of ferromagnetically aligned magnetic moments (up) and of the same chain with a magnetic moment with reversed magnetization (down).

By comparing the expression of the energy barrier for a SMM (Eq. 3.4) and the one just defined for a SCM, we can see that in both cases an energy barrier is responsible for the slow relaxation of the magnetization, although these barriers are associated to a different key parameter. In the case of SMMs and SIMs, it is the axial ZFS parameter D, whereas in SCMs it is the magnitude of the nearest-neighbor exchange interactions J (Launay and Verdaguer, 2014). Both depend on the magnitude of the ground spin state **S**. Since exchange coupling constants are commonly much higher than typical D values the result is that, in principle, higher energy barriers can be obtained with SCMs, thus extending the slow relaxation of the magnetization toward higher temperatures.

Finally, it is worth mentioning that SCMs can be prepared by linking SMMs in the same way as shown in Fig. 3.41 for the simpler mononuclear Co^{II} complexes. In such an arrangement, the slow relaxation of the magnetization of SMMs due to the ZFS is combined with the magnetic exchange interactions between them, as discovered for SCMs. This means that the temperature dependence of the relaxation time will be proportional to an energy barrier given by the sum of the two above-mentioned contributions. As a result, the relaxation times are expected to be longer. An archetype example of such compounds is the $[(Et)_4N]_2[Mn^{III}(5\text{-MeOsalen})]_2[Fe^{III}(CN)_6]$ (5-MeOsalen = 5-metoxysalycilethylenediamine) (Ferbinteanu et al., 2005) polymer, where [Mn(5-MeOsalen)] SMMs units are linked through paramagnetic $[Fe^{III}(CN)_6]^{3-}$ complexes (Fig. 3.43).

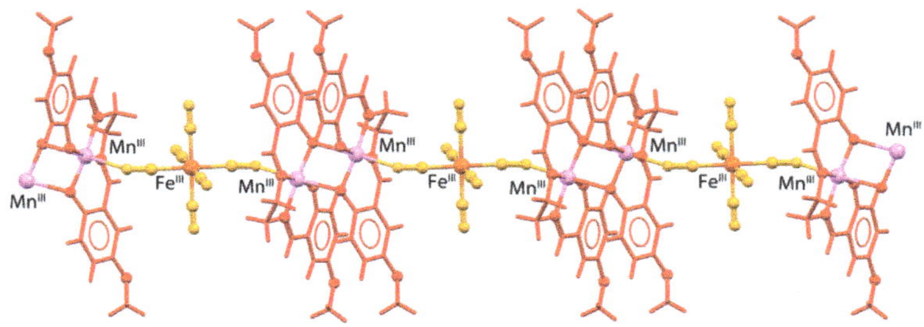

Figure 3.43 Portion of the molecular structure of the $[(Et)_4N]_2[Mn^{III}(5\text{-}MeOsalen)]_2[Fe^{III}(CN)_6]$ 1D coordination polymer, an archetype example of SCMs constructed by self-assembling of SMMs and paramagnetic anions. [Mn(5-MeOsalen)] and $[Fe(CN)_6]$ building blocks are reported in red and yellow, respectively. For graphical clarity tetraethylammonium cations are omitted and only one chain of the crystal structure is reported.

3.1.7.4 Molecular spin qubits

In the previous sections, molecular materials showing slow relaxation of the magnetization resulting from the presence of an energy barrier between two spin states have been presented and their properties described. All these materials have attracted great attention because of the occurrence of magnetic bistability at the molecular scale that can be, in principle, exploited to store information on quantum bases. Quantum storage is one of the branches of quantum information technology (QIT), that is, the use of quantum systems to develop a technology to store, retrieve, transmit, and manipulate data and information. Another branch of relevant interest for QIT, especially because it is at the basis of the realization of quantum computers, is the quantum information processing, that is, the use of quantum systems to perform quantum logic operations (Nielsen and Chuang, 2004). Whereas SMMs and SIMs are better suited to store information due to the large spin and the magnetic bistability, molecules with a smaller spin value that show slow relaxation of the magnetization can be exploited to perform quantum logic operations. These systems are called molecular spin qubits, that is, molecules in which a spin, either nuclear or electronic, can be exploited as a quantum bit (qubit) of information. Interest in this class of molecular materials has rapidly increased in the last few years as the realization of quantum computers is seen as one of the most fascinating challenges of this century and it would represent a real revolution in the field of information technology.

While a classical bit can assume only two states, $|0\rangle$ and $|1\rangle$, a qubit can assume any coherent superposition of these two states, as a linear combination: $|\psi\rangle = \alpha|0\rangle + \beta|1\rangle$ (Fig. 3.44) (Nielsen and Chuang, 2004). The assumed state must be long-lived to perform a quantum logic operation, and decoherence, that is, the decay of the superposition state, is one of the main limitations in the use of qubits.

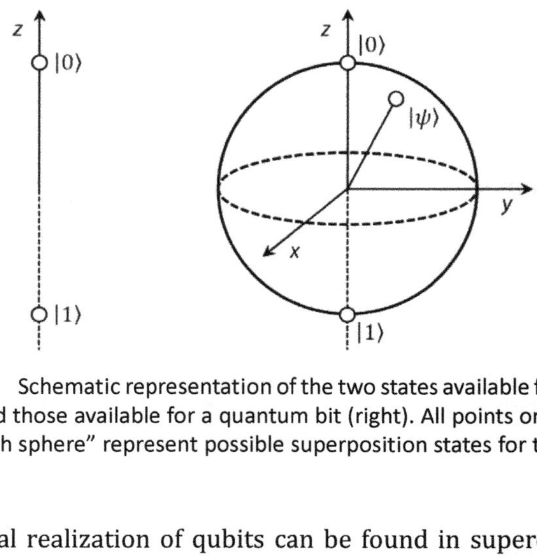

Figure 3.44 Schematic representation of the two states available for a classical bit (left) and those available for a quantum bit (right). All points on the surface of the "Bloch sphere" represent possible superposition states for the quantum bit.

Physical realization of qubits can be found in superconductive circuits and trapped ions and can be based on photons, spins, charges, or quantum dots. Among them, both nuclear and electronic spins are particularly interesting since the superposition of spin states can be realized by pulsed magnetic resonance techniques.

In the field of molecular materials, electronic spins are the most intuitive realization of qubits, and in particular, paramagnetic molecules with a spin $\mathbf{S} = ½$ guarantee two easy accessible quantum states, $m_S = +½$ and $m_S = -½$, which can be placed in a coherent superposition for a certain amount time. The time during which the superposition state is maintained is defined as phase memory time or quantum coherence time, T_m, and is governed by the spin-spin relaxation time, usually indicated as T_2 in magnetic resonance techniques (NMR and EPR). Spin-spin relaxation is in turn limited by the spin-lattice relaxation time, τ (indicated as T_1 in magnetic resonance techniques), which is the same physical phenomenon that is at the basis of the slow magnetic relaxation of SMMs and SIMs, although the origin is intrinsically different (vide infra). If the spin-lattice relaxation time is sufficiently long, slow relaxation of magnetization is observed, and relevant phase memory times can be detected. If the phase memory times are longer than the time required to perform a quantum logic operation (the simplest case is the reversal of the spin orientation in a NOT operation), the system can behave as a qubit. It must be highlighted that to perform more complex quantum logic operations longer phase memory times are required.

After this short overview on spin qubits and their characteristic time constants, we will first define the main structural and electronic features that a molecule should have to behave as a potential molecular qubit. Afterward we will analyze in more detail each aspect and the physical phenomena behind it.[14] Indeed, systems with the following features seem the most promising to prepare effective molecular qubits:

- A molecular system with a global spin $\mathbf{S}$ = ½ based on transition metal ions with small spin–orbit coupling
- An orbital in which the single unpaired electron resides well separated in energy from the other orbitals
- An electronic spin magnetically diluted in a diamagnetic matrix
- An organic ligand linked to the metal mostly based on atoms without a nuclear spin
- A molecular structure, and in particular the nearest spin environment, as rigid as possible

As far as the nature of the spin carrier is concerned, mononuclear complexes of V^{IV} or Cu^{II} seem the most performing candidates. Both are stable ions of the first-row transition metals series featuring two

[14]The reader should be aware that this class of molecular materials was still scarcely investigated by the time this book was written, and its study represents an actual and challenging research field for which many questions have yet to be answered.

electronic configurations, d^1 and d^9, respectively, in which a single unpaired electron is accommodated in a nondegenerate d orbital whose nature depends on the metal ion's coordination geometry. Heavier transition metal ions or lanthanides such as Ru^{III} (Graham et al., 2014) or Yb^{III} (Pedersen et al., 2016) have been also taken into consideration for this purpose, but they seem to suffer from spin–orbit coupling effects, which induce fast spin-lattice relaxation and thus short spin-spin relaxation times. Metal ions with spin values higher than **S** = ½ (e.g., Fe^{III} and Cr^{III}) also undergo rapid collapse of the superposition state (Graham et al., 2014). Polynuclear complexes in which several transition metal ions interact with each other through antiferromagnetic interactions to give a global spin state of **S** = ½ at low temperature have been investigated to achieve molecular qubits (Troiani et al., 2005). As polynuclear complexes, their electronic configuration is determined by a multitude of energy states, with the **S** = ½ state fully populated only at very low temperatures (1.9–5.0 K). As a result, they suffer from loss of appreciable phase memory times as the temperature increases because electronic states higher than the **S** = ½ state become populated (vide supra).

The coordination geometry can play a role in isolating the unpaired electron of the metal ion in a well-defined energy state. Indeed, by limiting the discussion to the case of the two most promising metal ions, V^{IV} and Cu^{II}, mentioned above, the square pyramidal and the square planar coordination geometries have to be preferred for the former and the latter ion, respectively, over the octahedral coordination geometry, which has partially filled degenerate orbitals (Fig. 3.45).

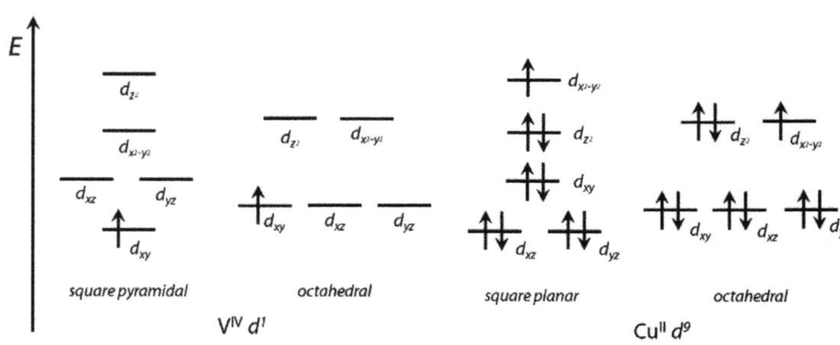

Figure 3.45 Approximate d orbital energy level diagrams for V^{IV} (d^1) in square pyramidal and octahedral coordination geometries and Cu^{II} (d^9) in square planar and octahedral coordination geometries.

Electronic and nuclear spins of the surrounding environment promote decoherence (vide supra) as they behave as local magnetic fields, which affect the spin state once placed in a state of coherent superposition. Interactions with other electronic spins can be limited by magnetically diluting the paramagnetic metal complex in a diamagnetic isostructural crystalline matrix, as for example by cocrystallizing a square planar Cu^{II} complex with a Ni^{II} diamagnetic analog (Bader et al., 2014) or a V^{IV} complex with its Ti^{IV} or Mo^{IV}-*ls* analogs (Atzori, Morra et al., 2016). Although very high dilutions of the paramagnetic component in the surrounding matrix have been initially investigated (0.01%–0.001%) (Bader et al., 2014), it has then been observed that good results can also be obtained with higher concentrations (10%–0.1%) (Atzori, Tesi et al., 2016). Alternatively, the paramagnetic metal complex can be solubilized in a proper mixture of solvents and the solution frozen to cryogenic temperatures to form an amorphous glass. Higher separation between electronic spins with respect to that provided by solid-state dilutions is achieved, and longer relaxation times are commonly observed, but, as an inconvenience, the investigable temperature range is limited by the melting point of the solvent.

Nuclear spins of the ligands, counterions, or solvent are one of the main sources of decoherence. Unfortunately, only a few chemical elements commonly constituting the organic ligands are essentially nuclear-spin-free (C, O, and S). The chemical tunability of the ligand seems thus highly compromised by this requirement, especially because of the ubiquitous presence of the nuclear spin-active H atoms in organic ligands (and counterions). Although some completely nuclear-spin-free ligands are available, good results have also been obtained in the presence of a limited number of nuclear-spin-active atoms (N and H) (vide infra).

We can anticipate that the mechanisms governing the spin relaxation in these systems involve spin-phonon coupling (cf. box titled "Spin-Lattice Relaxation Mechanisms"). Phonons are the quantum mechanical descriptions of local or reticular vibrational modes. This means that if the spin is involved in low-energy local vibrations, low-energy phonons can resonate with the spin, thereby inducing spin-lattice relaxation at a lower temperature with respect to the same metal ion embedded in a more rigid structural moiety, which requires higher-energy phonons to satisfy the resonance condition. In this latter instance, the spin-lattice relaxation times will not collapse so effectively as the temperature is increased. In this regard, the V^{IV} ion in a oxovanadium(IV) moiety, vanadyl, in a square pyramidal coordination geometry should be preferred to the octahedrally coordinated V^{IV} ion (Atzori, Morra et al., 2016).

With this view, some archetype examples of molecular qubits are the metal complexes reported in Fig. 3.46.

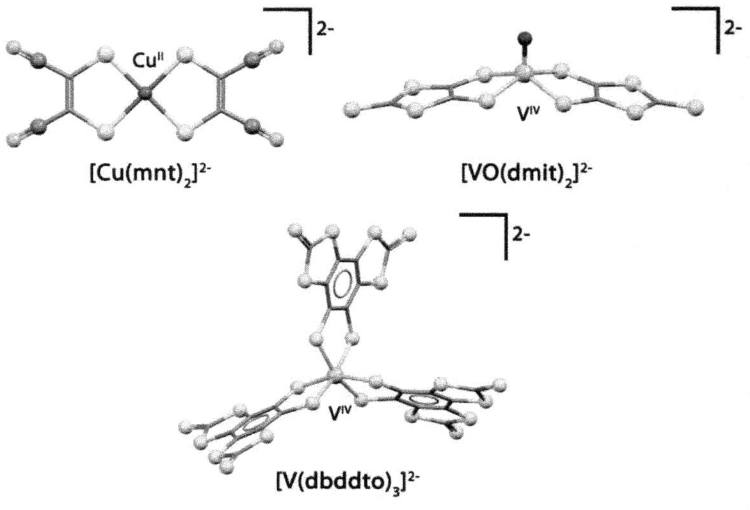

Figure 3.46 View of the molecular structures of the $[Cu(mnt)_2]^{2-}$, $[VO(dmit)_2]^{2-}$, and $[V(dbddto)_3]^{2-}$ metal complexes as archetype examples of molecular quantum bits. The organic counterions are omitted for graphical clarity.

The $[(Ph)_4P]_2[Cu(mnt)_2]$ (mnt = maleonitriledithiolate) complex diluted in a diamagnetic crystalline matrix of its $[(Ph)_4P]_2[Ni(mnt)_2]$ analog (Bader et al., 2014) represents the first reported potential molecular qubit in which coherence times of the order of 1 μs have been observed at room temperature, although at very high dilutions (0.001%). The reason for this good performance is that it satisfies most of the previously mentioned design criteria: it is a square planar Cu^{II} metal complex, with a single unpaired electron localized in a nondegenerate d orbital, and the ligand is mostly made of nuclear-spin-free atoms (C and S), although two nuclear-spin-active nitrogen atoms are present at the molecular periphery.

The $[(Ph)_4P]_2[VO(dmit)_2]$ (dmit = 1,3-dithiole-2-thione-4,5-dithiolate) complex diluted in a diamagnetic crystalline matrix of its $[(Ph)_4P]_2[MoO(dmit)_2]$ analog represents a case in which the molecular rigidity of the vanadyl moiety, when compared to the less rigid V^{IV} in an octahedral environment, has been demonstrated to be crucial to extend the observation of quantum coherence up to room temperature (Fig. 3.47) (Atzori, Morra et al., 2016).

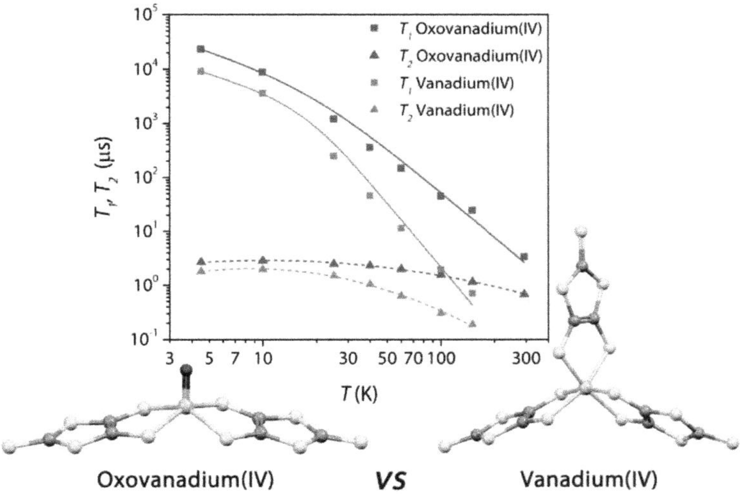

Figure 3.47 View of the thermal dependence of the spin-lattice (T_1) and spin-spin (T_2) relaxation times for the two molecular quantum bits reported below in the figure. Quantum coherence is observed for the oxovanadium(IV) complex up to room temperature. On the contrary, a more abrupt decrease for the octahedrally coordinated vanadium(IV) center induces the quantum coherence to collapse above 150 K. Adapted with permission from Atzori, Morra et al. (2016). Copyright (2016) American Chemical Society.

Here, the unpaired electron is well localized in a low-energy nondegenerate orbital provided by the square pyramidal coordination geometry. Record quantum coherence times at room temperature are observed in this highly concentrated material (ca. 5%) as a result of the benefit provided by the choice of a fully nuclear spin free ligand (the dmit ligand is exclusively made of C and S atoms).

Instead, the $[(Ph)_4P]_2[V(dbddto)_3]$ (dbddto = 2,5-dithioxobenzo(1,2-3,4)bis-(1,3)dithiolene-7,8-dithiolate) complex (Zadrozny et al., 2015) represents a case in which the extended nuclear-spin-free ligands and the dispersion in a nuclear-spin-free solvent (CS_2) as diamagnetic matrix are valuable conditions for the observation of record coherence times (of the order of 1 ms), at low temperature, which are comparable to those observed for atom-based inorganic qubits such as electron vacancies in silicon and silicon carbide.

In most of the studied examples of molecular qubits, the metal complex has an anionic nature, which requires the presence of a counterion. Unfortunately, the most commonly used organic countercations in coordination chemistry bear hydrogen atoms. As already mentioned, H is nuclear-spin-active and such nuclear spins are one of the main sources of decoherence in molecular qubits. One viable solution to overcome this drawback is represented by the replacement of the hydrogen atoms with deuterium ones, which have an integer nuclear spin and are less effective in promoting decoherence. In this way, higher relaxation times can be obtained.

On the other hand, another promising strategy to avoid the presence of hydrogen-rich counterions is the preparation of neutral metal complexes, which can best the performances of charged molecular complexes, even if the presence of nuclear-spin-active nuclei on the ligands cannot by completely excluded. One such example is the vanadyl phthalocyanine (VOPc) system dispersed into the isostructural diamagnetic matrix made of titanyl phthalocyanine (TiOPc) (Fig. 3.48) (Atzori, Tesi et al., 2016).

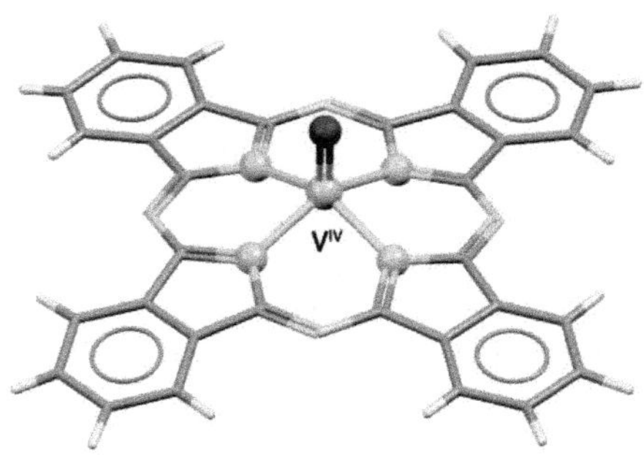

Figure 3.48 View of the molecular structure of vanadyl phthalocyanine (VOPc) metal complex.

This molecular material features the already described electronic/structural benefits provided by the vanadyl moiety, except for the fact that the ligand presents hydrogen atoms on its periphery and nitrogen atoms as donor sites. Beside this drawback, which limits the absolute values of the relaxation times, room temperature quantum coherence with T_m of the order of 1 μs has been observed. This value is comparable to that obtained for the $[(Ph)_4P]_2[Cu(mnt)_2]$ complex in higher dilutions (0.1% versus 0.001%), despite the fact that the Cu^{II} complex has less nuclear spin active nuclei on the ligands. This is worth mentioning because, although important, the absence of nuclear-spin-active nuclei in the metal environment is in the end not a strict condition to observe quantum coherence.

So far, we have analyzed in detail the electronic/structural features of those systems that are among the most performing candidates as potential molecular spin qubits. The reader should also know at this point that the observation of appreciable quantum coherence times is not the unique criterion that has to be satisfied by a molecule to behave as a qubit. A qubit is actually an object that needs to be manipulated, that is, it has to be placed in a state of coherent superposition. A proper sequence of EPR pulses allows the investigation of its manipulability, the outcome of which is the observation of the so-called Rabi oscillations. They can be viewed as the flipping motion of the spin between two opposite configurations (up and down) in a single-qubit quantum operation, that is, the NOT logic operation. After each cycle the spin returns in its original configuration and the flipping is repeated continuously until it loses the memory of the phase. The time constant between two opposite configurations can be regarded as the time required to perform one of the simplest quantum operations, the gate time (Fig. 3.49).

Although Rabi oscillations were observed in the past only at cryogenic temperatures, the VOPc/TiOPc system described above represents a recent example of their observation at room temperature. This means that even at this temperature, obviously required for future large-scale applications, coherent manipulations of the electronic spins can be performed.

So far, we have described molecular structures and related physical properties of molecules containing a single spin per molecule. Actually, to perform more complex quantum logic operations, manipulations on a single qubit are insufficient. In traditional logic gates, operations between two or more bits are commonly performed. In quantum logic gates the same must be realized through the interaction between two or more qubits. Coherent interactions between qubits are commonly described through the entanglement phenomenon. Without entering into details of the entanglement phenomenon and other conditions that qubits must satisfy to be considered viable qubits (as defined by the DiVincenzo criteria [DiVincenzo, 1995]), it is clear from the chemical design point of view that mononuclear metal complexes cannot satisfy this requirement. Indeed, a very challenging direction in this research field is the realization of molecular quantum logic gates through rational design of polynuclear metal complexes containing two or more distinguishable qubits. A recent example of such a molecular system is the dinuclear lanthanide-based metal complex reported in Fig. 3.50 (Aguilà et al., 2014).

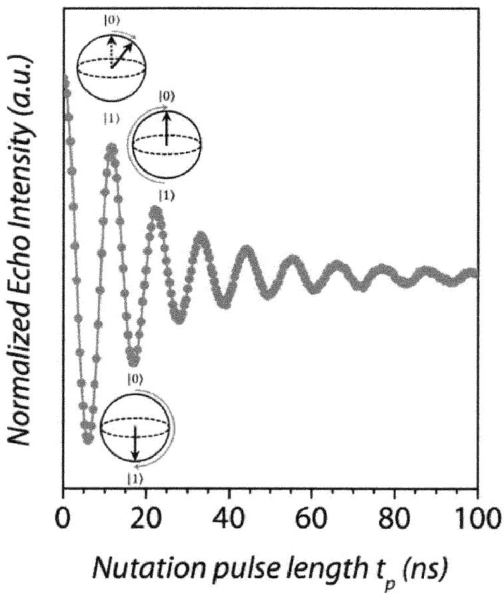

Figure 3.49 Rabi oscillations for a generic molecular spin qubit and relative spin state as a function of the nutation pulse length. In the initial state the spin is placed in the up configuration (maximum of the echo intensity), then it coherently goes to the opposite one (down, the deeper minimum), and subsequently it returns to the up state (second maximum). The time constant between maxima and minima is the gate time relative to the NOT logic operation. The echo intensity of the spin decreases as a function of time as a result of the loss of memory phase.

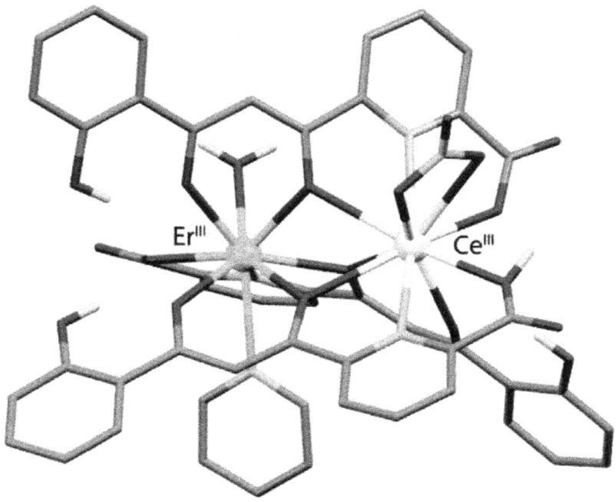

Figure 3.50 View of the molecular structure of the Er^{III}-Ce^{III} dinuclear compound, an archetype example of a molecular quantum gate.

Here, each metal center can behave as an electronic spin qubit because at low temperature both Ce and Er ions have a populated ground state doublet, thus satisfying the fundamental requirement for a system to behave as a qubit (vide supra). The two metal ions are, in principle, distinguishable through EPR techniques as they have very different $\mathbf{g}_J$ factors (cf. Table 2.2) and strong magnetic exchange interactions between them (which hinder the realization of an entangled state between qubits) are limited by the intrinsic inner nature of the f orbitals. Through a specific sequence of EPR pulses, the Control-NOT (C-NOT) logic operation[15] has been performed on this system, thus demonstrating the capability of molecular polynuclear complexes to behave as molecular quantum gates.

Before concluding this section it is important to highlight that the slow relaxation of magnetization in molecules that behave as potential spin qubits is of foremost importance for the observation of quantum coherence. Without a long spin-lattice relaxation time (τ or T_1), detectable spin-spin relaxation (T_2) cannot be observed (cf. Fig. 3.47). The origin of the slow relaxation of the magnetization in $\mathbf{S}$ = ½ spin systems cannot be related to the presence of an energy barrier resulting from intrinsic magnetic properties of the molecular system (magnetic anisotropy and ZFS) as found in SMMs,

[15]The Control-NOT logic operation consists of the flip of a qubit from $|0\rangle$ to $|1\rangle$ or from $|1\rangle$ to $|0\rangle$ if and only if the control qubit is $|1\rangle$.

SIMs, and SCMs. This is due to the fact that an **S** = ½ system does not have multiple spin substates and thus it cannot be subjected to the ZFS. The two available $m_S = +½$ and $-½$ spin substates can only be separated in energy by the presence of an external magnetic field, the Zeeman energy (cf. Section 2.1.9), so slow relaxation of the magnetization can only be observed in the presence of an external magnetic field. If no low-lying magnetic energy states are available to induce spin-phonon relaxation through the Orbach mechanism, only the direct and Raman processes can intervene to allow the magnetization to relax (see the box titled "Spin-Lattice Relaxation Mechanisms"). Among them, the direct process dominates at low temperatures (ca. <20 K) and high magnetic fields (ca. >3 T) as it has a T^1 and $\mathbf{H}^4$ dependence, whereas the Raman process is more relevant at higher temperatures (ca. >20 K) and low magnetic fields (ca. <3 T) due to its T^n (n = 3–6) and $\mathbf{H}^2$ dependence. This means that the thermal and field dependence of the spin-lattice relaxation in a wide temperature and magnetic field range can be described as the sum of these two contributions. Both these mechanisms are less effective than the Orbach process in promoting magnetic relaxation and are the cause of the very high temperatures (>40 K) and magnetic fields (up to 3–4 T) at which the relaxation time remains long (from microseconds to seconds) for **S** = ½ species even in concentrated samples. At such temperatures and magnetic fields, the slow relaxation of the magnetization of most SMMs, SIMs, and SCMs is completely suppressed along with the loss of appreciable magnetic properties. This feature is of tremendous importance and determines the actual relevance of molecular spin qubits.

3.2 Conducting Molecular Materials

We have seen in Chapter 2 that an ordered arrangement of atoms can afford a material showing conducting properties as a result of the superposition of atomic orbitals between neighboring atoms, which gives rise to the so-called conduction band. These properties can be also achieved in more sophisticated materials, such as organic polymers with extended π-delocalization (high aromaticity), which allows high electron mobility across the organic backbone. Remarkably, even discrete entities, like organic or metallorganic molecules (coordination compounds), when partially oxidized and properly organized into a crystal lattice, can behave like metal conductors (called metal-like conductors) and, in some cases, also as superconductors below a certain critical temperature. These materials are called organic conductors, and their discovery represented a real breakthrough in material science because, until then, these properties were considered to be achievable only with classic inorganic solids.

While a classification based on each magnetic property was suitable for magnetic molecular materials, as their physical properties arise from well-defined electronic and structural features, this approach is hard to follow with conducting molecular materials. This is related to the fact that the conducting properties of various classes of molecular materials, whose fundamental bases have been presented in Chapter 2, derive from common electronic and structural features that are hardly controllable. The extent of the interactions between molecular components, driven by the structural arrangement, determines the degree of charge transport, and this ultimately defines which class a certain material belongs to: semiconductors, metal-like conductors, or superconductors (cf. Section 2.2).

In this section we will present the main classes of conducting molecular materials with the aim of pointing out which electronic and structural features are required to prepare molecular components with good transport properties. However, it should be kept in mind that a prediction a priori of the resulting conducting properties of the synthesized material is hard to make when there is lack of precise structural data on its supramolecular arrangement.

A Bit of History

The pioneering work of Knop, who synthesized in 1842 the first molecular metal based on the tetracyanoplatinate anion, may be considered as the origin of this research field. Actually, only in 1968 was the stoichiometry of this system determined and its conducting properties investigated. It consists of a partially oxidized salt of the formula $K_2[Pt(CN)_4]Br_{0.3}\cdot 3H_2O$, whose structure is formed by stacks of $[Pt(CN)_4]$ units extended along the overlapping direction of the Pt^{II} d_{z^2} orbitals (see figure below).

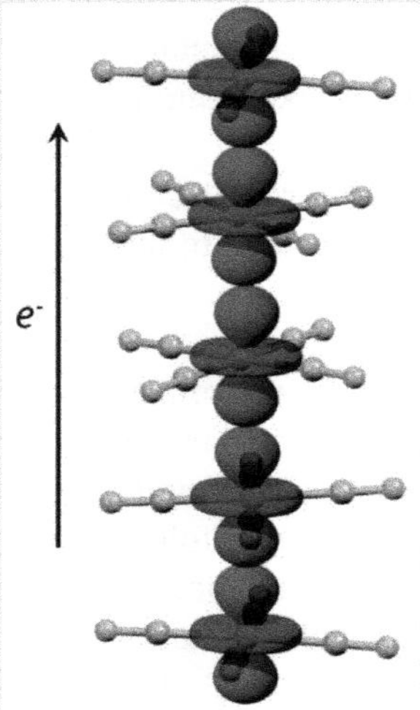

This high structural anisotropy reflects the monodimensional nature of the conductivity, which shows its maximum value along the stacking direction ($\sigma_{\parallel,RT}$ = 300 S cm^{-1}), whereas it is negligible perpendicularly to the stacks ($\sigma_{\perp,RT} \approx 10^{-5}$ S cm^{-1}).

It is interesting to note that although this first example was inorganic in nature, the development of inorganic conductors was nonetheless hindered by the strict electronic and structural requirements needed to observe these properties. The discovery in 1973 of the first organic metal, the tetrathiafulvalene-tetracyanoquinodimethane (TTF-TCNQ) system (see Section 3.1.2), gave a renewed impulse to this research field, and several organic-based systems were quickly developed due to their wider chemical tunability.

3.2.1 Charge Transfer Salts

Many of the molecular conductors known in the literature are charge transfer (CT) salts. These systems are obtained when two neutral molecular building units undergo the following chemical reaction:

$$m\mathrm{D}^0 + n\mathrm{A}^0 \rightarrow \mathrm{D}_m{}^{\rho+} + \mathrm{A}_n{}^{\rho-}$$

Here D, the donor, is the electron-donating species, and A, the acceptor, is the electron-accepting species, while the exponent ρ quantifies the CT degree between the two molecules. The CT degree can be complete (ρ = 1) or partial (0 < ρ < 1), depending on the difference between the oxidation potential of the donor and the reduction potential of the acceptor.

In Fig. 3.51 some of the most relevant molecular donors and acceptors employed for the preparation of CT salts are reported.

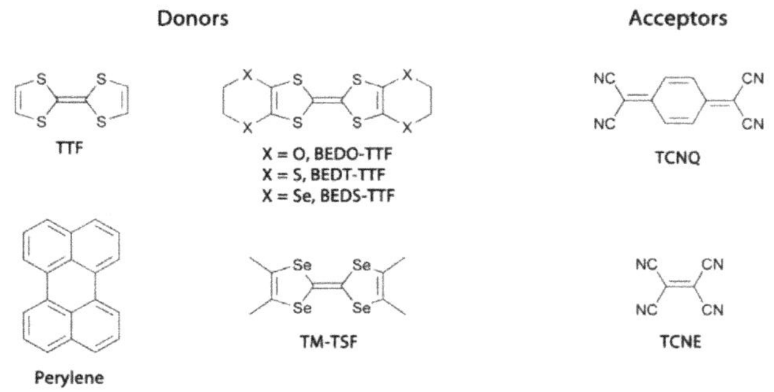

Figure 3.51 Molecular structures of some of the most important molecular donors and acceptors. TTF, tetrathiafulvalene; BEDO-TTF, bis(ethylenedioxo)-tetrathiafulavelene; BEDT-TTF, bis(ethylenedithio)-tetrathiafulavelene; BEDS-TTF, bis(ethylenediselena)-tetrathiafulavelene; TM-TSF, tetramethyl-tetraselenafulvalene; TCNQ, tetracyanoquinodimethane; and TCNE, tetracyanoetylene.

Generally, donors and acceptors are planar molecules with an extended π delocalization. These are two fundamental features for the formation of crystal structures where the molecules are stacked in columns, ensuring a good molecular orbital (MO) overlap, and for the stabilization of different and/or fractional oxidation states. One of the most-known examples of such compounds is the TTF-TCNQ system (Fig. 3.52), where molecules of the TTF donor and molecules of the TCNQ acceptor are arranged in segregated stacks along which the conductivity reaches its maximum value. This system, which can be considered an archetype example of this class of compounds, shows a room temperature electrical conductivity, σ_{RT}, of 500 S cm^{-1} and behaves as a metal-like conductor down to ca. 55 K (Ferraris et al., 1973).

Generally, two types of stacking can be recognized:

- Mixed stacks, where the donors and the acceptors alternate in a -D-A-D-A-D- sequence
- Segregated stacks, where the donors and the acceptors form separated -D-D-D-D-D- and -A-A-A-A-A- stacks

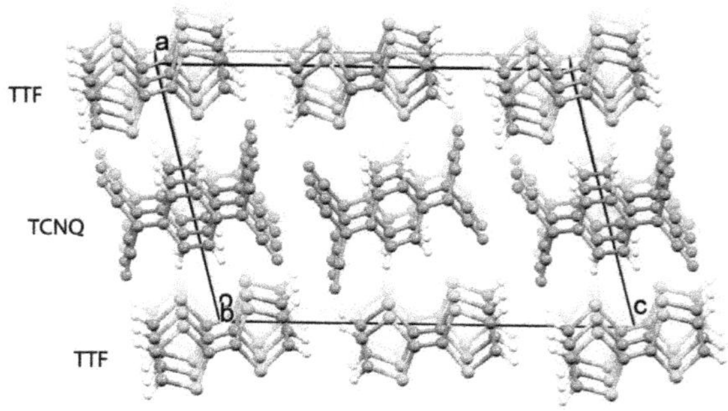

Figure 3.52 Segregated stacks of TTF and TCNQ molecules in the compound TTF-TCNQ.

It must be noted that the CT in these systems is a spontaneous phenomenon that arises, as mentioned above, from the reciprocal tendency of the two species to undergo partial or complete CT. In CT salts the two components have a fundamental role in the resulting conducting properties. The partially reduced acceptor is an electron-rich species that has more electrons than in the neutral state. These extra electrons populate the conduction band and have good mobility. On the contrary, the partially oxidized species has fewer electrons than in the neutral state, that is, electron holes in the valence band, and can be also responsible for good transport properties. These combined features make CT salts very efficient conductors when an electrical potential is applied at the ends of a crystal grown along the stacking direction. It must be noted that these materials are intrinsically structurally anisotropic and this is reflected in the high anisotropy of their physical properties. The conductivity of the material measured perpendicularly to the stack can be orders of magnitude lower than that measured along the stacking direction.

3.2.2 Radical Cation Salts

Another wide class of conducting molecular materials is represented by those systems that can be categorized as radical cation salts. These systems are obtained by combing an electron donor (as those reported in Fig. 3.51) in the neutral state with a nonredox anionic species. Here, the application of an electrical potential induces partial oxidation (fractional oxidation degree) of the redox active species (see the box titled "The Electrocrystallization Technique"), and the anionic species, already bearing an opposite charge, now act as counterions of the donor. An archetype example of this class of conducting materials is represented by the $[TM\text{-}TSF]_2[PF_6]$ system (Fig. 3.53), which was discovered in 1980 and is the first organic superconductor showing a metal–superconductor transition at 1.2 K under a pressure of 10 kbar (Jérome et al., 1980).

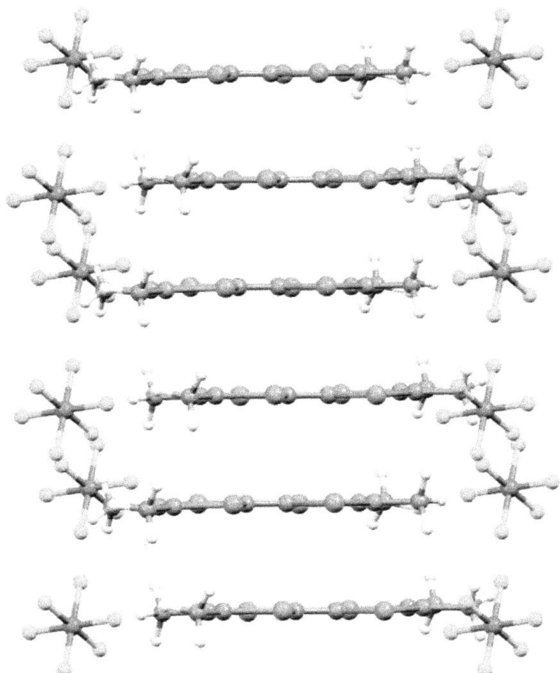

Figure 3.53 View of the crystal structure of $[TM\text{-}TSF]_2[PF_6]$ showing segregated stacks of the TM-TSF donor separated by PF_6^- counteranions.

Alternatively, such materials can be obtained through chemical oxidation, for example, by reacting a donor like the TTF with I_2, which once reduced acts as a counterion of the radical cation $TTF^{\cdot+}$ as I^- or I_3^-. Although practically simple and easy to handle, chemical methods offer fewer opportunities for chemical tunability since the number of species that can act as both oxidizing species and counterions is limited with respect to the number of nonredox active counterions that can be introduced through electrochemical methods.

The Electrocrystallization Technique

A molecular conductor can be obtained by chemical oxidation or electrochemical oxidation of an organic donor. Although very simple and easy to handle, the former approach yields in most cases powder materials, because their low solubility hampers slow crystallization. On the other hand, the electrochemical method, commonly called electrocrystallization, allows for a controlled and slow oxidation of the organic donor on the surface of a platinum wire electrode and, once the experimental conditions are optimized, provides high-quality single crystals.

Synthesis via electrocrystallization is commonly carried out in a special H-shape glass cell where two chambers are connected through a porous septum. The two chambers are called anodic and cathodic compartments. The figure below shows a schematic picture of an H-shape electrocrystallization cell equipped with a pair of electrodes.

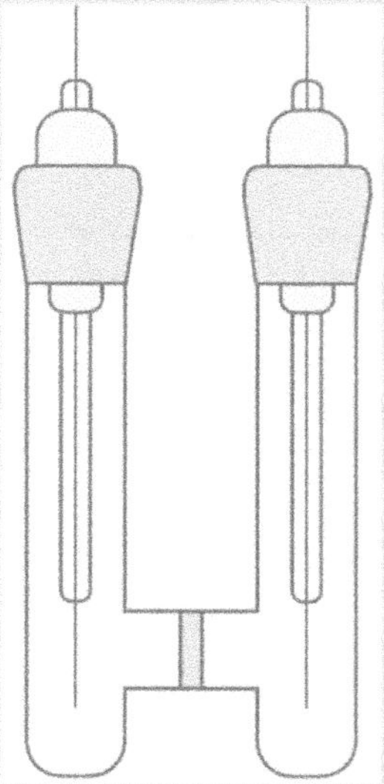

In the usual experimental procedure, a solution of the organic donor mixed with the counterion is usually placed in the anodic compartment, whereas the cathodic compartment contains a solution of the counterion only. Typically, the solutions are degassed before starting the experiment to avoid the presence of redox-active species, such as oxygen. The septum allows for a slow diffusion of the two solutions into one another, and, at the same time, permits ion-based electrical conductivity in the solution as the counterion also works as an electrolyte. The electrodes, usually two platinum wires, are placed in the two compartments and are connected to a potentiostat/galvanostat, depending on the choice of potentiostatic or galvanostatic conditions under which the experiment is performed. Two valves can be connected to the cell if an inert atmosphere is needed.

The electrosynthesis is usually carried out over a period of time that may vary from a few days to several weeks and is performed in a thermostatic bath under dark conditions and in an environment free of mechanical vibrations to ensure a good crystal growth.

During an electrocrystallization experiment, the following parameters have to be controlled and taken into account:

- **The solvent:** It can be pure or a mixture. It plays an important role in the solubility of the starting materials and the final product, in the redox potentials of the species in solution, and in the electrode/solution and crystal/solution surface tension, affecting the crystals' quality and, in some cases, the crystal phase.
- **The temperature:** It plays a role in the solubility of the starting materials and the final product and can affect the crystals' quality.
- **The current intensity**: The rate of growth of the crystals and, hence, the quality of the crystals, depends on its value. Galvanostatic conditions are usually preferred to potentiostatic conditions to promote regular crystal growth. To ensure reproducible experimental conditions, the value of the current density (current intensity/electrode surface ratio) should be taken into account rather than the current intensity.

Moreover, the experiment is affected by the following factors:

- The electrochemical properties of the counterion
- The purity and concentration of the reagents
- The purity of the solvent(s)
- The stability of the applied current intensity

Due to the high number of parameters involved in the outcome of an electrosynthesis, several experiments are often required to optimize the experimental conditions for obtaining high-quality single crystals.

The conducting properties of a certain material, either molecular or atom based, can be strongly influenced by the type and amount of impurities eventually included in the crystal structure. For this reason, in order to have reproducible results, great attention must be given to equipment cleaning, with particular attention to the electrodes' surface.

The electrodes' cleaning consists in the following steps:

1. **Chemical cleaning:** This is done with aqua regia (HNO_3/HCl 1:3 v/v) or piranha solution (H_2SO_4 95%–98%/H_2O_2 30% 3:1 v/v).
2. **Electrochemical cleaning (or depolarization):** It consists in placing the electrodes in an aqueous solution of diluted sulfuric acid (~1 M) and connecting them to a direct current (DC) generator. By periodically inverting the polarity of the electrodes, H_2 and O_2 are alternatively generated on their surface, allowing for the elimination by reduction or oxidation of all the undesired impurities.
3. **Rinsing:** This is done with distilled water and methanol.

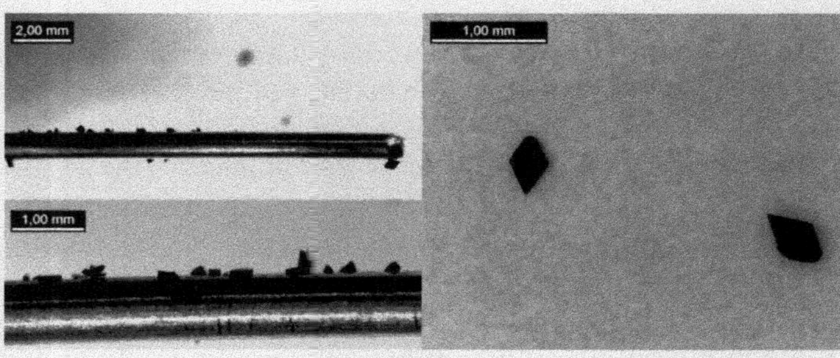

Electrocrystallization is the most employed technique for the preparation of crystalline conducting molecular materials. It presents some disadvantages, such as the meticulous preparation of the experiments, their duration, and the small quantity of obtained products (often of the order of a few milligrams). Nevertheless, it can afford materials as high-quality single crystals, with the desired morphological form to perform their full structural and physical characterization. The figure above shows some crystals of a molecular conductor grown on the surface of a platinum wire electrode (left) and a detailed view of two single crystals (right).

In the literature, these systems are also oftentimes recognized as CT salts, but it is important to note that there is a difference between these materials and those discussed in the previous paragraph. In the CT salts presented in Section 3.2.1 both the donor and the acceptor are responsible for the conducting properties of the resulting material. In radical cation salts, instead, there is only one redox-active building block, which is responsible for the conducting properties, while the counterion has the sole role to template the structure or, in case, bear other physical properties (vide infra).

With this view, the electronic and structural features that a molecule should have to behave as an effective building block for conducting molecular materials are:

- A high degree of π delocalization that allows for charge stabilization and reduces the electrostatic repulsions between molecules, leading to a more efficient overlap between MOs
- The presence of polarizable atoms at the molecule periphery, which play a key role in promoting inter- and intrastack contacts responsible for an increased dimensionality of the resulting material
- Good planarity, which allows for an effective π superposition of the MOs, and thereby, good conducting properties

TTF and its derivatives represent the most important family of organic donors showing these features. Nonetheless, TTF leads to conducting radical cation salts that, even if characterized by high room temperature conductivity values, are commonly affected by Peierls distortion (cf. Section 2.2.7) due to the low efficiency of this donor in establishing short contacts between stacks. To avoid or reduce the Peierls distortion, various structural modifications of this molecule have been considered and investigated. Specifically, the following strategies have been studied:

- The extension of the donor periphery by introducing O, S, and Se substituents, allowing the preservation of good planarity and the tendency to favor interstack contacts through noncovalent interactions
- The substitution of the core S atoms with atoms having more extended and polarizable orbitals, such as Se and Te

Among the various TTF derivatives, the most successful donors that have yielded the majority of known molecular superconductors and metals are bis(ethylenedithio)-tetrathiafulavelene (BEDT-TTF), bis(ethylenediselena)-tetrathiafulavelene (BEDS-TTF), and bis(ethylenedioxo)-tetrathiafulavelene (BEDO-TTF) (Fig. 3.51). In particular, the κ-$[BEDT\text{-}TTF]_2[CuN(CN)_2Br]$ system shows a metal–superconductor transition temperature of ca. 11.6 K, which represents the highest T_c value for this class of compounds (Kini et al., 1990).

The physical properties of molecular materials are strongly dependent on the nature of their building units as well as on the way these molecular bricks self-assemble in the crystal lattice. In fact, radical cation salts built with the same building units but showing completely different physical properties are well-known in the literature. This is mainly due to the different packing motifs exhibited by the donor molecules. Some of the most common molecular packings exhibited by TTF and its derivatives are reported in Fig. 3.54 (Mori, 1998, 1999; Mori et al., 1999).

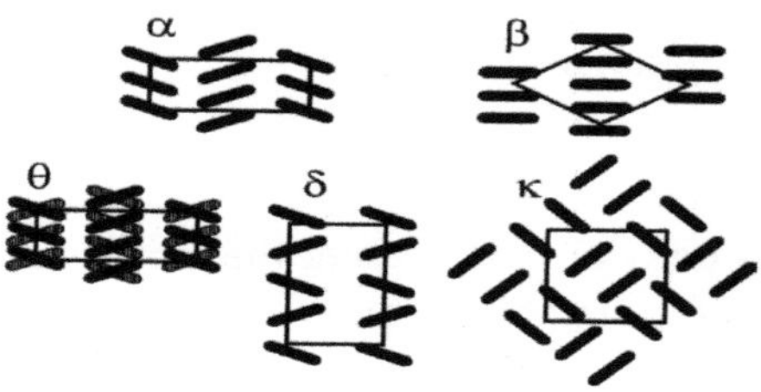

Figure 3.54 Schematic representation of some structural phases exhibited by TTF and its derivatives with classification symbols.

The molecular packing of the donor molecules is driven by several factors and, hence, is difficult to predict in advance. Moreover, it is not commonly easy to understand the reason why a specific packing would be favored with respect to the others, where no specific supramolecular interactions are present between the donors and the counterions.

In the case of radical cation salts where only one species undergoes partial oxidation or reduction process, the role of the counterion is of fundamental importance because it can influence the arrangement of the electroactive donor molecules in the crystal lattice, and thus the physical properties of the material, depending on:

- Its molecular geometry
- Its redox potential
- Its ability to establish intermolecular interactions with the electroactive donor molecules

One of the main advantages of radical cation salts with respect to CT salts is that good transport properties can be achieved with a single building block, whereas the counterion can be exploited as a carrier of an additional physical property, such as the magnetic or optical property, leading to multifunctional molecular materials. These materials will be the theme of Chapter 4.

3.2.3 Radical Anion Salts

Another class of conducting molecular materials is represented by the radical anion salts. In these systems, the building block that acts as a carrier of the conducting properties is a planar dianionic molecule with extended π delocalization that can be chemically or electrochemically oxidized to a stable radical anion. Here the charge is balanced by a countercation, similar to the case of radical cations where a counteranion is needed. Examples of molecules that satisfy such structural and electronic features are planar complexes of d^8 transition metal ions (Ni^{II}, Pd^{II}, and Pt^{II}) mostly with sulfur-based planar ligands, such as dithiolenes derivatives (Fig. 3.55).

Figure 3.55 Molecular structures of two of the most relevant square planar dithiolene-based metal complexes employed for the preparation of radical anion salts. dmit, 1,3-dithiole-2-thione-4,5-dithiolate; mnt, maleonitriledithiolate.

Such metal complexes can also be used in place of the purely organic acceptors presented above to obtain CT salts.

One relevant example is the $[TTF][Ni(dmit)_2]_2$ system, the first molecular superconductor based on a metal–dithiolene complex (Fig. 3.56) (Brossard et al., 1986). It shows a room temperature conductivity of 300 S cm^{-1} and behaves as a metal-like conductor down to 4 K. At that temperature the conductivity increases by over 10^5 S cm^{-1}.

The structural and electronic features that radical anion salts should have to show relevant conducting properties are the same as those reported above for CT and radical cation salts. Here it is interesting to note the role of the metal ion in the building block. Actually, its major functionality is to template the structure to planarity as a consequence of the tendency of d^8 metal ions to form square planar complexes. The metal ion has a paired electronic configuration as a result of the ligand field effects in such coordination geometry, but its orbitals well overlap those of the ligands to generate a fully delocalized highest occupied molecular orbital (HOMO) over the entire molecular structure. Such behavior will of course not be allowed with a metal ion imposing a different coordination geometry (e.g., tetrahedral with Zn^{II} as a metal ion), with concomitant loss of the resulting conducting properties. Once the metal complex is obtained, the total charge of these species will be dianionic since the oxidation state of the metal is M^{II} and each dithiolene ligand bears a charge of 2–. However, dithiolene ligands are redox-active and in such structural conformation, one-electron oxidation can be particularly favored due to the possibility of unpaired electron delocalization over the whole molecular structure. This means that metal dithiolenes show a great tendency to form radical anions with an open-shell configuration that satisfies the required electronic features to show interesting conducting properties (Fig. 3.57).

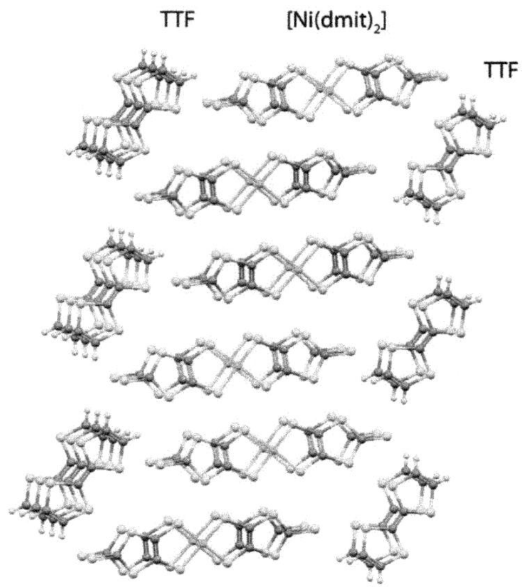

Figure 3.56 View of the molecular packing of the [TTF][Ni(dmit)$_2$]$_2$ system.

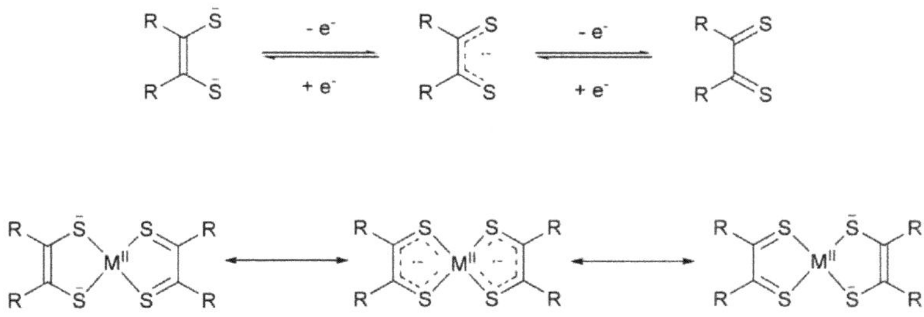

Figure 3.57 Redox equilibria for generic dithiolene ligands (above) and resonance structures describing the electronic charge delocalization of a generic neutral metal–dithiolene complex (below).

3.2.4 Single-Component Molecular Conductors

Single-component molecular conductors are a class of conducting molecular materials in which the starting closed-shell molecular building block is a monoanionic planar molecule that upon one-electron oxidation becomes neutral and has an open-shell configuration suitable for showing conducting properties. The term "single-component" here refers to the fact that these materials are built of only one building block. Although radical cation or radical anion salts can also, in principle, belong to this class as only one type of building block is responsible for the conducting properties (whereas the counterion only has a charge balancing and templating role), single-component molecular conductor neutral systems present only one type of chemical species in the crystal structure. Molecules that satisfy these structural and electronic features, that is, planarity and electron delocalization, are limited with respect to those of the other classes already presented. Most of them are represented by anionic closed-shell metal–dithiolene complexes of Au^{III} that, upon one-electron oxidation of the $[Au(dithiolene)_2]^-$ precursor, easily generate the neutral open-shell species. Other examples are extended metal–dithiolene complexes comprising one or two TTF units (susceptible to one or two electron oxidation) (Fig. 3.58).

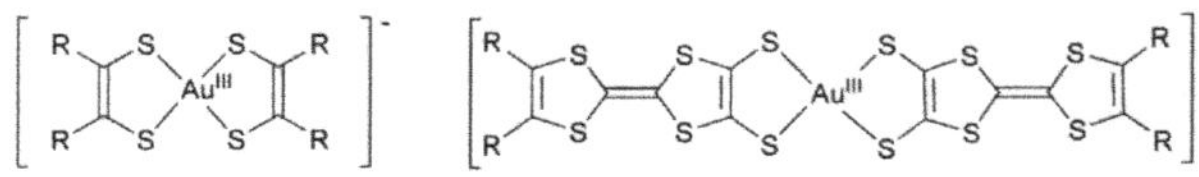

Figure 3.58 Molecular structures for two generic examples of square planar metal complexes employed for the preparation of single-component molecular conductors.

3.2.5 Chiral Molecular Conductors

Introduction of chirality into conducting molecular materials is a topic of special interest as chirality can modulate the structural disorder of achiral systems or express its influence through peculiar physical phenomena, like the electrical magnetochiral anisotropy (cf. Section 3.2.5.2, later in the text).

This section will point out the benefits that can be provided by the use of chiral building blocks (for simplicity only those based on the TTF moiety already introduced) to the conducting properties of the resulting materials.

3.2.5.1 Enhanced conductivity in enantiopure chiral conductors

From a structural point of view, it can be expected that a molecular material based on an enantiopure chiral donor will show some differences with respect to the same material based on a racemic mixture of the two enantiomers. This is related to the inherent source of structural disorder that a racemate can have since it is known that, sometimes, racemates can randomly accommodate opposite enantiomers on the same crystallographic site. Taking into account that the electrical conductivity is strongly influenced by the structural packing and consequently by the structural disorder, the presence of chiral centers could thus modulate the conducting properties.

This has been elegantly demonstrated through the preparation of a complete series of chiral donors of the TTF family, EDT-TTF-Me-oxazolines (ethylenedithio-TTF), and the corresponding radical cation salts with AsF_6^- as counterion in the racemic and both enantiomeric (*S* and *R*) forms. All systems crystallize in the triclinic space group (centrosymmetric for the racemate and noncentrosymmetric for the pure enantiomers) with the racemate that shows enantiomers of opposite chirality with a statistical 50:50 possibility to be accommodated in the same crystallographic site. This induces a considerable structural disorder in the racemic material, absent in the enantiopure forms, which in turn causes a huge variation of the conductivity value. For this reason, the conductivity at room temperature of the enantiopure form is 1 order of magnitude higher (σ_{RT} ca. 100 S cm^{-1}) than the racemic one (σ_{RT} ca. 10 S cm^{-1}) (Fig. 3.59) (Réthoré et al., 2005).

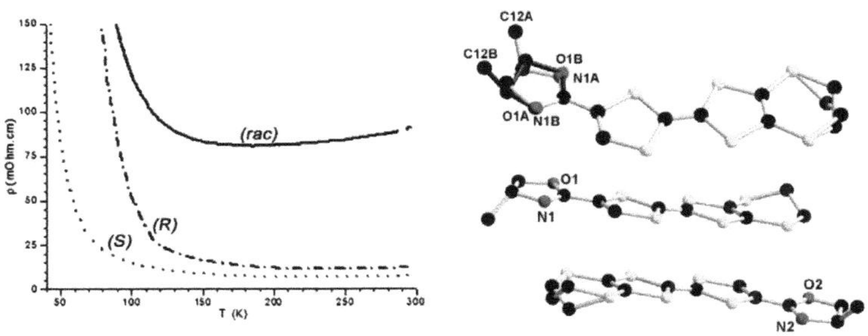

Figure 3.59 Thermal variation of the resistivity for the (*R*)-, (*S*)-, and (*rac*)-[EDT-TTF-methyl-oxazoline]$_2$AsF$_6$ systems where the difference in the conductivity values is highlighted (left), and molecular structures for the (*R*)-, (*S*)-, and (*rac*)-[EDT-TTF-methyl-oxazoline]$_2$AsF$_6$ (right). Adapted from Avarvari and Wallis (2009) with permission of The Royal Society of Chemistry.

3.2.5.2 Electrical magnetochiral anisotropy

The synergistic association of chirality and conductivity has also recently led to the observation in a molecular material of a new phenomenon, referred to as the electrical magnetochiral anisotropy (eMChA) effect, which was previously experimentally observed in the magnetotransport properties of chiral carbon nanotubes. This effect is related to the direct influence of the handedness of a chiral conductor on its conductivity measured under an applied magnetic field parallel to the direction of the current flow. Accordingly, small differences between the conductivities of the opposite enantiomers, and also the racemic compound, can be observed. Very recently, this has been experimentally demonstrated with the (*S,S*)- or (*R,R*)-[DM-EDT-TTF]$_2$ClO$_4$ molecular material. This system, where the two enantiomers show electrical magnetochiral effects with anisotropy factors opposite in sign, but equal in magnitude, field, and current dependencies, represents the most recent and significant advancement in the field of chiral molecular conductors (Fig. 3.60) (Pop et al., 2014).

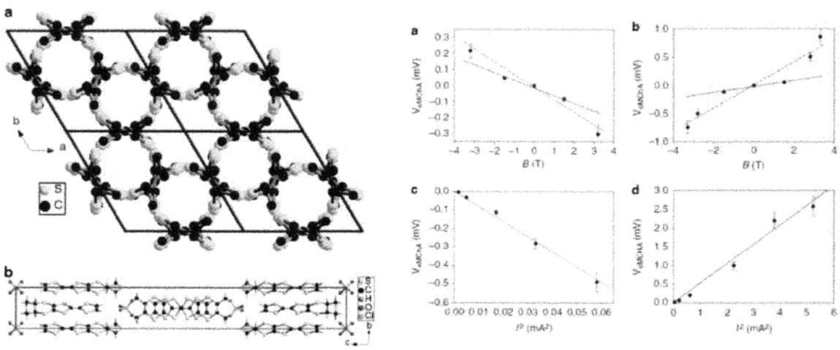

Figure 3.60 Views of the crystal structure (left) and eMChA results (right) for compounds (*S,S*)- and (*R,R*)-[DM-EDT-TTF]$_2$[ClO$_4$]. Reprinted by permission from Macmillan Publishers Ltd: [*Nature Communications*] (Pop et al., 2014), copyright (2014).

3.2.6 General Considerations

Planar organic molecules with good electron-donating and electron-accepting abilities, such as the derivatives of TTF and TCNQ, respectively, can be used to synthesize CT salts that behave as molecular conductors and superconductors. In this class, both components are responsible for the conducting properties of the resulting material.

We have also explained that intriguing conducting properties can be achieved even if only one of these two types of molecules is oxidized or reduced (radical anion or radical cation salts). In the case of TTF and related compounds, the additional positive charge generated upon oxidation has to be compensated for by a counteranion. The first experiments dedicated to the preparation of such materials were performed by using as counterions simple monoanions such as Cl^-, Br^-, I_3^-, PF_6^-, AsF_6^-, BF_4^-, ClO_4^-, and NO_3^-, which are essentially not functional (Coronado and Day, 2004). However, the need for a charge-compensating entity offers, contrary to CT salts and single-component molecular conductors, the opportunity to incorporate additional properties into the same material by using functional counterions. Inorganic and coordination chemistry provide a wide choice of metal complexes of various nuclearities and dimensionalities that can be used as charge-compensating counterions for conducting radical-cation salts. By limiting the examples to the sole field of magnetic molecular materials, they range from simple mononuclear complexes—such as tetrahalometallates, $[MX_4]^{n-}$ (M = Fe^{III} and Cu^{II}; X = Cl and Br); trisoxalatometallates, $[M(C_2O_4)_3]^{3-}$ (M = Fe^{III} and Cr^{III}); and hexacyanidometallates $[M(CN)_6]^{3-}$ (M = Fe^{III})—to cluster-type complexes, such as polyoxometalates; square-planar complexes, such as dithiolene complexes of the type $[M(mnt)_2]^-$ (mnt = maleonitriledithiolate; M = Ni, Pt, and Pd); and more sophisticated layered structures, such as the bimetallic oxalate complexes $[M^{II}M^{III}(C_2O_4)_3]^-$ (M^{II} = Mn, Co, Ni, Fe, and Cu; M^{III} = Fe and Cr), most of them already described in this chapter. Indeed, most of the examples of hybrid molecular materials showing a combination of conducting and magnetic components so far reported in the literature are based on the above organic/inorganic combinations.

With the same view, anionic conducting building blocks can be combined with functional magnetic cations to achieve a combination of magnetic and conducting properties. This is the case, for example, of conducting metal–dithiolene complexes and cationic Fe^{III} complexes exhibiting spin-crossover (SCO) behavior or cationic Mn^{II}/Mn^{III} complexes showing slow relaxation of magnetization.

In Chapter 4 some of the most instructive examples of such materials are described to provide hints to the reader on how to design, prepare, and understand the structure and properties of such conducting/magnetic multifunctional molecular materials.

In conclusion, among the four presented classes of conducting molecular materials, we can say that all of them can provide materials with good transport properties but only the radical anion and radical cation salts offer an easy way to achieve multifunctional molecular materials through charge compensation.

3.3 Optical Molecular Materials

A first general classification of optical molecular materials can be made on the basis of their optical properties related either to phenomena occurring in the linear optical regime or to the realm of nonlinear optics (NLOs). This makes this class of molecular materials extremely wide and varied as it comprises chromophores, luminophores, chemically photoactive species, NLOphores, etc. The various optical properties shown by molecular compounds can be suitably exploited and modulated on the basis of the applicative purposes that these molecules are intended for, such as emitters and amplifiers, light converters, optical switches, sensors, light-harvesting dyes, and photocatalysts. One of the most remarkable advantages of this class of molecular materials is perhaps that, more than the other previously described classes, it offers the opportunity to deliver "ready-to-use" compounds, which can find immediate application in several fields, ranging from optics to molecular biomedicine. This is mainly related to the fact that, in general, optical properties are typically associated with single molecules, so they are not limited by chemical restraints necessary to ensure physical cooperativity, as in the case of conducting materials, and, at the same time, unlike magnetic materials, their functionality is not precluded under a wide range of temperatures.

Aside from a general discrimination of linear and nonlinear optical phenomena, a rigorous classification of the optical properties that can be found in molecular materials can be quite complicated in view of the different, and most often closely associated, photophysical processes (e.g., absorption, emission, energy transfer [ET], and photochemical activity) that a single molecule can undergo. However, in the field of linear optics, it is usually a common practice to think of the class of molecular materials showing optical properties mostly as molecules displaying radiative emission, that is, luminophores. Nonetheless, the characteristics of the emitted radiation strictly depend on the origin and type of the associated electronic transition, that is, the nature of the electronic energy levels involved, and hence the type of related MOs. Whereas the properties of organic luminophores can be simply interpreted on the basis of the general rules presented in Chapter 2, in the case of metal complexes the radiative processes that can be displayed are certainly more sophisticated. This is due to the contribution of both the organic moiety (ligands) and the metal ion to the energy levels involved in the photocycle. With this view, the luminophores based on metal coordination compounds mostly encountered in the field of molecular materials will be herein classified as follows:

- **Metal complexes with emissive ligands:** The electronic transitions are localized on ligands' orbitals, and the metal plays a structural role.
- **Open-shell transition d-metal complexes:** The electronic transitions involve both ligands' and the metal's orbitals.
- **Lanthanide complexes:** Emission arises from the metal's f orbitals.

Prior to a specific description of the aforementioned three general types of linear luminophores and of molecular NLOphores, some introductory general considerations regarding the optical properties of metal complexes will be addressed in the following paragraphs.

3.3.1 Electronic Transitions and Light Absorption in Metal Complexes

In Chapter 2, a general introduction to the main photophysical processes giving rise to photocycles in generic molecular entities and the electronic transitions taking place in d- and f-metal ions was given. Before entering more deeply into the discussion of functional luminophores based on metal complexes, it is useful to provide a more detailed view of the MOs and the electronic transitions in these compounds, especially with reference to those related to the first step of a photocycle, that is, light absorption.

The sequence of electronic energy states in transition metal complexes is determined by the contribution of both the organic ligands and the metal orbitals. In the first instance, a simplified energy level diagram reporting the sequence of frontier MOs in a typical octahedral d-metal complex can help to identify the main types of the electronic transitions leading to light absorption in such molecular compounds, as reported in Fig. 3.61.[16] Depending on the localization of the MOs involved, these transitions can be

[16]The reader should be once again reminded, to avoid any confusion, that a mere sequence of MOs does not represent the energy states of the molecule, as it lacks information on the whole electronic configuration of the ground and excited states (cf. box in Section 2.3.2.3).

metal centered (MC) or ligand centered (LC) or they can either accompany an electron displacement or CT from an orbital with major metal contribution to one prevalently localized on the ligand usually having π or π^* character (metal-to-ligand charge transfer, MLCT) or the inverse (ligand-to-metal charge transfer, LMCT). Other peculiar transitions may involve electron transfer between orbitals mainly localized on two different moieties of the same ligand or between two different ligands of the same complex, usually denoted as intraligand charge transfer (ILCT) and ligand-to-ligand charge transfer (LLCT), respectively. A further classification of the types of CT transitions observed in transition metal complexes is often encountered in the literature as a consequence of the complexity of metal–ligand interactions and orbital mix, so labels such as metal-to-metal-to-ligand charge transfer (MMLCT), ligand-to-metal-to-metal charge transfer (LMMCT), and mixed metal–ligand-to-ligand charge transfer (MMLLCT) are also often used. However, this labeling scheme is often far than rigorous and may generate some confusion. Therefore, in this book, transition label usage will be restricted to the simplest cases described above so long as they are reliable enough to describe the considered phenomena.

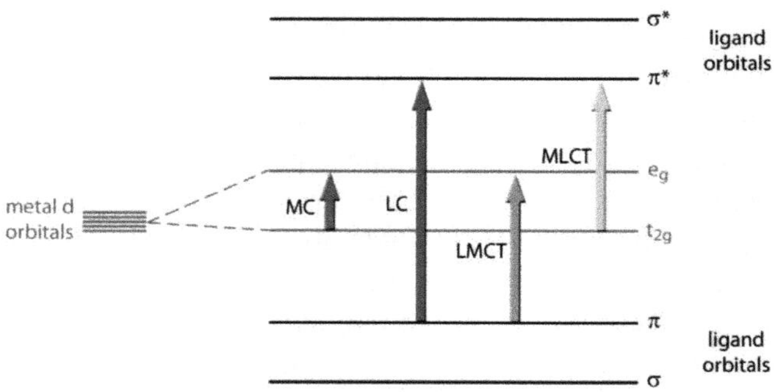

Figure 3.61 Simplified energy sequence of frontier MOs in an octahedral complex of a first-row transition d-metal ion and the main lowest-energy absorption transitions.

General considerations about the types of transitions displayed by metal complexes must be made before dealing in more detail with the different classes of luminophores. The following discussion will, for simplicity and clarity, focus on absorption transitions. However, it is clear that the intrinsic characteristics of each type of transition will be also reflected in the case of emissive phenomena. Therefore, the same information given herein provides nonetheless a useful guide to describe the luminescence properties of a metal complex when simply taking into consideration that emission is expected from the lowest-energy state, normally corresponding to the lowest-energy absorption transition (Kasha's rule, cf. Section 2.3.3.2).

3.3.1.1 Structural/optical properties relationship

While for p-metal and lanthanide complexes the poor covalent character of the ligand–metal bonding allows an apparently simple interpretation of the emission phenomena as only related to LC and/or MC (for f-metal compounds) transitions (vide infra), as if the two constituting entities were virtually isolated, in the case of transition d-metal complexes, where metal and ligand orbitals significantly mix, more sophisticated considerations must be made. The type of transition that can be observed in a d-metal complex depends on a multitude of features, including the electronic configuration of the metal ion, its oxidation state, the electronic properties of the ligands, and the molecular geometry.

Making accurate predictions about the nature of the electronic transitions expected in metal complexes is often a very challenging task. However, some very general prospects can be drawn from chemical considerations and practical experience, especially when comparing similar compounds. These empirical rules can be summarized as follows:

- As commonly found, LC transitions are the most significant in complexes of closed-shell metal ions having completely filled or empty d (or p) orbitals, especially when the metal–ligand bonding is largely noncovalent in character.
- CT bands involving both components of a metal complex, that is, the ligands and the metal center, are typically exhibited by open-shell d-metal ions, where the contribution of metal orbitals to the states involved in the emissive process becomes relevant. As a general rule, it can be inferred that MLCT transitions are displayed when the metal ion is in a low oxidation state and can be readily oxidized, especially when coordinated with π acceptor ligands, whereas LMCT

transitions are favored in coordination compounds bearing metals with a high positive charge and π donor ligands.

- MC d–d transitions are usually displayed by metal ions with open-shell configurations in complexes coordinated with strong organic σ donors.

On the basis of these observations, it is easily understandable that subtle chemical changes can dramatically influence the spectroscopic properties of d-metal complexes, as schematically summarized with the help of a few typical examples in Fig. 3.62.

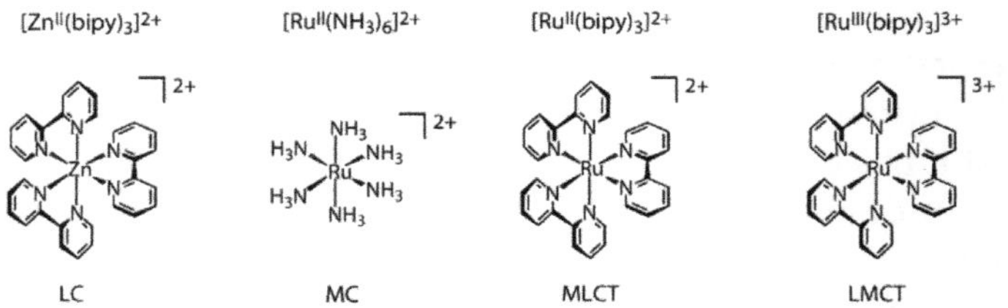

Figure 3.62 Examples of similar coordination complexes showing different absorption transitions depending on the metal electronic configuration (Zn^{II}, closed shell; Ru^{II}/Ru^{III}, open shell), its oxidation state (Ru^{II}; Ru^{III}) and the bonding properties of the ligand (bipy, 2,2′-bypiridine, π donor/acceptor; NH_3, σ donor).

The tris-bipyridyl complex of the closed-shell Zn^{II} ion, $[Zn^{II}(bipy)_3]^{2+}$, is therefore characterized by LC transitions, whereas analogous octahedral complexes of the open-shell Ru^{II} and Ru^{III} ions display transitions that are determined by relevant metal contribution. Among the series of ruthenium complexes shown in Fig. 3.92, MLCT or LMCT transitions are observed when the electron-rich Ru^{II} or the electron-poor Ru^{III} ion is coordinated with the π donor or π acceptor bipyridine (bipy) ligand, respectively. Instead, σ donor ligands such as NH_3 favor MC transitions, as observed in the $[Ru^{II}(NH_3)_6]^{2+}$ complex.

3.3.1.2 Energy

The type of transition favored in a d-metal complex determines its chromophoric properties, that is, the shape of the spectral absorption. The energy of MC d–d transitions is governed by several factors (including the geometry of the complex, the nature of the ligands, the periodic position of the metal ion, and its charge) determining the entity of the energy separation (Δ) between the sets of d orbitals in the molecular complex (cf. Section 2.3.7.1). In octahedral complexes of first-row transition metal ions, d–d transitions typically give rise to light absorption falling in the visible region, whereas second- and third-row d-metal ions absorb at higher energies as a consequence of an increased Δ_O. Differently to the case of d-metal ions, whose absorption properties can be chemically modulated, f–f transitions are hardly influenced by the ligand field (cf. Section 2.3.7.2), so the typical MC absorption bands in lanthanide complexes practically depend solely on the nature of the metal itself and they can span from ultraviolet (UV) to near-infrared (NIR) regions (refer to the Dieke diagram in Fig. 2.55, Chapter 2). The lowest-energy LC transitions are usually of the type π–π^* or n–π^* and, depending on the extension of the ligand aromaticity, the associated light absorption can fall in the near-UV up to the visible–NIR region, where also CT bands typically appear. LC σ–σ^* transitions are instead highly energetic (usually require excitation in the far-UV region) and may lead to bond breaking; unlike the other types of transitions, they are not much affected by the presence of the metal. An interesting characteristic of CT bands, which makes them easily distinguishable from the other types of transitions, is their solution solvatochromism, that is, a significant shift of the absorption wavelength in solvents of different polarity. This feature arises from the very different dipole moments associated with the ground and excited states, related to the displacement of the electron density between two different regions of the molecule, and hence the higher stabilization of the ground state upon solvent interaction with respect to the excited state.[17]

Figure 3.63 gives a schematic representation of the energy separation of orbitals sets related to the various types of absorption transitions in metal complexes.

[17]Solvatochromism is actually a sort of kinetic effect, as the excited state usually does not survive long enough to allow surrounding solvent molecules to rearrange.

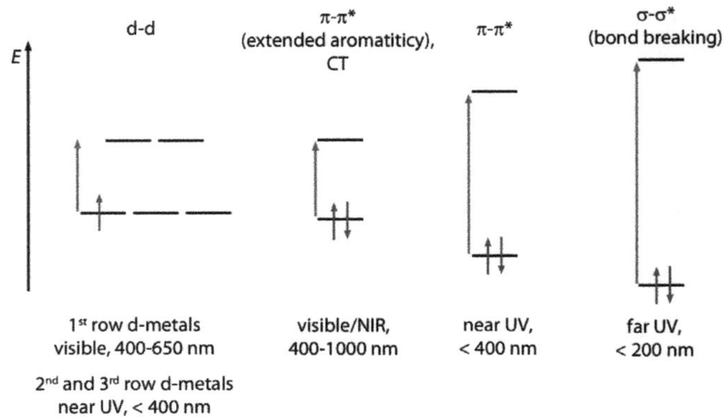

Figure 3.63 Schematic depiction of the typical energy separations between orbitals sets involved in the main electronic absorption transitions (red arrows) in transition d-metal octahedral complexes.

3.3.1.3 Intensity

Different absorption intensities are typically associated with the above-mentioned electronic transitions, depending on how strictly selection rules apply and on the value of the associated transition moment integral (cf. Section 2.3.2.3). LC transitions are usually fully allowed, giving rise to intense absorption (molar extinction coefficient, $\varepsilon = 10^4$–10^5 mol^{-1} cm^2). On the contrary, as we have already seen in Chapter 2, MC d–d and f–f transitions are formally prohibited by the angular momentum conservation rule ($\Delta \mathbf{l} = \pm 1$) and Laporte's rule, which can however be strictly applied only for electric dipole (ED) transitions in centrosymmetric complexes. Although MC absorption can become faintly allowed, thanks to vibronic coupling, related spectral bands remain very weak, with molar extinction coefficients in the 200–300 mol^{-1} cm^2 range for noncentrosymmetric d-metal complexes and below 100 mol^{-1} cm^2 for centrosymmetric ones. ED f–f transitions, which are hardly affected by the chemical environment, give rise to even weaker absorption lines with molar extinction coefficients of the order of tens of mol^{-1} cm^2.[18] On the other hand, transitions involving both

[18]Magnetic dipole (MD) transitions in lanthanide ions are formally allowed but intrinsically weak in intensity (cf. Section 2.3.7.2).

metal and ligand orbitals, that is, LMCT and MLCT, imply electron transfer between orbitals of prevalent d and p character so they are not subjected to spectroscopic selection rules other than the spin rule. Therefore, such transitions, which also involve a remarkable dipole moment variation, give rise to rather intense absorption bands ($\varepsilon = 10^3–10^4$ mol^{-1} cm^2) that are only slightly weaker than $\pi–\pi^*$ transitions due to the lower orbital overlap between the ground and excited states.

3.3.2 Luminophores with a Structural Role of the Metal

It has been already briefly stated that LC transitions take on a major role in coordination complexes of closed-shell metal ions and, in such a case, they usually show spectroscopic features similar to those of the free ligand. This means that whenever a luminescent organic entity is used as a ligand, its emissive properties can be mostly retained upon metal coordination while taking advantage of several benefits provided by a metal complex, such as:

- Improved thermal stability
- Improved emission quantum yield, thanks to the increased rigidity of the system (cf. Section 2.3.3.4)
- Molecular design in terms of modulation of the structural arrangement and geometry of the system

The latter point is of remarkable importance in the field of molecular materials as it would allow the introduction of emissive properties otherwise not observable in coordination compounds with open-shell d-metal ions, as a consequence of orbital mixing. This is the case, for example, of specifically designed classes of molecular building blocks, such as the anilate-based compounds already described in Section 3.1.4. In fact, the Al^{III} complex with the chlorocyananilate derivative, belonging to the class of the anilate ligands depicted in Fig. 3.18, displays similar spectral features with respect to those of the free ligand, thanks to a predominant noncovalent/ionic character of the bonding, which only poorly affects the ligand's frontier MOs. As a result, the Al^{III} complex retains the red emission of the free ligand while preserving the tris-chelated octahedral structural arrangement, ideal to form more sophisticated architectures, such as 2D honeycomb layers, of the analogous chlorocyananilato complexes of Cr^{III} and Fe^{III}, where emission is instead completely extinguished (vide infra), as shown in Fig. 3.64 (Atzori et al., 2015).

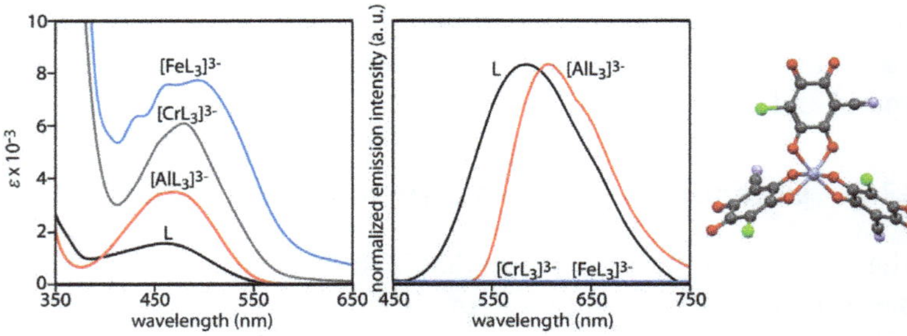

Figure 3.64 Absorption (left) and emission (center) spectra of the free L = chlorocyananilate ligand, $ClCNC_6O_4^{2-}$, and its isostructural complexes with Cr^{III}, Fe^{III}, and Al^{III}, and molecular structure of the tris-chelated complexes (right). Color codes: light blue, Al^{III}, Cr^{III}, or Fe^{III}; red, O; blue, N; green, Cl; gray, C.

The most archetypal example of a molecular complex showing improved emission properties arising from LC transitions is perhaps the tris-8-quinolinolato Al^{III} complex, commonly indicated as AlQ_3, which is the first example of a coordination compound used as an optically active center in electroluminescent devices (Tang and Van Slyke, 1987). In this compound, a green emission arises from an ILCT transition located on the highly aromatic 8- quinolinolate ligand, where the HOMO is mainly localized on the phenoxydic ring and the lowest unoccupied molecular orbital (LUMO) is mostly contributed by the pyridinic moiety (cf. Fig. 2.31), as schematically illustrated in Fig. 3.65. The increased structural rigidity and thermal stability provided by the coordination with a metal ion in an octahedral fashion with respect to the free Q ligand improves the emission quantum yield and allows the processing of this luminescent material through suitable techniques, such as vacuum deposition, for the fabrication of optical devices such as organic light-emitting diodes (OLEDs, see Chapter 5). In this regard, it is also worth anticipating herein that one of the major advantages of such molecular emitters with organic moieties, as will be better clarified in Chapter 4, is related to the high ligand aromaticity, which bestows these materials with good conducting properties suitable for exploitation in semiconducting devices and, at the same time, allows electrical excitation (electroluminescence, EL) as a consequence of the formation of a rather stable exciton.[19]

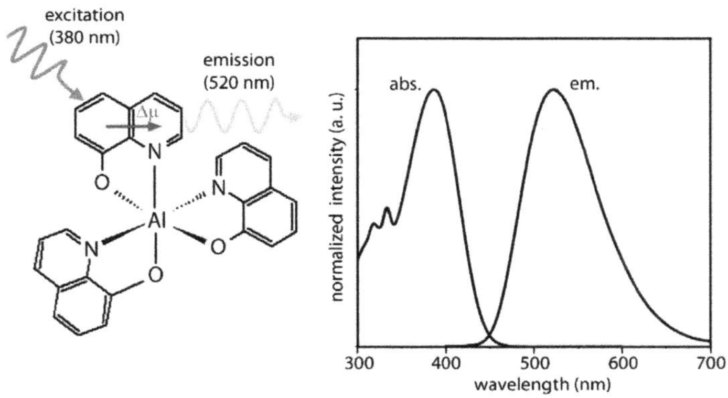

Figure 3.65 Structure of the AlQ_3 complex and schematic depiction of the photoluminescent mechanism giving rise to green emission after photoexcitation in the near-UV region (left) and absorption and normalized emission spectra of AlQ_3 in solution (right).

In light of the above-described examples, some general guidelines can be drawn for the design of this class of molecular emitters:

- The organic ligand must possess suitable frontier MOs to display fluorescent emission.
- The organic ligand must be stable and must retain emissive properties in the coordinated form (e.g., as an anion).
- The coordinated metal ion should have, in principle, a closed-shell configuration, although an open-shell configuration does not necessarily preclude pure LC emission.

In addition, the versatility of coordination chemistry allows extraordinary potential for fine modulation of the optical properties of this class of compounds. This can be achieved:

---

[19]An exciton is an electron–hole pair where the electron lies in the conduction band and the hole is located in the valence band of a semiconducting species as a consequence of light or electrical excitation. EL occurs when the exciton corresponds to the excited state of a molecular emitter (see Section 4.9).

- Through change of the metal ion, which can anyway have a significant influence on the ligand electronic energy levels
- Through subtle ligand alterations, for example, by introducing suitable substituents
- By taking advantage of solid-state effects due to supramolecular interactions

All these effects will be herein described.

3.3.2.1 Effect of the metal

As mentioned at the beginning of this section, the nature of the coordinated metal ion has a dramatic effect on the luminescent properties displayed by the complex. The limit case is the total quenching of LC emission upon coordination with an open-shell d-metal ion, where metal–ligand orbital mixing causes a modification in the energy level sequence so transitions giving rise to radiative emission from ligand-centered states are suppressed. This is, for example, the case of the Fe^{III} and Cr^{III} anilate complexes discussed in the previous paragraph and reported in Fig. 3.64. However, fine-tuning of the absorption and emission spectral features, that is, the energy of related electronic transitions between the HOMO and the LUMO, is achievable taking into account some of the characteristics of the metal ion:

- The hard/soft character, which affects the degree of covalency
- The electronic configuration (closed shell/open shell), which determines the entity of the orbital mix
- The atomic number

The first two parameters can have opposite effects depending on the nature of the ligand's frontier MOs. As a clarifying example, in the class of quinolinolato complexes, where the highest degree of covalency is associated with the metal–nitrogen bonding on the pyridinic ligand moiety (where the LUMO lies), an increased covalent character of the bond leads to stabilization of the LUMO and then to a red shift of the absorption/emission spectral features. This effect is of course larger and more sensitive to the metal in the case of open-shell configurations, as clearly evidenced in the absorption spectra of a series of quinolinolato complexes with both closed- and open-shell metal ions, shown in Fig. 3.66.

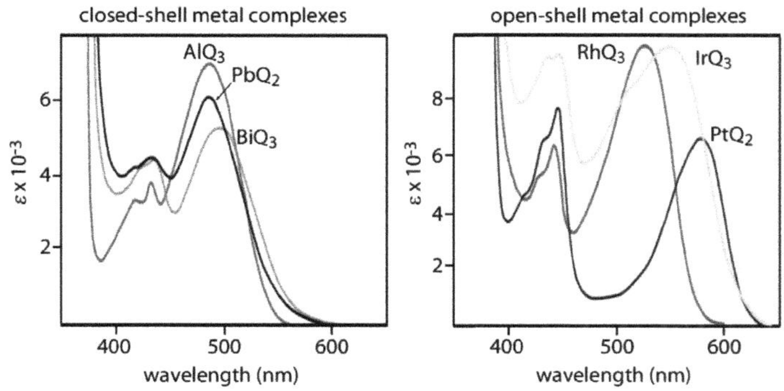

Figure 3.66 Absorption spectra of quinolinolato complexes of closed-shell (left) and open-shell (right) metal ions. Reprinted with permission from Ballardini et al. (1986). Copyright (1986) American Chemical Society.

Analogous effects related to the soft/hard character and the electronic configuration of the metal are also expected to affect the emission features of the complex. However, the most dramatic influence of the metal ion on the luminescence properties of a coordination compound is associated with its atomic number by means of the heavy-atom effect, which can favor intersystem crossing (ISC) between excited ligand states of different spin multiplicities (most frequently a singlet-triplet conversion). This would eventually result in a metal-induced LC phosphorescent emission that is significantly red-shifted with respect to the free-ligand fluorescence (cf. Section 2.3.3.3). This effect becomes evident, for instance, in the case of the quinolinolato compounds cited before, which can display green fluorescence and/or red phosphorescence depending on the nature of the coordinated metal, as shown in Fig. 3.67. These two types of emission originate from competitive mechanisms, that is, singlet decay and ISC followed by triplet decay, which can even coexist in the same molecular complex, giving rise to dual luminescence in two different spectral regions upon single-wavelength excitation, as, for example, in the case of the BiQ_3 and PbQ_2 complexes shown in Fig. 3.67. The atomic number effect of the metal has also important consequences for the temporal evolution of the complex luminescence output as phosphorescence is considerably longer-lived than fluorescence and can be exploited for different application purposes.

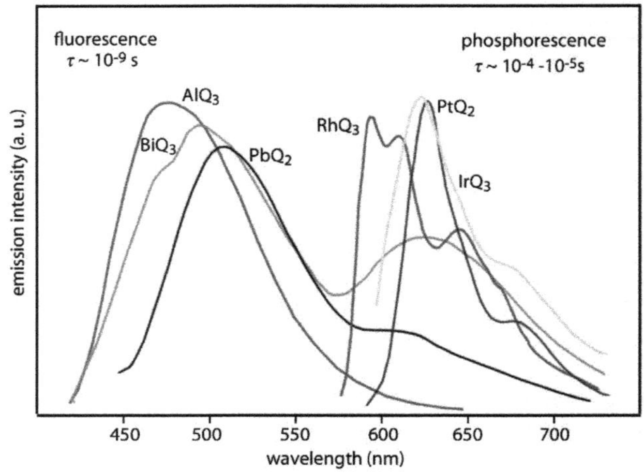

Figure 3.67 Emission spectra for a series of quinolinolato complexes. Reprinted with permission from Ballardini et al. (1986). Copyright (1986) American Chemical Society.

3.3.2.2 Effect of the ligand

Fine-tuning of the emission properties of organic luminophores can be achieved by the introduction of suitable substituents having an electron-donating or electron-withdrawing character, able to significantly influence the frontier MOs of the organic moiety. As a general empirical rule, we can infer that an electron-donating group is able to stabilize the filled HOMO and destabilize the empty LUMO, whereas the effect is reversed in the case of electron-withdrawing substituents. In particular, the influence of a ligand's substituents on its emitting properties is usually more pronounced and, at the same time, would allow a larger tuning range whenever the organic emitter gives rise to ILCT transitions between HOMOs and LUMOs located in different molecular regions. For example, in the typical case of the quinolinolate ligand, different substitution on the phenoxide ring (typically in the 5 and 7 positions) or on the pyridinic moiety (possible in the 2 position) would allow the independent control of the stabilization or destabilization of the HOMO or the LUMO, respectively. This can allow excellent fine-tuning of the emission color of the organic emitter upon very subtle chemical changes. A significant example of this effect is reported in Fig. 3.68 (Montes et al., 2004), where the emission colors, ranging from blue to orange-red, of a series of Al^{III} complexes with quinolinolate derivatives bearing different substituents in the 5 position of the ring are shown. In these compounds, a red shift of the emission is observed upon the introduction of electron-donating groups, whereas electron-withdrawing substituents induce a blue shift with respect to the green-emitting AlQ_3 complex.

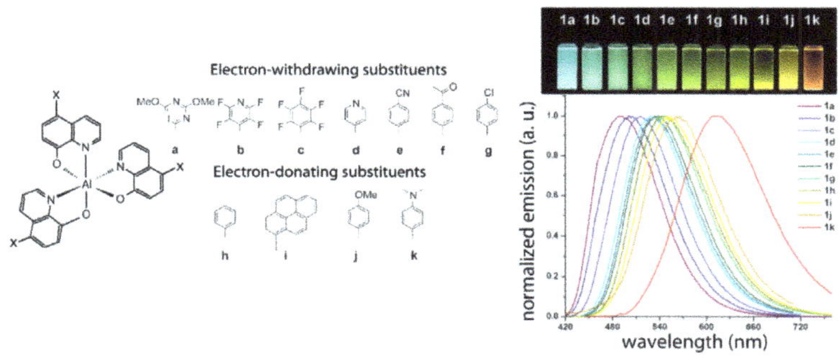

Figure 3.68 Molecular structure of Al^{III} quinolinolato derivatives with different substituents in the 5 position (left) and corresponding emission colors and spectra (right). Adapted with permission from Montes et al. (2004). Copyright 2004 Wiley-VCH.

3.3.2.3 Supramolecular effects

Supramolecular interactions can significantly affect the optical properties of a molecular emitter, in particular inducing a remarkable red shift in the emission wavelength. In this case this aggregation-induced emission is often defined as excimer-like, as it can be somehow compared to the radiative emission of an excimer. This entity is a supramolecular dimer held together by some kind of bonding interactions established when one constituting molecule is in its excited state (thus having semifilled HOMOs and LUMOs) and the other one is in the ground state (with completely filled and empty orbitals).[20] This excited dimeric species is then a chemically different object with respect to the originating free molecules. On

[20]The dimeric species is formally called excimer when it originates from the interaction of two identical monomers, while it takes the name exciplex (short for excited complex) whenever it is formed by one excited and one unexcited molecule of a different nature.

relaxing, the excimer returns to a ground state corresponding to the monomer species in their dissociated form. This decay may be accompanied by radiative emission at a longer wavelength with respect to that expected from the isolated excited monomers, giving rise to a broader spectrum as a consequence of the dissociative nature of the ground state (see Fig. 3.69).

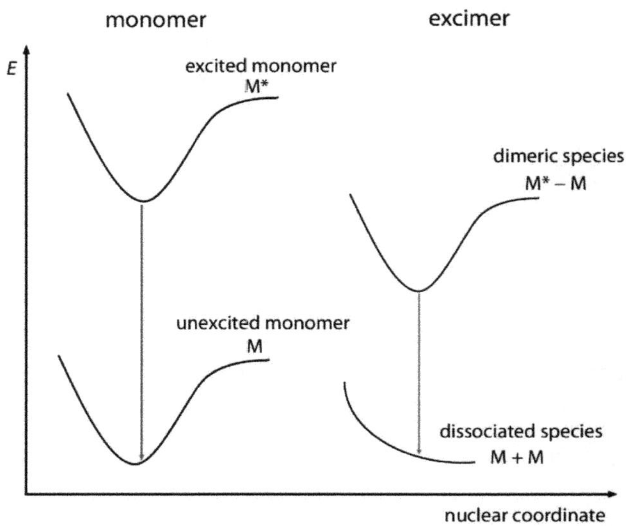

Figure 3.69 Comparison of the ground and excited states of a monomeric emitter (M) and of the corresponding excimer. Vibrational levels are omitted for clarity.

The formation of excimers is, strictly speaking, a diffusional process, which can also occur at the atomic level, where the excited and unexcited monomers come into contact after random kinetic collisions.[21] In the case of molecular materials, especially for coordination complexes, dimerization phenomena may take place, thanks to a whole range of possible chemical interactions between specifically designed molecules, especially π-stacking interactions between highly aromatic organic ligands. This effect is usually observed in crystalline compounds, where the interactions between molecules are well established and the metal coordination templates

[21]For example, the formation of the excimer Xe*-Xe in the gas phase is responsible of the broad fluorescence emitted by commercial Xe lamps.

suitable supramolecular arrangements. In Fig. 3.70 a meaningful example of excimer-like emission showing substantial red shift with respect to the isolated emitting complexes, as a consequence of an extensive π-stacking between aromatic ligands of adjacent molecules forming a dimeric unit, is shown (Artizzu et al., 2008).

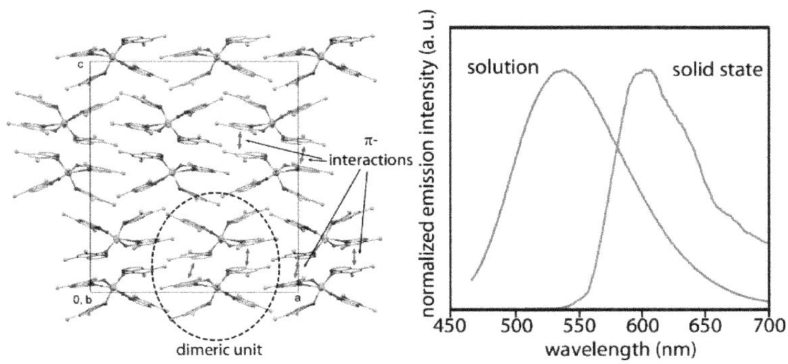

Figure 3.70 Crystal packing (left) and dilute solution and solid-state excimer-like LC emission spectra (right) of a Gd^{III} complex with the 5,7-dichloro-quinolinolate ligand showing extensive π-stacking between dimeric units.

The templating effect of the metal in favoring supramolecular interactions between the coordinated ligands can be so effective that excimer-like emission can be even observed in the diluted solution. This can be the case when flexible aromatic ligands wrap around the metal ion, forming cluster aggregates or "cage" polynuclear complexes. Typical examples of these molecular architectures, where several π-interactions are established between adjacent multidentate Schiff-base ligands assembled together through coordination with multiple Zn^{II} ions, are shown in Fig. 3.71 (Al-Rasbi et al., 2006).

It is well understood in chemistry that aggregation or clustering phenomena, both in the solid state and in solution, can be often induced by appropriate counterions by means of a variety of supramolecular interactions, including π-stacking, hydrogen bonding, and halogen bonding. Therefore, aggregation-induced emission is particularly useful for the development of luminescent sensors for the recognition of specific chemical species.

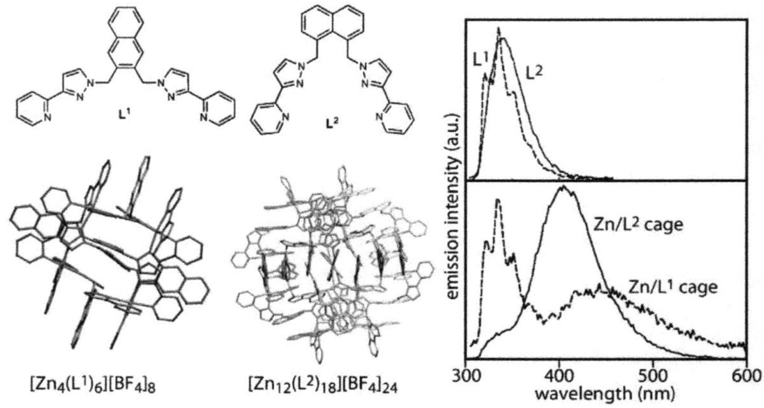

Figure 3.71 Molecular structures (left) and solution emission spectra (right) of aromatic flexible luminescent ligands and their Zn^{II} "cage" complexes. Hydrogen atoms are omitted for clarity. Adapted from Al-Rasbi et al. (2006) with permission of The Royal Society of Chemistry.

3.3.3 Luminophores Based on Open-Shell d-Metal Complexes

As mentioned in the introductory paragraphs of this section, the complexity of the chemical, electronic, and physical features of d-metal complexes, the variability of metal–ligand interactions, the nature of frontier MOs, the numerous parameters affecting the energy of the excited states, and many other factors make the emissive properties of these compounds not easily predictable. Therefore, an exhaustive account of all the possible kinds of luminophores based on open-shell d-metal complexes can be quite complicated or dispersive. For this reason in this book we will limit the discussion to the most studied examples and related classes of luminescent d-metal complexes, taking into account that the fundamental principles described herein for model compounds can be oftentimes applied to a wider class of materials.

As a first general consideration, it can be pointed out that the vast majority of luminescent transition d-metal complexes are investigated and developed in view of their phosphorescent properties, whereas fluorescent materials do not usually garner the same attention. The reasons behind this mainly lie in the applicative purposes that these kinds of compounds are tailored for. In particular, phosphorescent emission is greatly advantageous for the following applications:

- Emissive centers in electroluminescent OLED devices where excitons are mostly harvested in the triplet state (cf. Chapter 5)
- Phosphorescent sensors whose long-lived emission is well distinguished from background light
- Long-lived excited species for photochemical reactions, photocatalysis, and sensing
- Photoredox processes involving triplet state species, such as oxygen quenching by ground state $^3O_2$[22]

A rough subclassification of emissive materials based on transition d-metal complexes can be quite satisfactorily made simply on the basis of the periodic position of the coordinated metal, which mostly determines the ordering of the energy levels in the molecular complex depending on the crystal/ligand field strength (which increases along a group). This can be pictured through an energy state diagram as the one reported in Fig. 3.72 for the case of VIIIB group's divalent metal ions in an octahedral field. Such diagrams are a sort of transposition of the Jablonski diagram (cf. Section 2.3.3.5) for d-metal complexes, where the various energy states are named after the associated electronic absorption transitions (e.g., MC, LC, and MLCT), which contain information on the involved orbitals. Although this labeling may seem not entirely orthodox, as it, strictly speaking, refers to the types of transitions not to actual molecular energy states, it still is a convenient way to construct energy level diagrams for d-metal complexes.

From the diagram of Fig. 3.72 and according to Kasha's statement, which predicts emission from the lowest-lying excited state, it is easy to conclude that, as a general rule, radiative emission is expected to be MC in nature for first-row transition metal ions and MLCT for second- and third-row d-ions.

[22]Ground-state triplet oxygen $^3O_2$ can easily undergo energy transfer (ET) from an excited species in its triplet state as a consequence of diffusional processes (dynamic quenching, cf. Section 2.3.6.1) to become the highly reactive singlet $^1O_2$ oxygen. As a result, phosphorescent emission from the donor species is quenched. For this reason, triplet emitters are often used for oxygen sensing.

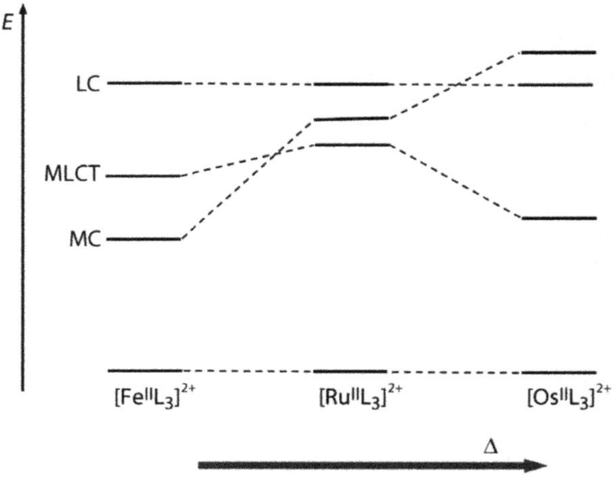

Figure 3.72 Sequence of electronic energy states in a series of isostructural tris-chelated octahedral complexes of the VIIB group's divalent metal ions with a generic ligand L.

3.3.3.1 First-row transition metal complexes

Overall, complexes of first-row transition metal ions displaying emissive properties are rather unusual, that is, d–d transitions rarely yield emissive states. This is quite understandable since in transition metal complexes electronic states are in general much more densely packed than in typical organic systems and, in most cases, the energy gap between d orbitals (and related energy states) induced by crystal field effects is so small that rapid internal conversion processes are largely favored with respect to radiative emission. In purely octahedral complexes, where d orbitals are well separated into two sets, the promotion of one electron from a nonbonding (or bonding) t_{2g} orbital to an antibonding e_g one weakens the metal-to-ligand bonding and induces a distortion in the molecular structure that results in a spatial displacement of the excited state with respect to the ground state, hence favoring nonradiative decay (cf. Section 2.3.3.4). However, for suitable d^n electronic configurations where not all the excited states necessarily imply further population of e_g antibonding orbitals but rather a variation in the spin multiplicity, the relatively long-lived phosphorescence, favored by the heavy coordinated metal ion, can be more easily observed than the highly quenched fluorescence.

All these observations can be easily explained by taking into account the most typical example of MC d–d phosphorescence found in Cr^{III} compounds. The diagram of potential energy curves for the electronic states of the Cr^{III} (d^3 configuration) in an octahedral complex[23] provides an immediate explanation for the optical properties of these compounds (Fig. 3.73). In fact, it can be observed that, whereas the first excited, spin-allowed, quartet state $^4T_{2g}$ is usually significantly displaced with respect to the ground $^4A_{2g}$ state, the doublet ^2E_g state, where all the three electrons lie in t_{2g} orbitals, maintains the nuclear molecular coordinates of the unexcited complex. At the same time, the $^4T_{2g}$ state is "nested" inside the ^2E_g potential energy curve, allowing overlap of vibrational levels and enhancing the ISC process. Therefore, several Cr^{III} compounds display a red (~755 nm) long-lived phosphorescent emission due to the $^2E_g \rightarrow {}^4T_{2g}$ transition.

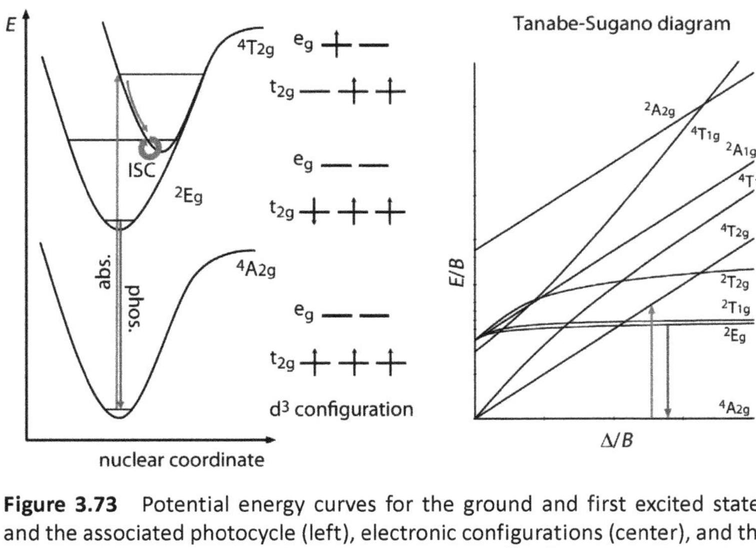

Figure 3.73 Potential energy curves for the ground and first excited states and the associated photocycle (left), electronic configurations (center), and the simplified Tanabe–Sugano diagram (right) for a Cr^{III} (d^3) octahedral complex. ISC, intersystem crossing; abs., absorption; and phos., phosphorescence.

[23]The reader can be reminded that the d orbitals of a metal ion in a coordination complex are directly involved in bond formation; therefore the use of Morse potential energy curves is in this case suitable for representing the energy states associated with the configuration of atomic orbitals (which anyway have a significant degree of orbital mix).

However, except some few additional examples, such as some Mn^{II} d^5 *hs* compounds whose emission is related to a d–d $^4T_1(G) \rightarrow {}^6A_1$ transition, the luminescent properties that can be detected in complexes with first-row transition metals are seldom attributable to "pure" MC transitions as a consequence of significant ligand orbital mix. For example, Cu^I or Ni^0 complexes often show weak emission properties related to MLCT transitions.

It is, nonetheless, generally concluded that efficient emissive properties are rarely observed for first-row transition metal complexes, also in view of the relatively weak spin–orbit coupling induced by the "lightweight" metal ions. As a consequence, this class of compounds is rarely investigated for optical applications. However, we can briefly summarize the main requirements for achieving MC d–d emission in first-row transition metal complexes:

- Metal orbital sets must be far enough apart in energy to favor radiative decay, so octahedral complexes with a strong crystal/ligand field are the most promising.
- The metal should have a suitable d^n configuration, so the lowest excited state only differs for spin multiplicity with respect to the ground state and phosphorescence is the main decay path.

3.3.3.2 Second- and third-row transition metal complexes

Second- and third-row transition metal complexes are by far the most widely studied in view of their emissive properties. The stronger spin–orbit coupling induced by the heavier 4d and 5d metal ions with respect to 3d ones and the relevant metal–ligand orbital mix once coordinated, easily favor the ISC mechanism,[24] leading to phosphorescent states and quite efficient emission quantum yields with long lifetimes, up to the microsecond range. Moreover, the larger crystal/ligand field strength (and consequently the larger Δ_O) typical of these complexes determines a sequence of energy levels suitable to lead to the population of excited MLCT states (see Fig. 3.72) that are better susceptible to radiatively decay to the ground

[24]Due to the high crystal/ligand field strength in coordination complexes, 4d and 5d metal ions normally are in a low-spin configuration. Therefore the ground state is usually associated with a lower-spin multiplicity than the excited state.

state with respect to, usually nonemissive, MC states,[25] which are raised in energy and are less prone to becoming populated. These CT transitions are fully allowed by selection rules and generally accompanied by a significant degree of dipole moment variation; thus they can give rise to a strong emission whose intensity is, of course, largely determined by the energy gap law (cf. Section 2.3.3.4). In light of these considerations, it appears logical that the molecular photophysics and the photochemistry of transition metal complexes are essentially dominated by systems, particularly complexes of platinum group's metals (Ru, Os, Rh, Ir, Re, and Pt), giving rise to emission (especially phosphorescence) from MLCT states. With this view, we will discuss the main features of luminescent second- and third-row metal complexes on the basis of some model compounds and their relatives, with particular reference to the following two classes:

- Octahedral d^6-metal complexes
- Square-planar d^8-metal complexes

3.3.3.2.1 *Octahedral d^6-metal complexes*

The field of luminescent transition metal complexes has been largely dominated since the 1970s by Ru^{II} polypyridyl compounds, following the report of the most archetypal example Ru^{II} tris-bipyridyl, $[Ru(bipy)_3]^{2+}$, complex (Paris and Brandt, 1959), shown in Fig. 3.62. $[Ru(bipy)_3]^{2+}$ is currently a widely used luminescence reference standard (cf. the box titles "Spectroscopic Evaluation of Emission Quantum Yield," in Section 2.3.4.1) emitting in the red region of the visible spectrum ($\lambda_{em} \approx 615$ nm).

From the inspection of the MO diagram of $[Ru(bipy)_3]^{2+}$ for a Ru^{II} ion in a low-spin (*ls*) d^6 configuration, reported in Fig. 3.74, it can be easily deducted that the lowest absorption band for this complex would correspond to a singlet state ^{1}MLCT transition. This transition implies an internal redox process for which an electron localized on the metal is transferred to a ligand orbital of antibonding character ($\pi^* \leftarrow t_{2g}$):

[25]As already explained in Section 3.3.3.1, spin-allowed d–d transitions imply the population of an antibonding orbital, leading to an excited (MC) state, which is spatially displaced with respect to the ground state and therefore more prone to deactivation via internal conversion.

$$[Ru^{II}(bipy)_3]^{2+} \xrightarrow{h\upsilon} [Ru^{III}(bipy)_3(bipy^{-})]^{2+*}$$

The displacement of an electron from the inner metal ion to the peripheral part of the complex facilitates further electron exchange from the excited complex to an external species/matrix, allowing for a number of applications, such as artificial photosynthesis, water splitting, hydrogen production, and dye-sensitized solar cells. Some of these applications will be discussed in Chapter 5.

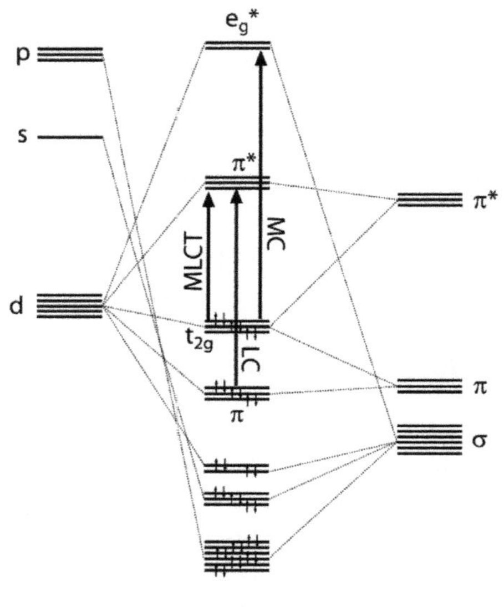

Figure 3.74 Molecular orbital diagram of the $[Ru(bipy)_3]^{2+}$ complex.

From the ^{1}MLCT-excited singlet state ISC is then likely to occur as a consequence of the heavy-atom effect induced by the metal ion, leading to the population of a triplet ^{3}MLCT state according to the photocycle depicted in Fig. 3.75, where the Jablonski diagram for the $[Ru^{II}(bipy)_3]^{2+}$ complex (which can be also taken as generally valid for related compounds) is reported. Radiative decay from the excited ^{3}MLCT state brings the complex back to its original ground state, giving rise to phosphorescent emission.

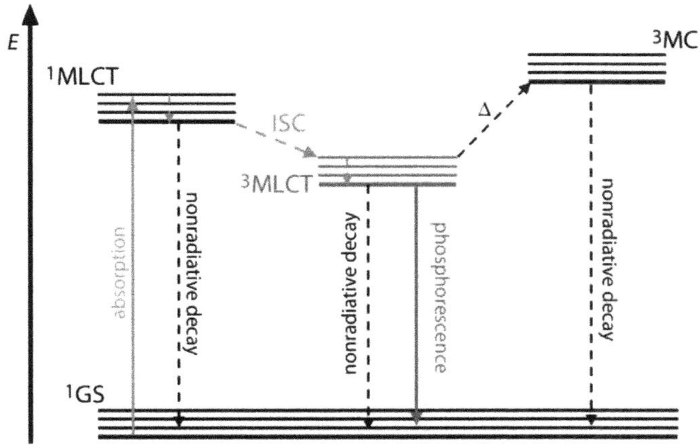

Figure 3.75 Simplified Jablonski diagram for a generic luminescent second- or third-row transition metal ion complex with a singlet ground state (^{1}GS) and photocycle leading to phosphorescent emission. Dashed arrows represent nonradiative processes. Δ is the extra thermal energy necessary to achieve the population of the higher-lying ^{3}MC state.

The nature of the radiative emission as arising from a triplet state is confirmed by the observation that the quantum yield is significantly affected by oxygen quenching, that is, excited triplet deactivation due to the interaction with ground state triplet oxygen $^3O_2$ (cf. footnote 22). It is worth remarking that phosphorescence originating from a lower-lying triplet state implies a relevant Stokes shift, which is greatly advantageous for efficient emission since the negligible overlap of the absorption and emission bands (see Fig. 3.76) allows for the minimization of self-absorption and consequent concentration quenching. Another interesting characteristic of ^{3}MLCT emission is that it is particularly sensitive to the chemical environment, especially in regard to the solvent polarity. In fact, the triplet excited state survives long enough ($\tau = 10^{-6}$–10^{-7} s) to allow solvent molecules to rearrange around the excited complex, which bears an excess of charge density on a peripheral ligand. As a result, nonpolar solvents tend to push up the energy of the excited state, causing an ipsochromic shift of the spectral features as well as an increase in the radiative lifetime and quantum yield according to the energy gap law, whereas polar solvents stabilize the excited state and exert an opposite effect.

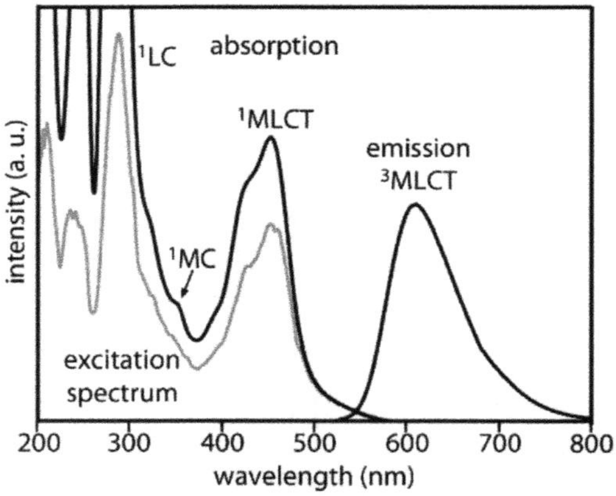

Figure 3.76 Absorption, emission, and excitation (gray line) spectra for the $[Ru^{II}(bipy)_3]^{2+}$ complex in solution. The good overlap of the excitation spectrum with the absorption spectral features indicates that the lower-lying ^{1}MLCT state can be also populated by nonradiative internal conversion (IC) from superior energy levels (^{1}MC and ^{1}LC).

Besides these considerations, the emission quantum yield and lifetime of $[Ru^{II}(bipy)_3]^{2+}$ and its analogs are largely determined by nonradiative mechanisms competing with the photocycle described above. In the absence of photoredox processes, and apart from external fluorescence and phosphorescence quenching phenomena, a third competing process is related to thermal activation leading to the population of the higher-lying nonradiative ^{3}MC state (see Fig. 3.75). As a consequence, emission efficiency is highly temperature dependent, and in the case of $[Ru^{II}(bipy)_3]^{2+}$, fast equilibrium between the emissive ^{3}MLCT and the nonemissive ^{3}MC state occurs even at room temperature due to the small energy gap between them. The nature of the ligand bonding and the strength of the crystal/ligand field, which can be controlled by chemical modifications, for example, through the introduction of suitable substituents on the ligands, allow for the fine-tuning of the spectral features of $[Ru^{II}(bipy)_3]^{2+}$ related compounds, but at the same time, crucially determine whether a complex is luminescent or not, regardless of the temperature. The population of the ^{3}MC state is, in fact, considered responsible for the deactivation of radiative emission in $[Ru^{II}(bipy)_3]^{3+}$-related complexes. This is likely to occur in complexes where the crystal/ligand field is too weak and/or the ligand's π^*

orbitals are too high in energy, as a result of which an inversion of the sequence of π^* and e_g^* MOs (cf. Fig. 3.74) is expected, leading to nonemissive complexes due to the lowering of the MC state below the MLCT one. Therefore, one of the most effective strategies to improve the luminescent properties of $[Ru^{II}(bipy)_3]^{3+}$-related complexes relies either on the use of strong σ donor ligands in order to push up in energy the e_g^* orbitals, and hence the associated MC state, or, conversely, on the integration of strong π acceptor ligands to stabilize the π^* MO and lower the energy of the MLCT state. However, these measures may even be insufficient to guarantee efficient luminescent output, as many additional parameters, such as structural tension and distortion, may play a crucial role. For example, the Ru^{II} bis-terpyridyl complex, $[Ru^{II}(terpy)_2]^{2+}$, is essentially nonluminescent due to the geometry distortion induced by the sterical tension of the tridentate terpyridine ligand and the consequent weakening of the crystal/ligand field strength, which lowers the energy of the MC state. As a further case, the nonemissive behavior of *cis*-$[Ru^{II}(bipy)_2(py)_2]^{2+}$, is attributed to the photodegradation caused by the replacement of one monodentate pyridine ligand, more labile than the bidentate bipy, with a solvent molecule. In this case the weakening of metal–ligand bonding after photoexcitation is a direct consequence of the promotion of one electron in an antibonding e_g^* orbital. Since solvent molecules usually are weak σ donors, this results, once again, in a lowering of the MC state below the emissive MLCT one. These examples are schematically depicted in Fig. 3.77.

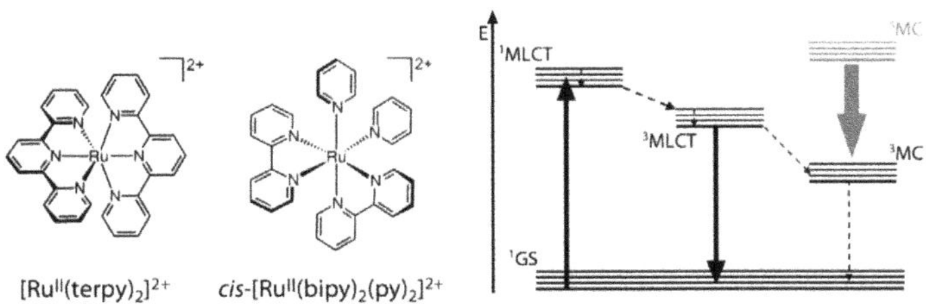

Figure 3.77 Molecular structure (left) and Jablonski diagram (right) for the nonluminescent $[Ru(terpy)_2]^{2+}$ and *cis*-$[Ru(bipy)_2(py)_2]^{2+}$ complexes.

Aside from chemical modifications of the coordinated ligands, the substitution of Ru^{II} with an isoelectronic (d^6) metal ion belonging to the third row of the transition block represents another effective approach to raise the energy of the MC state as a consequence of the increased crystal/ligand field strength, as exemplified in Fig. 3.72. The most immediate example is represented by Os^{II} compounds, since osmium can be considered a chemical vicariant of ruthenium. However, differently to its ruthenium analog, which is intensely emissive, only a weak luminescence is detected for $[Os(bipy)_3]^{2+}$ at room temperature. This is ascribed to the much easier oxidation of Os^{II} with respect to Ru^{II}, which stabilizes the MLCT state so that the energy gap with the ground state becomes smaller and, according to the energy gap law, nonradiative deactivation becomes more competitive. On the contrary, the less electron rich Ir^{III} is much more difficult to oxidize with respect to Os^{II} and Ru^{II}. Consequently, in Ir^{III} complexes, the MLCT state is raised in energy, and, at the same time, the stronger crystal/ligand field places the MC state well above both the MLCT and the LC ones so that it can be hardly populated (cf. Fig. 3.72). Moreover, the stronger spin–orbit coupling induced by the higher atomic number of Ir^{III} increases the efficiency of ISC and the formation of excited triplets. Therefore, Ir^{III} compounds generally are strong luminophores with long emission lifetimes over a wide range of temperatures and offer the additional advantage to allow considerable tuning of the emission color upon chemical modifications and consequent change in the ordering of the excited levels. In fact, Ir^{III}–based luminophores can be roughly subdivided into two categories:

- Complexes with anionic ligands able to stabilize the MLCT state by balancing the charge loss on the metal and giving rise to ^{3}MLCT emission
- Complexes with neutral ligands displaying ^{3}LC emission

For example, whereas the $[Ir^{III}(ppy)_3]$ complex with the anionic phenylpyridine (ppy^-) ligand displays ^{3}MLCT emission, in the case of $[Ir^{III}(terpy)_3]^{3+}$, bearing the neutral terpyridine (terpy) ligand, phosphorescence is attributed to a ^{3}LC transition, as schematically exemplified in Fig. 3.78.

It should, however, be taken into account that in many cases a mixing of MLCT and LC states is more appropriate to describe the electronic transitions occurring in these systems, as even transitions that are mostly LC in character can have a significant metal contribution, depending on the mixing of the metal's and ligands' orbitals. The extent of orbital mixing and the energy of the excited states, in turn, depend on a variety of factors, including the nature of the external matrix and the temperature. For instance, the emission observed in $[Ir^{III}(bipy)_3]^{3+}$ can be attributed to a ligand-centered transition for the complex in a glassy solvent at 77 K but it nonetheless shows some MLCT character in solution at room temperature. The fact that emission can occur from both LC and MLCT states means that significant color tuning can be achieved through ligand modifications (vide infra).

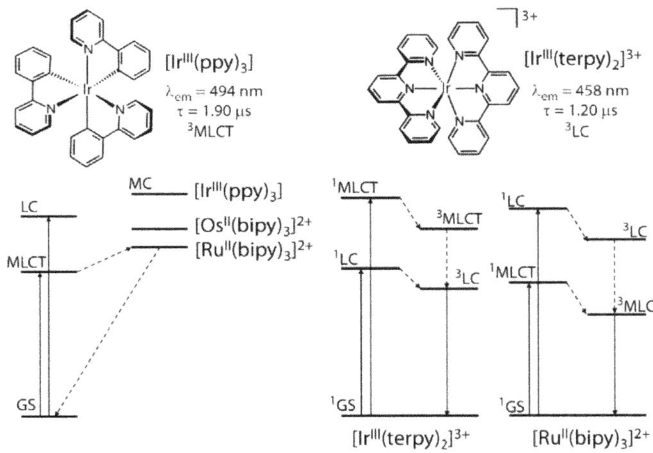

Figure 3.78 Structure of $[Ir^{III}(ppy)_3]$ (upper left) and a simplified energy level scheme for Ru^{II}, Os^{II}, and Ir^{III} octahedral complexes (lower left). Structure of $[Ir(terpy)_2]^{3+}$ (upper right) and a comparison of the energy level diagrams and photocycles of $[Ir(terpy)_2]^{3+}$ and $[Ru(bipy)_3]^{2+}$ (lower right). Solid arrows represent absorption and emission transitions, and dashed arrows denote nonradiative processes.

In general, lower quantum yields and lifetimes are associated to ^{3}LC emission with respect to that originating from ^{3}MLCT states. The reason behind this is the fact that the efficiency of the spin–orbit coupling favoring ISC and subsequent population of excited triplet states also depends on the degree of the contribution of the metal orbitals to the energy states involved in the transition. This observation, together with the difficulties encountered in the synthesis of Ir^{III} complexes with neutral organic ligands as a consequence of the kinetic inertness of this *ls* d^6 metal ion, makes its coordination chemistry largely dominated by compounds with anionic ligands. It is also worth remarking that the use of anionic ligands also offers the additional advantage to allow neutral complexes, which can be more conveniently processed for the fabrication of optical devices such as OLEDs. Among this class of compounds, cyclometallated carbon complexes, especially those with N,C-cyclometallating ligands, such as the archetypal [Ir(ppy)$_3$], are largely studied in view of their good luminescent properties, their stability, and the opportunity to easily tune the emission color by chemical modifications of the ligand. In fact, on increasing the number of anionic carbon donors by replacing more electronegative, uncharged nitrogen atoms of the six total coordinating sites in the octahedral complexes, both the metal d and the ligand π orbitals are raised in energy. As a result, a whole series of analogous complexes can be obtained showing emission originating from different types of transitions whose spectral region depends on the energy gap between the involved levels, as shown for Ir^{III} bis-terpyridine derivatives in Fig. 3.79.

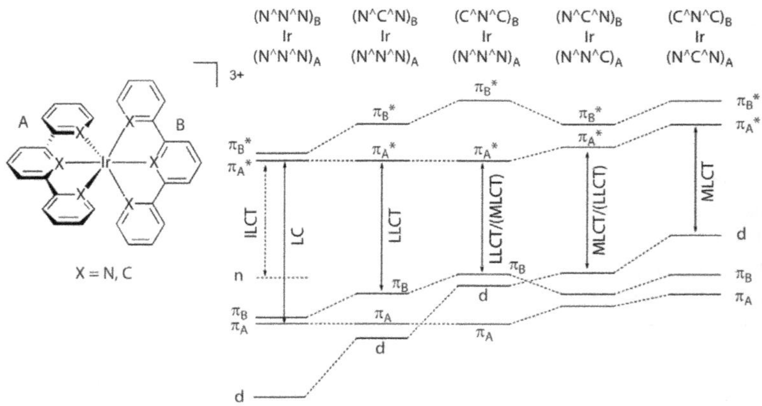

Figure 3.79 Molecular structure of Ir^{III} bis-terpyridine derivatives (left) and simplified schematic energy level diagram (right) showing the influence of cyclometallation on the frontier orbitals in the bis-tridentate complexes, that leads to excited states of a different character. The dashed n orbital represents an electron-rich n or p orbital localized on a remote pendant group. Adapted from Gareth Williams et al. (2008) with permission of The Royal Society of Chemistry.

It is also worth remarking that a further opportunity to tune the emissive features of such compounds is also achieved through the chemical modification of the ligand periphery, for instance, by introducing suitable electron-rich "pendant arms" or ancillary units whose MOs do not directly intervene in the coordination with the metal and are thus unaffected by the metal–ligand bonding. In other words, these units provide nonbonding orbitals whose energy is higher than that of the ligand bonding orbitals, as in the first example reported in Fig. 3.79. In this case, ILCT emission can be observed as if the ligand were coordinated to a closed-shell metal ion, similar to the class of compounds discussed in Section 3.3.2.

3.3.3.2.2 *Square-planar d⁸-metal complexes*

Another widely studied class of third-row transition metal luminophores is represented by Pt^{II} ($5d^8$) square-planar complexes, which mostly display phosphorescent emission, thanks to the high spin–orbit coupling exerted by the heavy-metal ion. Among the Ni^{II}, Pd^{II}, and Pt^{II} triad, the platinum ion is the only one susceptible to inducing a crystal/ligand field strong enough to raise the energy of ^{3}MC states, so they are not thermally accessible at room temperature. However, the ordering of energy levels in Pt^{II} compounds is particularly sensitive to the characteristics of the coordinated ligands, so only complexes with good σ donor or π acceptor ligands, causing a strong ligand field, are found to display significant luminescence properties. Therefore, it appears evident that the nature of the displayed emission, which can have both ^{3}LC and ^{3}MLCT character, can be tuned by changing the features of the coordinated ligand and similar considerations to those already made for Ir^{III} complexes can be applied. However, unlike the prototypical octahedral d^6 complexes that we have considered so far, the open structural arrangement of square planar Pt^{II} d^8 complexes allows the formation of molecular aggregates, dimers, or even chained supramolecular assemblies, which can lead to remarkable changes in the optical properties of the compounds with respect to the isolated monomeric molecules. These aggregation phenomena, typically observed in "dense" materials such as solids or in concentrated solutions, can occur either through ligand stacking interactions and/or through the formation of metal–metal bonding. The effect of supramolecular interactions through ligand moieties has been already described in Section 3.3.2.3, and the same principles discussed therein for the case of metal complexes giving rise to LC emission can be also applied for open-shell platinum complexes. In addition, metal–metal bonding interactions occurring between adjacent square-planar Pt^{II} complexes can be established through the superposition of the out-of-plane d_{z^2} and even the empty p_z, metal orbitals. The combination of these orbitals gives rise to a bonding orbital and an antibonding MO, usually labeled dσ and dσ* (or pσ and pσ*), respectively, which hold together the supramolecular assembly, as illustrated in Fig. 3.80 for the case of the $[Pt(bipy)_2CN_2]$ complex. This newly generated set of MOs determines a modification of the sequence of energy levels so that the dσ*–π transition, labeled MMLCT, becomes the lowest-energy transition and is susceptible to give rise to an emissive state. In simple terms, this process can be described as the transfer of one electron from a metal orbital of an excited molecule (dσ*) to the ligand orbital of the second unexcited monomer (π), in a way similar to the excimer-like emission mechanism already described in Section 3.3.2.3. From the MO scheme for the dimeric species reported in Fig. 3.80, it can also be noticed that the π-stacking between the aromatic bipy ligands of the two monomers in the "eclipsed" configuration creates a combination of four π-type MOs of different energies so that the intermolecular, excimer-like, π–π* transition in the dimer is of a lower energy, that is, is red-shifted, with respect to the intramolecular LC (π–π*) transition in the isolated molecule.

Thanks to the reversibility of the dimerization/polymerization phenomenon, which determines the optical properties of the material, the class of square-planar Pt^{II} complexes offers the intriguing opportunity to design efficient luminescent switches that can respond to external stimuli such as temperature (dimer/polymer formation may occur upon cooling whereas heating restores the monomeric phase) or sonication (which favors the dissociation). These properties may be enhanced when a specific ligand design is addressed to extend the intermolecular interactions between the monomer units, for example, by introducing side chains, up to the point that even a phase change, particularly the formation of gels, can accompany the variation of the optical properties. These kinds of molecular "devices" are currently attracting considerable interest in the field of molecular materials in view of their wide potential applications related to physical/chemical changes that can be followed simply by the naked eye. One of the related practical applications, and perhaps the most investigated, concerns the development of luminescent sensors for the recognition of specific analytes. In fact, supramolecular interactions can also occur between an open-structure Pt^{II} complex and a metal ion of a different nature, typically group Ib closed-shell d^{10} metal ions, Cu^{I}, Ag^{I}, or Au^{I}. These ions can bind to the complex via metallophilic interactions $Pt^{II} \rightarrow M^{I}$ established between the Pt 5d orbitals and the empty outer s and p orbitals of the closed-shell d^{10} metal ion and/or through the donor atoms of the coordinated ligands, especially in the case of soft donors. This circumstance can induce a perturbation in the distribution of the frontier MOs of the molecular complex, which in turn can lead to emissive states and resulting optical properties not recognizable in the isolated molecule, as in the case of the tetranuclear heterometallic complex based on the Pt_2Ag_2 cluster reported in Fig. 3.81.

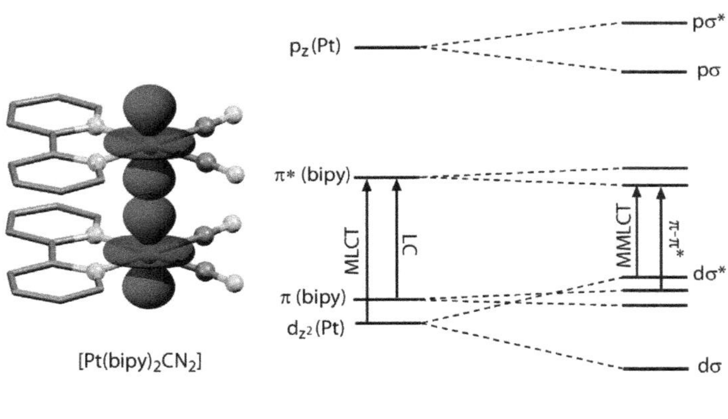

Figure 3.80 Molecular structure of a dimeric unit of [Pt(bipy)$_2$CN$_2$)] showing the overlap of d_z^2 orbitals (left) and a corresponding molecular orbital diagram for the monomeric and dimeric units (right). The arrows represent the lowest-energy-absorption transitions in the two species. Hydrogen atoms are omitted for clarity (Miskowski and Houlding, 1991).

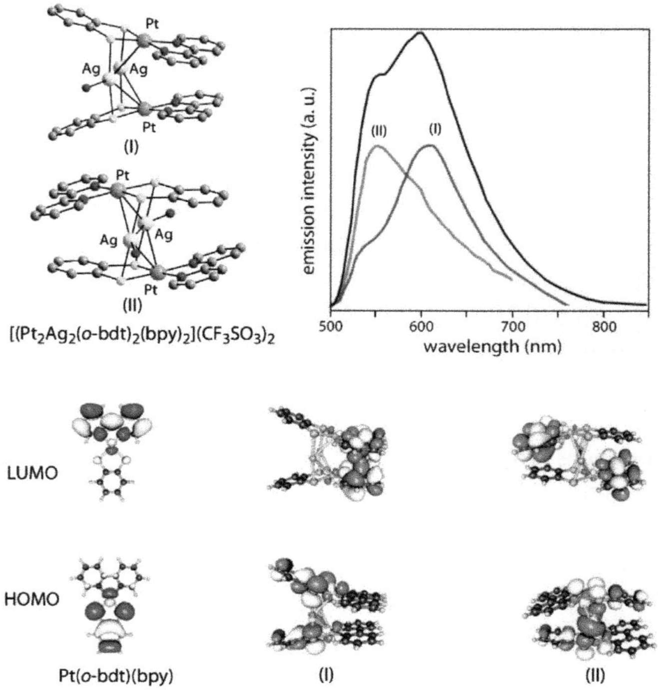

Figure 3.81 Molecular structure of the "eclipsed" (I) and the "staggered" (II) dimeric units constituting the $[(Pt_2Ag_2(o\text{-bdt})_2(bpy)_2](CF_3SO_3)_2$ assembly, their deconvolved emission spectra (right), and their calculated frontier MOs compared with those of the [Pt(o-bdt)(bpy)] complex (below). Adapted from Moussa et al. (2016) with permission of The Royal Society of Chemistry.

3.3.3.2.3 *General considerations*

Second- and third-row transition metal complexes undoubtedly represent the most extensively studied class of molecular materials for the preparation of efficient luminophores in a wide range of spectral regions. The discussion in this paragraph has highlighted that the number of variables and parameters to be taken into account in the design of these luminophores can be extremely ample and varied and results are not always easily predictable. However, a few general guidelines can be drawn:

- The crystal/ligand field must be strong enough to raise the energy of the nonemissive MC state so that its population is not thermally accessible.
- The complex should not bear tensed ligands that would cause structural distortion and the consequent weakening of the crystal/ligand field.
- The atomic weight of the metal should be as high as possible to favor ISC leading to phosphorescent emission.

The emission efficiency, chemical versatility, and optical tenability of this class of luminophores, accompanied by other valuable characteristics typical of d-metal ions, such as the ability of undergoing redox processes and their intrinsic magnetic properties, offer an extraordinary potentiality toward their employment in several applicative fields, ranging from optoelectronics to artificial photosynthesis.

3.3.4 Luminophores Based on Lanthanide Complexes

Before starting an in-depth discussion on the optical properties of lanthanide complexes, it is useful to recall very briefly the main chemical aspects and electronic features of these compounds as they do not obey the same common rules that apply to transition metal complexes. This is principally ascribed to the inner core nature of the 4f electrons, which are well shielded from the outer 5s and 5p closed shells and mostly do not participate in bonding. The chemistry of lanthanide (Ln) ions is therefore related to a strong Lewis acid behavior, implying a high ionic character of the bonding, lack of orbital directionality, and tendency to reach high coordination numbers. As a result, the formation of lanthanide complexes is mainly governed by entropy and steric factors and their structural arrangement is not easily predictable. Molecular architectures may easily vary from simple mononuclear complexes with variable coordination numbers to discrete polynuclear assemblies and infinite polymeric networks depending on subtle chemical changes.

The electronic features of lanthanide complexes are extremely different from those of transition metal complexes and more closely resemble those of closed-shell metal ions. This is due to the poor metal–ligand orbital mixing, which virtually rules out interconfigurational CT transitions and allows the optical properties of both the ligand and the metal to be practically preserved with respect to the free components. This aspect provides the remarkable benefit of allowing the association of the versatility of coordination chemistry with the unique photophysical properties of lanthanide ions, which can display emission that spans from the UV to the near-infrared (NIR) spectral windows. With this view, lanthanide-based materials are currently finding numerous applications in many different fields, ranging from optoelectronics to biology, in particular as emitters for lighting industry, electroluminescent and/or photoluminescent materials for display technology and telecommunications, and functional probes for biological assays and medical imaging purposes.

The actual reason behind the remarkable interest in lanthanide-based emitters is their particularly long-lived and narrow-band emission related to the forbidden nature of lanthanide intrashell f–f transitions (cf. Section 2.3.7.2). The downside of this peculiarity is that the absorption cross section of the ED transitions of lanthanide ions is very low (molar extinction coefficients $\varepsilon < 10\ M^{-1}\ cm^{-1}$) and consequently the direct optical excitation of these metal emitters is very inefficient, leading to weak luminosity ($L = \varepsilon \times \Phi$). Moreover, the narrow absorption lines of lanthanide ions require very stable, wavelength-tunable excitation sources. A way to overcome these drawbacks, first proposed by Weissman, is to encapsulate the metal emitter with a suitable organic light-harvesting ligand (chromophore), which can allow indirect excitation of the lanthanide ion by populating its higher levels by means of a ligand-to-metal energy transfer (ET) mechanism (sensitized emission, cf. Section 2.3.5) (Weissman, 1942). This strategy takes advantage of the fact that lanthanide ions' optical features are quite insensitive to changes in the chemical environment, so their spectral properties are preserved upon coordination. At the same time, this two-step excitation process can afford a large excited-state population using pump light fluences (J/cm^2) 4 to 5 orders of magnitude lower than those required for bare lanthanide ions and allows overcoming their weak absorptivity. In the early 1990s, the interest in lanthanide complexes as luminescent materials was renewed, when Lehn proposed that such compounds could be seen as light conversion molecular devices (LCMDs) for photoluminescence (PL) applications, coining the term "antenna effect" to denote the absorption-ET-emission sequence involving distinct absorbing (the ligand chromophoric unit) and emitting (the lanthanide ion) components (Lehn, 1990). Since the emitting system consists of two well-separated units, a large induced Stokes shift[26] between absorbed and emitted light results. Therefore, through this indirect excitation process narrow-band visible or NIR radiation can be obtained by exciting the "antenna" (sensitizer) with high efficiency using convenient light sources such as readily available lasers in the UV–visible regions, as schematically depicted in Fig. 3.82.

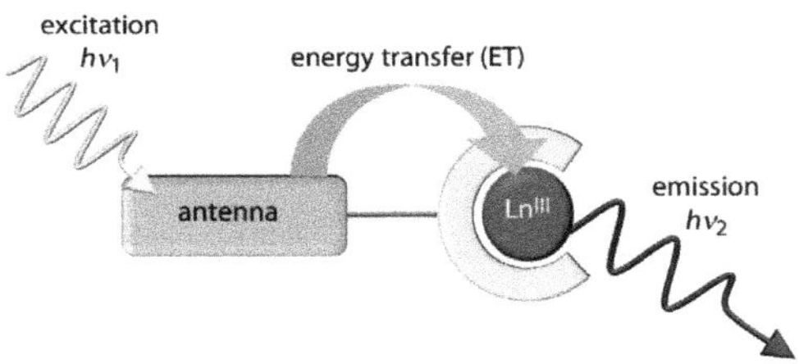

Figure 3.82 Simplified scheme of a lanthanide complex working as a LCMD through photoexcitation of an organic antenna unit.

One of the most archetypal examples of this class of molecular converters is undoubtedly represented by the erbium complex of the 8-quinolinolate (Q) ligand, which has been widely studied as a NIR emitter. The molecular structure of this compound, which adopts a trinuclear architecture formulated as Er_3Q_9, and its spectral properties are shown in Fig. 3.83 (Artizzu et al., 2005).

In addition to light converting properties, this class of compounds offers the unique and extremely important advantage of allowing the direct transformation of electricity into electromagnetic radiation. In fact, suitable organic ligands having high aromaticity and electrical/conducting properties can allow for efficient excitation by means of an electric pulse (electroluminescence, EL). Following this property, lanthanide complexes have been proposed as luminescent materials for electroluminescent devices such as OLEDs (Kido, 1990), prompting a growing interest of the scientific community as well as

[26]It must be underlined that this induced Stokes shift has a different physical origin with respect to the regular Stokes shift related to the energy dissipation in closely spaced energy levels, as described in Section 2.3.3.1.

the industry in this class of compounds. Electroluminescent devices will be discussed in more detail in Chapter 5, whereas we will herein focus more on the PL properties of lanthanide-based compounds.

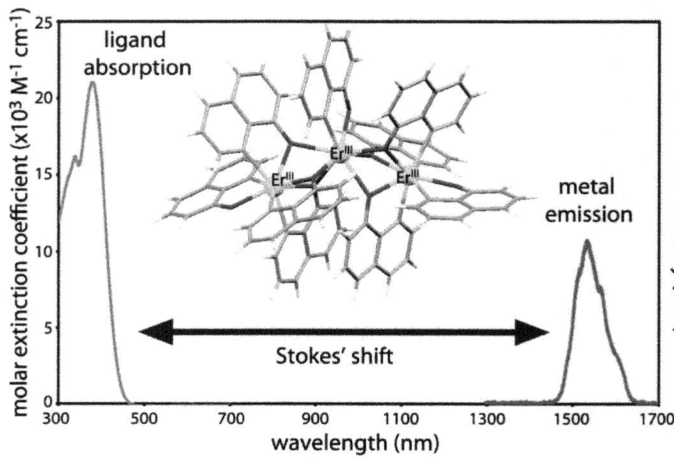

Figure 3.83 Structure and spectral properties of the Er_3Q_9 complex, showing a large induced Stokes shift between ligand absorption in the UV-visible and NIR erbium emission peaked at 1530 nm.

Figure 3.84 shows a simplified energy level scheme illustrating the multistep photocycle typical of most lanthanide complexes with antenna ligands. First, optical excitation in the near-ultraviolet (NUV)–visible regions populates ligand singlet excited states S_1, which can decay either to the ground state S_0 or to triplet states T_n through an ISC mechanism enhanced by the heavy-atom effect (cf. Section 2.3.2.3). Excited triplets can subsequently populate the upper levels of the lanthanide ion via resonance energy transfer (RET), which can occur either via Dexter's (cf. Section 2.3.5.1) or Förster's (cf. Section 2.3.5.2) mechanism (vide infra). Once the Ln^{III} ion becomes excited, a nonradiative "cascade" decay between closely spaced levels may lead to the population of an emitting level, usually the lowest one of the upper-energy-level manifold (cf. Section 2.3.7.2). Following a fast excited-state ion relaxation, the radiative decay ${}^{(2S'+1)}\Gamma'_{J'} \rightarrow {}^{(2S+1)}\Gamma_J$ yields f–f intrashell narrow emission whose wavelength depends on the nature of the lanthanide ion, as explained in Chapter 2.

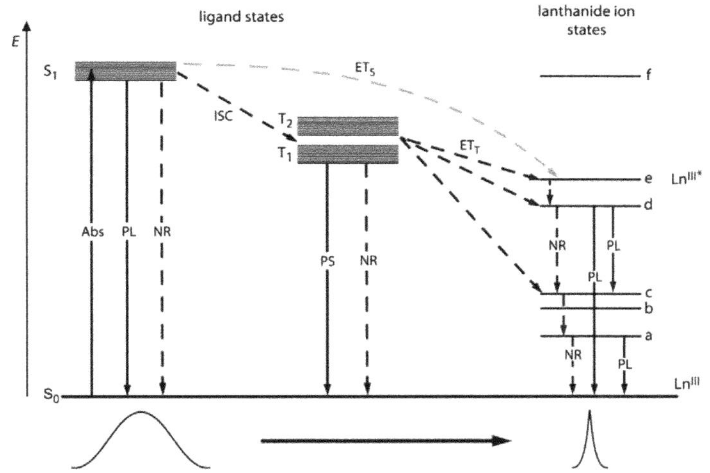

Figure 3.84 Jablonski diagram depicting the multistep photocycle in luminescent lanthanide complexes with antenna ligands. Dashed arrows represent nonradiative processes, and solid arrows indicate photon absorption or emission. Secondary processes, such as excited ligand decay, are also reported. ABS, absorption; PL, luminescence; PS, phosphorescence; NR, nonradiative decay; ISC, intersystem crossing; ET, energy transfer; a–f, Ln^{III} energy levels.

The mechanism described above is the most commonly accredited in antenna-sensitized luminescent lanthanide complexes and should be taken as a general scheme although, in principle, different pathways may be possible for this photocycle. Nonetheless, some statements of general validity can be drawn herein.

The overall (or absolute) luminescence quantum yield Φ_{tot} of sensitized luminescence from a lanthanide complex depends on several factors pertaining to the diverse steps of the photocycle schematized in Fig. 3.84. In a simplified model, a basic equation summarizing the aforementioned parameters, which will be discussed in more detail in the following paragraphs, can be written as follows:

$$\Phi_{tot} = \eta_{sens}\Phi_{Ln} \quad (3.7)$$

According to this equation, the efficiency of the emissive process in lanthanide complexes can be seen as the product of the intrinsic lanthanide quantum yield Φ_{Ln} (defined in Eq. 2.85), related to the emissive process undergone by the metal ion, and the effectiveness of the ligand-to-metal ET, that is, the sensitization efficiency η_{sens}. It must be underlined that only Φ_{tot} can be directly measured by means of specific instrumental equipment or spectroscopic methods (cf. Section 2.3.4.1), whereas careful consideration must be made for η_{sens} and Φ_{Ln}, as will be clarified further in this section.

In summary, the light output of luminescent materials based on lanthanide complexes essentially depends on:

- The absorption properties of the antenna unit
- The efficiency of the sensitization process
- The luminescence efficiency of the metal ion, affecting the emission lifetime

Whereas the first parameter can be rather easily controlled by selecting intensely absorbing organic chromophores, usually highly aromatic molecules, the remaining two issues require judicious considerations. These points will be developed in the following paragraphs according to a general subdivision of the photocycle leading to lanthanide emission in coordination complexes into two main critical steps: the ligand-to-metal sensitization and the lanthanide-centered emission. These aspects will be henceforth discussed by taking into account the simplest molecular system where a luminescent lanthanide ion is directly coordinated with an antenna ligand.

3.3.4.1 Ligand-to-metal sensitization

Ligand-to-metal sensitization efficiency, η_{sens}, is a key parameter for the design of luminescent lanthanide complexes since it critically influences the overall emission quantum yield. Nonetheless, its evaluation may be quite complex as many factors are to be carefully taken into account. In fact, several kinetic steps are involved in the sensitization of the lanthanide ion, and this main process competes with other external relaxation mechanisms, such as ligand internal decay, triplet oxygen quenching (cf. footnote 22), and internal intraligand CT. Therefore, to obtain high lanthanide luminescence overall quantum yields upon excitation of an organic antenna it is then necessary to rationally design and select suitable ligands, ensuring high sensitization efficiencies. This implies in the first instance the assessment of ligand energy states and the prediction of the mechanisms that would be involved in the sensitization process. Only then an overall evaluation of the factors governing the efficiency of ligand-to-metal ET can be drawn out.

3.3.4.1.1 *The nature of the ligand-feeding states and ET mechanisms*

The general description of the photocycle occurring in luminescent lanthanide complexes, provided in the introduction to this section, pointed out the role of ligand triplet states in mediating the ET to the metal emitter. However, inspection of Fig. 3.83 brings to attention that two energy-feeding channels from the ligand to the lanthanide ion are actually available, either via excited triplet (ET_T) or singlet (ET_S) states. To settle on the most reliable energy pathway, one has to keep in mind that the first requirement to be fulfilled to allow for ligand-to-metal ET is that the ligand excited state must survive long enough to be able to relax via the channel that feeds the upper energy levels of the lanthanide ion rather than by internal decay. This indeed explains one of the main reasons long-lived excited triplet states of the ligand are mostly held accountable for metal sensitization rather than excited singlets. Secondly, the effectiveness of the ET mechanism depends on the energy match between the donor and acceptor levels (resonance condition) so lower-lying triplet levels are generally better suited to fulfill this requirement. However, the role of singlet ET cannot be a priori ruled out and it may be crucial in some circumstances. For instance, when a visible emissive lanthanide ion such as Tb^{III} is involved, ligand excited triplet states may lie below the high-energy receiving levels of the metal (cf. Fig. 2.57), so the only available transfer channel is via the excited singlet states of the ligand. Another aspect to be considered to determine the spin nature of the ligand donor states and, at the same time, the type of mechanism involved is related to the selection rules for ligand-to-metal ET. As briefly mentioned in the previous paragraph, ET from the organic ligand can, in principle, occur either via Dexter's or via Förster's mechanism, depending on the total angular momentum variation ($\Delta\mathbf{J}$) undergone by the lanthanide ion. According to Malta and coworkers (de Sá et al., 2000), the selection rules for ET from the excited singlet or triplet state of the sensitizer to the $^{(2S+1)}\Gamma_J$ levels of lanthanide ions are as follows: $\Delta\mathbf{J} = 0$ and ± 1 ($\mathbf{J} = \mathbf{J}' = 0$ forbidden) for Dexter's (exchange) mechanism, and $\Delta\mathbf{J} = \pm 2$, ± 4, and ± 6 for Förster's (multipolar) mechanism. It is often stated in literature that triplet states usually populate the upper energy levels of the lanthanide ion through the exchange mechanism whereas excited singlet states transfer energy by means of the multipolar mechanism.

Nevertheless, the reader must be warned that although these theoretical considerations may be useful as general guidelines, judgment must be exercised. In fact, under real, practical circumstances, several factors concur in lowering the validity of the mentioned selection rules, especially when considering that the quantum treatment of lanthanide energy levels is a simplified approximation (cf. Section 2.3.7.2), and that mixing of states, including **J** mixing, or 4f–5d orbital mixing, is likely to occur. As a matter of fact, experimental evidence[27] allow us to infer that ligand-to-metal ET via ligand triplet states can be elected as the main mechanism responsible for lanthanide sensitization whereas the direct path via excited singlets can be reliably neglected in most cases. For the sake of completeness, it is also worth mentioning that another type of sensitization mechanism via LMCT state exists, which is not a nonradiative process but implies the transfer of one electron from the organic moiety to the lanthanide ion. However, this sensitization pathway is only limited to those cases where the ligand is susceptible to oxidation and the lanthanide ion can achieve a stable half-filled $4f^7$ or filled $4f^{14}$ configuration by reduction (Sm^{III}, Eu^{III}, and Yb^{III}). Therefore, we will hereafter focus the discussion on ligand-to-metal ET via ligand triplet states. In this regard, it is also important to point out that all the following considerations will refer to the simple case of a triplet state on a single donor ligand. This model works well for the majority of lanthanide complexes in view of the weak covalent character of the metal ligand bonding and the poor orbital mix. Theoretical calculations in fact usually show that triplet energy states in LnL_n (Ln = lanthanide ion; L = ligand) model complexes consist of a set of n degenerate levels corresponding to ligand-centered singly-occupied HOMOs and LUMOs that are

[27]Sophisticated time-resolved measurements, namely excited-state absorption (ESA) experiments performed through the pump and probe method, allow the determination of the different steps of the photocycle by assessing their temporal dynamics, that is, the evolution of the excited states. The involvement of ligand triplet states has been therefore well established for luminescent lanthanide complexes, whereas only indirect hints could be retrieved for ET from singlet states.

typically π and π* in nature, respectively, and closely resemble those of the free ligand, as shown in Fig. 3.85 for the Er_3Q_9 complex as an example.

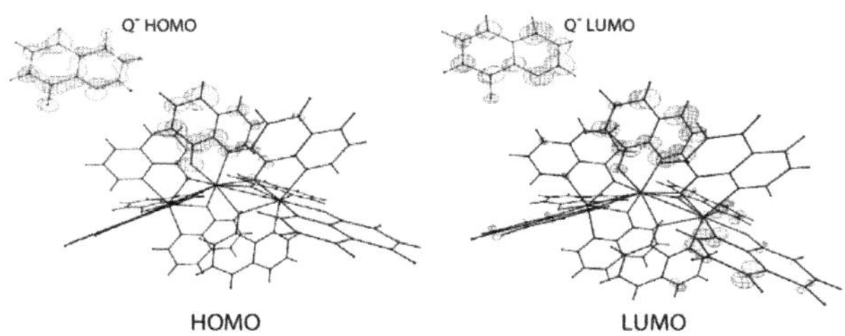

Figure 3.85 Structure and frontier molecular orbitals[28] for the Er_3Q_9 complex and the ligand Q^- (8-quinolinolate).

Each coordinated ligand has then independent triplet states and can individually work as antenna chromophore. This is also valid in heteroleptic molecular complexes, where only one ligand is suitable to act as a sensitizing unit. However, it should not be forgotten that the whole system of coordinated ligands should be, in principle, held accountable for indirect lanthanide excitation, especially if interligand interactions or ET is likely to occur.

The determination of the type of mechanism, either Dexter's or Förster's, as the one responsible for the ligand-to-metal ET, takes on a limited practical importance in simple complexes where the antenna donor is directly bound to the metal ion. It is, however, essential to remind that both mechanisms are governed by two parameters that should be always taken into consideration when designing lanthanide-based LCMDs (cf. Section 2.3.5):

- The spatial distance between the donor and the acceptor, which must be kept short
- The spectral overlap between the emission band of the donor and the absorption band of the acceptor, which should be maximized

[28]MOs in Fig. 3.75 were calculated with the approximate extended Hückel method, which provides in this case reliable results, although more sophisticate approaches exist in the framework of density functional theory (DFT). The reader should refer to specialized texts on MOs' theoretical calculations for details.

Whereas the first issue can be controlled by chemical design, and it is somehow fixed by the metal–ligand bond distance, the second one concerns the choice of an organic ligand with suitable electronic characteristics and related spectral features. This latter point is crucial in determining the efficiency of lanthanide sensitization and will be better clarified in the following paragraph.

3.3.4.1.2 *Efficiency of ligand-to-metal sensitization*

According to the most reliable and acceptable general model, described by the scheme $S^* \rightarrow T^* \rightarrow Ln^*$, ligand-to-metal sensitization proceeds via two determining steps:

1. Intraligand ISC
2. RET from ligand excited triplets to metal upper levels

Consequently, the sensitization efficiency can be expressed as the product of two parameters:

$$\eta_{\text{sens}} = \eta_{\text{ISC}}\eta_{\text{RET}} \tag{3.8}$$

where η_{ISC} and η_{RET} are the efficiencies of the ISC and RET processes, respectively.

The first step of the ligand-to-metal ET mechanism is of course enhanced by the spin–orbit coupling effect induced by the lanthanide ions, which makes triplet population from ligand excited singlets very efficient. Experimental time constants of this process, determined in a series of lanthanide complexes of differently substituted quinolinolato ligands similar to the one reported in Fig. 3.70, fall in the timescale of tens of picoseconds, leading to ISC efficiencies by far exceeding even those that would be expected on the sole basis of the bare heavy-atom effect. This observation points out that the ISC mechanism in lanthanide complexes can be considered as a fully efficient process independently of the ligand composition and the presence of heavy substituents.

On the other hand, RET from a ligand triplet to a lanthanide ion's levels is a much more subtle issue to be considered. First and foremost, it has to be kept in mind that the fundamental requirement for this mechanism is the energy match between the donor and the acceptor levels, and the resulting spectral overlap between the donor emission and the acceptor absorption. This means, in the first place, that the receiving levels of the lanthanide ion must lie below the ligand triplet level. For this reason, antenna ligands must be carefully designed and selected to achieve sensitization of the lanthanide ion in accordance with its emissive properties. As a general and practical rule, visible emitting lanthanide ions such as Eu^{III}, Tb^{III}, Dy^{III}, and Sm^{III} can be usually only sensitized by organic antennas absorbing in the UV region and hence having excited levels high enough in energy to lie above the metal receiving levels, while Gd^{III} levels are too high in energy to be populated by RET from an organic molecule, as exemplified in Fig. 3.86.[29]

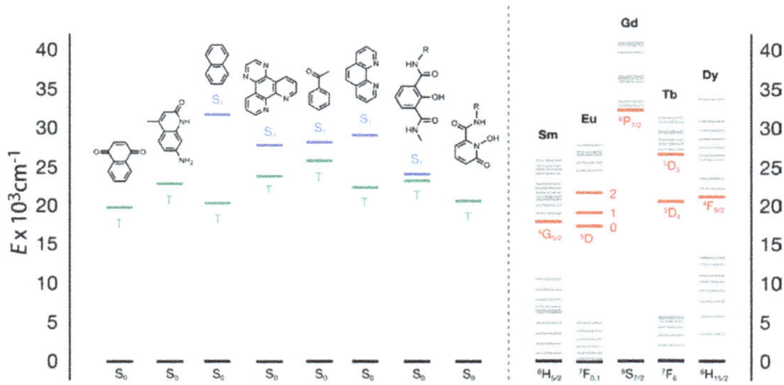

Figure 3.86 Energy diagram of emissive levels of some lanthanide cations (right) and chromophores (left); singlet states and triplet states of chromophores are evidenced in blue and green, respectively, while relevant emitting states of some lanthanides are drawn in red. Reprinted from Armelao et al. (2010), Copyright 2010, with permission from Elsevier.

On the other hand, NIR emitters can, in general, be more easily excited also by ET from a visible antenna chromophore, and the wealth of absorption lines of most of these metal ions ensures a good spectral overlap with the donor emission band. In this regard, Yb^{III} represents a special case in the class of NIR emitters in view of the simple two-level manifold energy diagram of this ion, giving rise to a single absorption line at 1 μm from the $^2F_{5/2}$–$^2F_{7/2}$ transition (cf. Fig. 2.57), as illustrated in Fig. 3.87, where a comparison of the absorption spectra of different lanthanide ions is reported.

[29]The reader must be reminded that although most lanthanide ions possess a wealth of closely spaced energy levels, not all of them are suited to become sensitized by RET. Quantum mechanics considerations, not discussed herein, can be applied to retrieve selection rules for this mechanism, although, as already discussed, they are largely relaxed under experimental conditions.

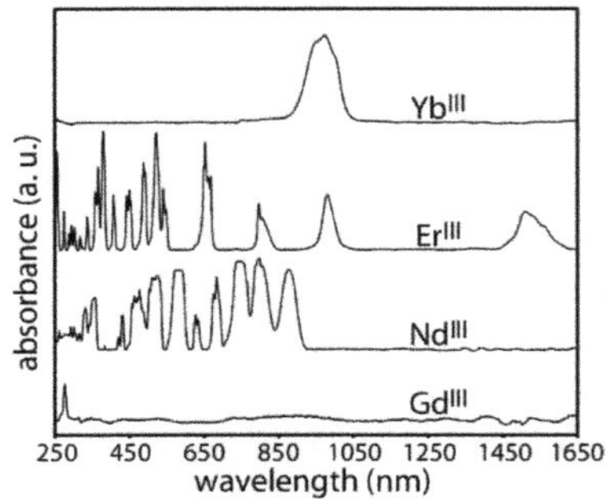

Figure 3.87 Absorption spectra of some lanthanide trichlorides.

Efficient sensitization of Yb^{III} therefore requires an accurate choice of the antenna ligand with suitable spectral properties in the visible region. However, this does not necessarily mean that the selection of the antenna has to be restricted solely to NIR-emitting dyes. In fact, for aromatic molecules, phosphorescent bands are usually broad enough to satisfy the requirement of spectral overlap with the 1 μm absorption band of Yb^{III}, so examples of good sensitizing efficiencies in complexes with this metal ion are not as rare.

As long as it is granted that ligand triplet levels lie above the metal ion's receiving levels, it is important to take into account that the involved energy gap should be not too high, in order to minimize energy losses due to thermal relaxation, and not so low that back ET from the excited lanthanide ion to the ligand would compete effectively with metal emission. Ideally, an energy gap of 1500 cm^{-1} would be recommended to prevent this detrimental phenomenon, which is often invoked to account for the low quantum yield in complexes of visible emitting lanthanide ions. In addition, potential low-lying ILCT or interligand CT states can provide a de-excitation channel for the excited metal, as back ET, or for excited ligand nonradiative decay. A well-designed antenna system that ensures a good energy match between the triplet donor levels and the lanthanide accepting levels would also allow for fast metal-to-ligand ET kinetically favored with respect to other triplet deactivation pathways, such as triplet-triplet annihilation, especially by oxygen, which is ubiquitous in the environment.

On the basis of all these considerations, the organic antenna ligand must be judiciously selected in order to afford suitable optical and electronic properties for achieving optimal lanthanide sensitization. Although, in principle, it can be very difficult to predict a priori the electronic features of an organic chromophore, and only limited changes in the spectral properties can be achieved by chemical modification (cf. Section 3.3.2.2), the implementation of appropriate theoretical models on the basis of structural data can be helpful in the design of efficient antenna ligands.

The direct experimental evaluations of ligand triplet energy levels and, ultimately, of the η_{sens} parameter can be rather tricky tasks, and they will be shortly addressed in the box titled "Evaluation of Ligand Triplets and Ligand-to-Metal Sensitization Efficiency (η_{sens})." Otherwise, a quantification of the η_{sens} parameter can be possible only in an indirect way, by retrieving its value from Eq. 3.6. This implies that the other terms of the mentioned relation, that is, the overall and the intrinsic quantum yield, must be measured or experimentally evaluated, the latter being discussed in the next paragraph. However, beyond these issues, some general guidelines can be drawn in regard to the design of efficient antenna ligands as summarized hereafter:

- Intensely absorbing chromophores, basically highly aromatic ligands, should be selected to improve the overall luminosity.
- Long-lived excited states are crucial to allow for ligand-to-metal ET channel; therefore luminescent organic ligands are often the best choice.
- The antenna ligand should have accessible and long-lived triplet states.
- Bearing in mind that the ideal energy gap between a ligand donor and a lanthanide ion's accepting levels is 1500 cm^{-1}, near-UV chromophores are usually better suited for visible emitters while visible light absorbing units can better sensitize NIR luminescent lanthanide ions.
- The ligand system should not give rise to low-lying CT states that can potentially deactivate metal sensitization through radiationless decay.

Evaluation of Ligand Triplets and Ligand-to-Metal Sensitization Efficiency (η_{sens})

Apart from sophisticated theoretical calculations, ligand triplet energies can be experimentally determined in complexes with the optically silent Gd^{III} ion. Since Gd^{III} cannot be sensitized by the ligand but is still able to favor triplet state population, the low-temperature (77 K)[30] emission spectrum of a Gd^{III} complex can be attributed to ligand phosphorescence. The high-energy tail of the phosphorescent band is granted to correspond to the onset of the triplet-energy-level manifold.

In view of these characteristics, gadolinium complexes are useful references for assessing the optical properties of isostructural lanthanide analogous compounds, especially in regard to the assessment of the efficiency of the double step sensitization process, through the comparison of the temporal dynamics of the ligand excited states (investigated through advanced time-resolved photophysical studies). This is schematically depicted in the following figure where the excited singlet and triplet dynamics in complexes of the generic ligand L, with Gd^{III}, Al^{III}, and an optically sensitized lanthanide ion (Ln), are reported.

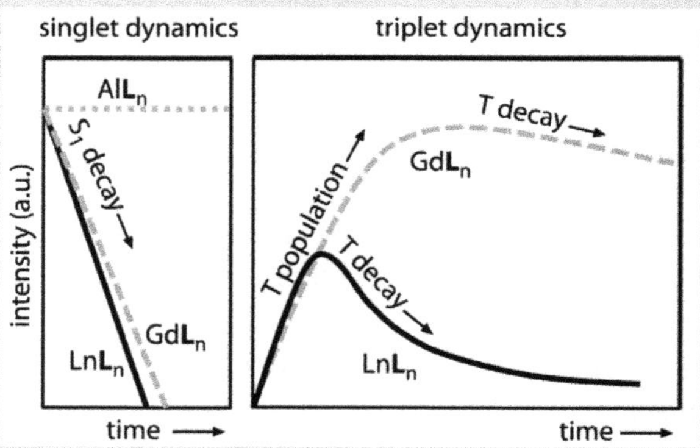

The panel on the left shows the excited singlet (S1) decay (reported on a logarithmic scale), which is much faster in complexes of the heavy lanthanide ions with respect to the light Al^{III} ion, indicating efficient ISC. The right panel, instead, illustrates the dynamics of the ligand triplet (T) state, which is more quickly depopulated in complexes with the sensitized Ln^{III} ions with respect to the Gd^{III} analog, due to

[30]Cryogenic temperatures are often necessary to observe phosphorescence since long-lived triplet states easily undergo thermal deactivation.

efficient resonance ET. The comparison of the time constants for the different kinetic steps of the photocycle in the Gd^{III} complex, with respect to analogous isostructural complexes of the same ligand with optically sensitized lanthanide ions, provides a reliable direct idea of the efficiency of the whole ligand-to-metal sensitization process.

3.3.4.2 Sensitized lanthanide emission

The intrinsic quantum yield of lanthanide emission, Φ_{Ln}, is defined according to Eq. 2.85 as the ratio between the observed emission lifetime τ and the radiative lifetime τ_r of the emitting lanthanide ion:

$$\Phi_{Ln} = \frac{\tau}{\tau_r} \tag{3.9}$$

These two parameters provide different pieces of information that are crucial to achieving a full understanding of the factors that govern the quantum efficiency of lanthanide complexes. While the measurement of τ can indeed deliver a direct feedback on the role of the chemical environment in affecting the luminescence output, its absolute value takes on a limited meaning if not compared to τ_r, that is, the inverse of the intrinsic oscillator strength ($\kappa_r = 1/\tau_r$) of the lanthanide emitter, which is determined by the specific coordination surroundings. In other words, the efficiency of the lanthanide emission can be only assessed by comparing how effectively the lanthanide ion does emit in respect to how much it can potentially emit in that specific chemical environment under ideal conditions, that is, without quenching phenomena.

A careful evaluation of both these parameters is therefore central to defining the factors affecting the performances of the emitter and its intrinsic potentialities, as will be described in more detail in the following paragraphs.

3.3.4.2.1 *Lanthanides' oscillator strength*

The radiative lifetime of a lanthanide ion acting as an emitting oscillator represents its tendency to decay radiatively to the ground state, that is, the higher it is, the lesser the ion will be prone to emitting light and vice versa. Therefore higher intrinsic quantum yields can be achieved for those ions that are characterized by lower values of τ_r, where radiative emission is less prohibited. It is clear that different lanthanide ions have different oscillator strengths, by virtue of the different natures of their energy level configurations. In simple terms, they can be more or less luminescent with respect to each other. However, although scientific literature often disregards this issue, it is of utmost importance to take into account that this parameter is not a constant but critically depends on the coordination environment of the metal emitter, other than the external matrix into which it is inserted.

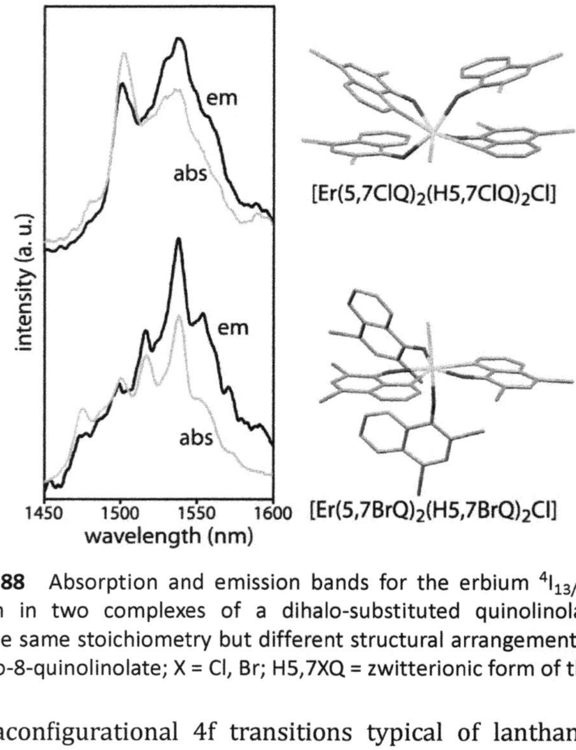

Figure 3.88 Absorption and emission bands for the erbium $^4I_{13/2} \leftrightarrow {}^4I_{15/2}$ transition in two complexes of a dihalo-substituted quinolinolate ligand having the same stoichiometry but different structural arrangements (5,7XQ = 5,7,dihalo-8-quinolinolate; X = Cl, Br; H5,7XQ = zwitterionic form of the ligand).

Intraconfigurational 4f transitions typical of lanthanide ions have been briefly described in Chapter 2, where it was underlined that the ligand/crystal field of a molecular complex slightly allows f–f transitions. As a consequence, the spectral features arising from induced electric dipolar (ED) transitions are sensitive to the coordination environment, in particular to the coupling of the oscillating EDs, generated in the ligand system by the electromagnetic radiation, with the 4f electron cloud. The ligand/crystal field affects both the intensity and the shape of lanthanide ions' spectral bands. The latter phenomenon is related to a partial removal of the degeneracy of 4f levels into 2**J** + 1 terms, which induces a broadening and partial resolution (Stark lines) of lanthanide absorption and emission bands. The fine structure arising from this Stark splitting and the variation of intensity of ED responsive bands upon changes in the symmetry of the complex can be used as a tool to probe the metal ion's environment. Noteworthy is the fact that the spectral shape becomes broader as the symmetry lowers, and this is of particular interest since the spectroscopic breadth is decisive for achieving high gain bandwidth for optical amplification in laser systems. A significant example of the influence of the ligand system on the spectral lines of lanthanide ions is provided in Fig. 3.88, where the 1.5 μm emission and absorption spectra of two erbium complexes with the 5,7-dihalo-quinolinolate ligand (5,7XQ; X = Cl and Br) having the same stoichiometry but different structural arrangements are reported (Artizzu et al., 2007).

A substantial amount of literature exists on the correlation of the molecular structure with the spectral features of lanthanide ions, especially for some widely studied ions such as Eu^{III}. As already briefly mentioned in Chapter 2, a fine model for predicting the intensities of lanthanide ions' spectral lines has been extensively implemented since the work of Judd and Ofelt in the early 1960s. A thorough description of this topic is beyond the scope of this book, but it is nonetheless of utmost importance to underline the critical role of the chemical design in dramatically influencing the spectroscopic features and, ultimately, the oscillator strength of lanthanide emitters. This is the main reason why the radiative lifetime should be experimentally evaluated for each specific case under investigation whenever the intrinsic quantum yield of lanthanide emission is addressed. Working out a reliable experimental value of the radiative lifetime can certainly be a rather delicate task, and accurate spectroscopic measurements must be made to retrieve consistent data. Two models are usually taken into account for this purpose in the case of lanthanide emitters, both dependent on the absorption spectrum and on an inverse power of the refractive index of the medium. Leaving the quite sophisticated formal mathematical treatment of the first model to other literature reports, it can be pointed out that it is based on Einstein's equation for spontaneous emission and takes into account the spectroscopic parameters experimentally retrieved in the framework of the Judd–Ofelt theory. While this model is perhaps the most rigorous, it often leads to large errors mainly arising from the difficulties in applying the Judd–Ofelt theory to the experimental case. A simpler and probably more reliable approach for assessing the radiative lifetime makes use of the Strickler–Berg equation (Eq. 2.86, Chapter 2), under the restriction that the emission and the absorption transitions share the same set of levels, that is, the absorption spectrum corresponding to the emission band must be known,[31] which is however not applicable to all the lanthanide emitters, such as for the case of Nd^{III}.

In any case, it should be kept in mind that limited tuning of the radiative lifetime, addressed to reduce its value in order to maximize the intrinsic quantum yield, can be achieved either by controlling the coordination environment or by varying the refractive index of the medium into which the emitter is immersed. Whereas the latter issue is rather straightforward, since the incorporation of the molecule into matrices of high refractive indexes is noticeably effective in shortening the τ_r value, the former is much more elusive, and scientific literature still lacks a sufficient amount of data to provide a full understanding of the factors correlating the nature of the coordination sphere to the oscillator strength of lanthanide emitters. In principle, the stronger the interaction of the ligand field with the 4f electron cloud, the more allowed intraconfigurational 4f transitions will be and the shorter the radiative lifetime. Therefore, as a general guideline, strong and polarizable bonds should be implemented in the design of luminescent lanthanide complexes. Following this rule of thumb, it can be underlined that radiative lifetime values for lanthanide ions in molecular complexes may vary rather widely and are in general much shorter than those retrieved for the same ions immersed in a purely inorganic environment. This may appear quite obvious if you recall that a highly asymmetric ligand/crystal field is quite effective in relaxing selection rules and makes f–f transitions more allowed. As an indicative example, experimentally determined τ_r values for the ${}^2F_{5/2} \rightarrow {}^2F_{7/2}$ transition

[31]As a practical consideration, it is pointed out that the accurate measurements of lanthanide ions' absorption cross section can be a tricky task in view of the weak absorptivity of these ions. This can be especially true for NIR emitters as the background signal due to the chemical surroundings is particularly strong in that region (vide infra) and accurate baseline subtraction is necessary.

of Yb^{III} in coordination compounds with organic ligands fall in the range 0.5–1.3 ms and are noticeably smaller than the often misused "literature value" of 2 ms, indicating a different coordination environment.

3.3.4.2.2 *Luminescence lifetime*

Several nonradiative deactivation processes (quenching effects) can severely limit the intrinsic quantum yield of lanthanide complexes, leading to observed luminescent lifetimes τ, which are usually significantly shortened with respect to radiative lifetimes. Among them, quenching via vibrational deactivation by chemical species present in the molecular surroundings and acting as high-energy oscillators is by far the most effective phenomenon. As was briefly explained in Section 2.3.6.1, vibrational quenching occurs through a Förster mechanism of radiationless ET induced by the resonant coupling of the vibrational modes of a chemical species, such as water molecules, with the emitting lanthanide ion acting as an oscillating dipole, as schematically depicted in Fig. 3.89.

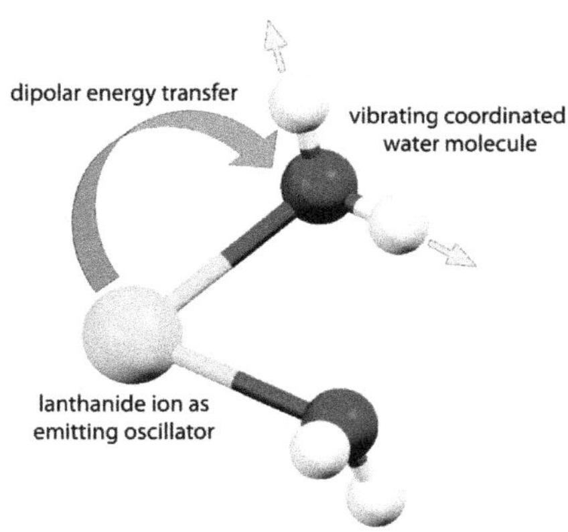

Figure 3.89 Schematic depiction of quenching via vibrational deactivation occurring through Förster's dipolar ET between the emitting lanthanide ion, acting as an oscillating dipole, and a coordinated water molecule undergoing normal vibrational modes (symmetric stretching is represented here as an example).

The extent of nonradiative vibrational deactivation undergone by an emitting lanthanide ion essentially depends on the energy gap (ΔE_g) between the lowest-lying excited (emissive) state of the metal ion and the highest sublevel of its ground multiplet. The smaller this gap, the easier is its de-excitation by high-energy vibrations of small molecules and organic groups such as water, OH, and CH groups. This rule of thumb is often referred to as the "energy gap law,"[32] and it finds its logic in the donor acceptor energy match (resonance condition) required to allow for Förster's resonance energy transfer (FRET, cf. Section 2.3.5). This condition is met by the acceptor species through the superior harmonics or combinations[33] of its normal vibrational modes, which are energetically close to the lanthanide emitting levels (the donor), as shown in Fig. 3.90 for some examples of luminescent lanthanide ions and typical quenching organic groups.

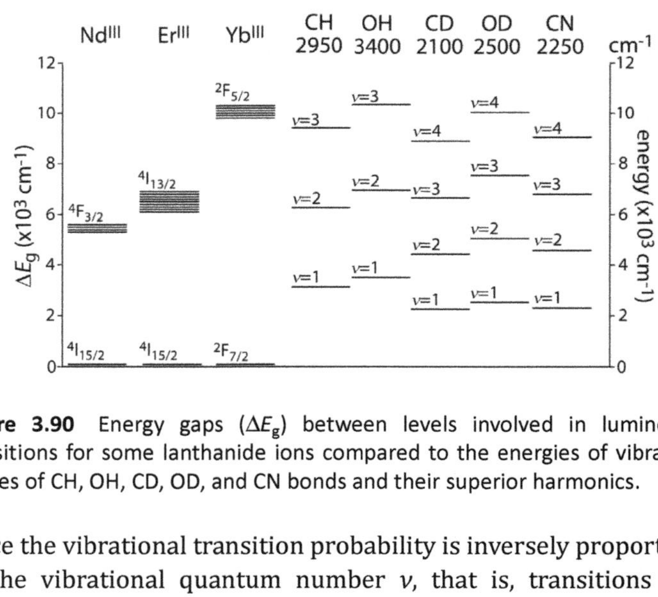

Figure 3.90 Energy gaps (ΔE_g) between levels involved in luminescent transitions for some lanthanide ions compared to the energies of vibrational modes of CH, OH, CD, OD, and CN bonds and their superior harmonics.

Since the vibrational transition probability is inversely proportional to the vibrational quantum number v, that is, transitions from the $v = 0$ level to levels of higher v become progressively weaker approximately by a factor of 10, quenching via vibrational excitation drastically decreases for superior harmonics of deactivating oscillators. Therefore, the higher the number of phonons necessary

[32]It has to be pointed out that the energy gap law described here for a lanthanide ion's emission has a different origin with respect to the general rule discussed in Section 2.3.3.4, which refers to molecular systems.

[33]For example, the combination of symmetric (v_1) and antisymmetric (v_3) stretching modes of water molecules gives rise to a $v_1 + v_3$ band in the NIR at ca. 1420 nm (vide infra).

to bridge the gap between the excited and ground levels of a lanthanide ion, the lower the extent of vibrational deactivation. As a result, NIR-infrared luminescent lanthanide ions are by far more sensitive to this type of quenching than visible-emitting lanthanide ions, and this is mirrored by the observed emissive lifetimes, which usually fall in the microsecond or nanosecond timescale for the first class of luminophores while they can easily exceed the millisecond timescale for higher-frequency emitting species such as Eu^{III} and Tb^{III}. Table 3.3 reports the number of phonons of OH and OD vibrations necessary to reach resonance with the emitting levels of different lanthanide ions.

Table 3.3 Energy gaps and lifetimes of luminescent lanthanide ions in H_2O and D_2O

Lanthanide ion	**ΔE_g (cm^{-1})**	**No. of phonons**		**τ (μs)#**	
		OH	**OD**	**H_2O**	**D_2O**
Gd^{III} ($^6P_{7/2}$)	32,100	9	15	2300	NA
Tb^{III} ($^5D_4$)	14,800	4	7	467	3800
Eu^{III} ($^5D_0$)	12,200	3–4	5–6	108	4100
Yb^{III} ($^2F_{5/2}$)	10,300	3	4–5	0.17	3.95
Dy^{III} ($^4F_{9/2}$)	7850	2–3	3–4	2.6	42
Sm^{III} ($^4G_{5/2}$)	7400	2	3	2.7	60
Er^{III} ($^4I_{11/2}$)	6600	2	3	NA	0.37
Nd^{III} ($^4F_{3/2}$)	5400	1–2	2–3	0.031	0.14

Source: Reprinted from Bünzli (2015), Copyright (2015), with permission from Elsevier.

From the above considerations and the data of Table 3.3, it becomes evident that high-energy oscillators, especially OH groups and water molecules, are undoubtedly the most powerful quenchers for lanthanide emission and have a remarkable shortening effect also on the lifetimes of visible emitters. This effect is so ubiquitous and assumes such a crucial importance for luminescent lanthanide complexes that a model correlating the hydration number to the observed emission lifetimes was conveniently established as described in the box titled "Hydration Number in Luminescent Lanthanide Complexes."

Hydration Number in Luminescent Lanthanide Complexes

The effect of water molecules in significantly shortening the luminescence lifetimes of lanthanide complexes has been known since the very first studies on this class of functional molecular materials. This detrimental phenomenon can originate both from molecules present in the first coordination sphere of the lanthanide emitter and in the immediate molecular environment when water is used as a solvent. In the latter case, the high number of available quenching species surrounding the emitter leads to dramatic quenching effects. Following the work by Horrocks and Sudnick (1979) an empirical model was developed to retrieve the number of water molecules coordinated to a lanthanide emitter (q) by comparing the observed lifetimes of the molecular complex in the H_2O and D_2O solution:

$$q = A(\Delta\kappa - B)$$

where

$$\Delta\kappa = \kappa_{H_2O} - \kappa_{D_2O} = \frac{1}{\tau_{H_2O}} - \frac{1}{\tau_{D_2O}}$$

In this equation κ_{H_2O}, κ_{D_2O}, τ_{H_2O}, and τ_{D_2O} are the observed rate constants and lifetimes in water and deuterated water, whereas A and B are empirical parameters determined from structurally known reference complexes that refer to the quenching contribution of the inner (coordinated water) and the outer (solvent molecules) coordination spheres, respectively. This model is obviously based on the assumption that vibrational quenching by water molecules is by far the most important deactivating phenomenon and all the other possible ET channels are the same in both solvents so their contribution is elided when comparing the data in the two systems.

Despite the fact that this equation is commonly accepted and widely used, it is just an empirical model and several exceptions to it have been found. Therefore, it has to be used with caution and we cannot assume that it will help us make fully reliable predictions.

Vibrational quenching can be rather easily quantified and its origin identified in the case of NIR emitters, thanks to the fact that vibrational overtones and combination bands of organic groups and small molecules are easily spectroscopically detected in this spectral region. Recalling the resonance condition for nonradiative ET (cf. Section 2.3.5), the overlap integral between the donor (the lanthanide ion) emission band and the acceptor absorption, which can be effortlessly calculated from experimental data, will provide significant evidence of the extent of vibrational quenching and the nature of the species that are mainly responsible for this phenomenon. An explicative example of spectroscopic evidence of the contribution of organic groups and water molecules in quenching NIR luminescence of Er^{III} is illustrated in Fig. 3.91.

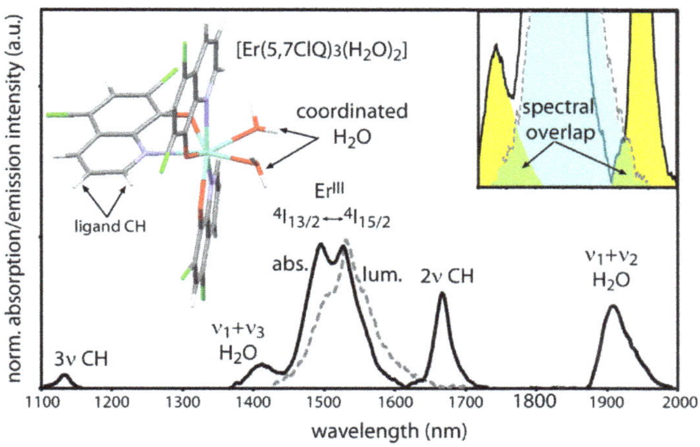

Figure 3.91 Solid-state absorption and emission spectra of the depicted erbium complex with the 5,7ClQ ligand. Overtones of ligand CH groups (2ν and 3ν second and third stretching overtones) and combination bands of water molecules (ν_1, ν_3 symmetric and antisymmetric stretching, ν_2 bending) are indicated. In the inset a close-up of the spectral overlap between the erbium $^4I_{13/2} \rightarrow {}^4I_{15/2}$ emission band (light-blue-shaded area) and the H_2O $\nu_1 + \nu_3$ and CH 2ν absorption bands (yellow-shaded areas) is evidenced.

Inspection of Fig. 3.91 points out that both water molecules and ligand CH groups are resonant with erbium emission and are responsible for vibrational quenching. In particular, the role of water is highlighted by the significant shortening of the observed erbium lifetime of more than 1 order of magnitude ($\tau = 0.1$ μs versus $\tau = 2$ μs) for the complex reported in Fig. 3.91 with respect to the water-free complex with the same ligand[34] previously depicted in Fig. 3.88.

[34]As a consequence of the peculiar coordination chemistry of lanthanide ions, briefly described at the beginning of this section, obtaining complexes of the same metal and ligand with different stoichiometry and structural arrangements is not a rare case and is mainly dependent on the conditions used in the synthetic procedure.

The quantification of the extent of vibrational quenching can be therefore evaluated in the framework of FRET on the basis of experimental spectroscopic and structural data, that is, the spectral overlap described above and the number and orientation of quenching species in the molecule, by implementing Eq. 2.95. Förster's model works rather well in the case of NIR-luminescent complexes, accounting for over 90% of quenching phenomena in this class of emitters. Useful guidelines can be then drawn from Förster's model of RET to significantly reduce quenching via vibrational deactivation, on the basis of the observation that its extent depends on the number and distance from the metal emitter of OH and CH groups (above others):

- The complex should be water-free, that is, no water molecules should be directly coordinated to the lanthanide ion.
- The lanthanide ion coordination sphere should be fully saturated, and ligands should sterically shield the ion from the environment to prevent water or solvent molecules from entering the immediate surroundings of the metal emitter.
- Organic ligands should not bear CH groups or at least they should be at a distance no shorter than 7–8 Å from the lanthanide ion to reach 50% of emission efficiency.[35]

In accordance with these conclusions, it has been in fact observed that fluorination of organic antenna ligands, and the consequent replacement of the high-energy CH oscillators with low-energy CF ones, is one of the most promising strategies for achieving significant enhancement of emission efficiency in lanthanide complexes, particularly for NIR emitters. A remarkably long luminescence lifetime (τ = 1.1 ms), of more than half the value of the radiative lifetime (taken as 2 ms), was found for the Yb^{III} complex with the anion of the fully fluorinated N-{P,P-di-(pentafluorophinoyl)}-P,P-di pentafluorophenylphosphinimidic acid ($HF_{20}tpip$) ligand, shown in Fig. 3.92, in a solution of deuterated acetonitrile (Glover et al., 2007).

However, a significant margin of deviation of the FRET model from the observed data and its inapplicability to visible emitters,

[35]Neglecting quenching phenomena other than vibrational deactivation.

for which only predictions based on empirical observations can be made, indicate the necessity of a deeper understanding of the factors limiting the luminescence efficiency in lanthanide complexes.[36]

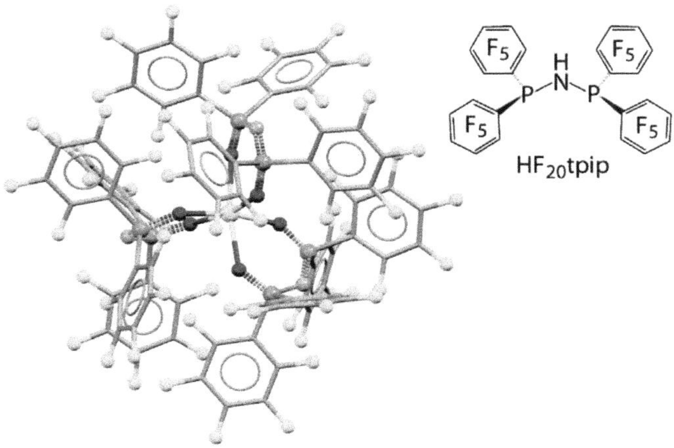

Figure 3.92 Molecular structure of the complex $[Yb(F_{20}tpip)_3]$ with the anion of the fully fluorinated ligand $HF_{20}tpip$ also reported in the figure.

Nevertheless, quenching via vibrational excitation, although most effective, especially for NIR emitters, is not the only mechanism responsible for lanthanide radiationless decay. Other deactivation processes could occur through ET between two neighboring lanthanide complexes. These nonradiative processes can be easily induced by diffusional collisions of complex molecules, especially in a liquid medium. Among them, the cross-relaxation mechanism consists of energy sharing between two lanthanide ions, one of them initially in an excited state, that ends up with both ions in a lower-lying nonemissive state. Instead, the excitation migration mechanism is a rapid exchange of energy between close-contact lanthanide ions that does not imply itself a radiationless deactivation but may increase the quenching probability by favoring the transfer of excitation

[36]Surprisingly and inexplicably, some of the best-performing visible-emitting lanthanide complexes do actually bear water molecules in the first coordination sphere.

where more quenching processes may take place.[37] It is evident that both mechanisms, schematically depicted in Fig. 3.93, are greatly enhanced by concentration and the reduced luminescent lifetimes observed in highly concentrated media are generally ascribed to the so-called concentration quenching effect.

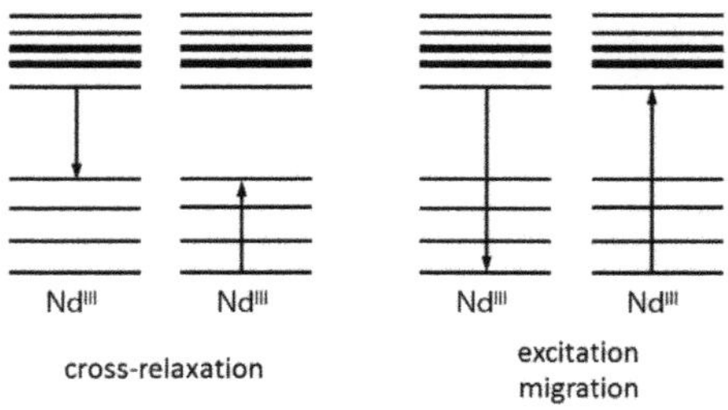

Figure 3.93 Diffusional quenching mechanisms for Nd^{III} complexes.

3.3.4.3 Advanced molecular design and general considerations

We have so far only considered the general case where lanthanide emission is sensitized via an organic antenna chromophore directly bound to the metal. Nonetheless, more sophisticated molecular designs can be accessed to achieve lanthanide luminescence sensitization as Förster's mechanism of ET is not limited by orbital overlap and close contact between the donor and the acceptor, insofar as the two components are close enough to communicate efficiently. This allows antenna units to be placed in the molecular periphery and be therefore available for interactions with external stimuli (such as on/off property switch through chemical binding with external entities), while the immediate coordination environment of the lanthanide ion can be appropriately tailored to ensure molecular thermodynamic stability and/or to provide low-energy-phonon surroundings for the metal emitter. For example, visible-emitting complexes of Eu^{III}, Tb^{III}, and Pr^{III} with the polydentate ligand L^3 bearing quinoline units as pendant arms reported in Fig.

[37]Excitation migration is not usually considered important within metal centers in polynuclear complexes since, as a result of the short distances between the connected communicating units, it is spatially contained and usually so fast that it does not influence the overall kinetics of the photocycle.

3.94 have been successfully studied for DNA binding, sensing, and photocleavage (Singh et al., 2015).

Besides organic units, lanthanide ions can also be sensitized through ET from a variety of metal-containing entities in different environments, such as:

- Transition metal complexes, such as $[Ru(bipy)_3]^{2+}$, in a supramolecular arrangement or in a concentrated solution of the two donor acceptor entities
- Suitable d-metal ions such as Cr^{III}, whenever the two interacting metal ions are embedded in the same molecular architecture, through direct d–f ET
- Different lanthanide ions having upper energy levels that are resonant with the emitter's receiving levels in a heterolanthanide polynuclear complex, through direct f–f ET.

The strategy suggested in the latter two points is particularly fascinating as additional, peculiar properties can be achieved by combining the features of d- and f-metals, as will be discussed in more detail in Chapter 4. However, besides potential "bonus" features, heterometallic systems can be very promising for improving the emission properties of lanthanide emitters. For example, apparent lanthanide NIR emission lifetimes up to the microsecond timescale can be afforded in systems where the intramolecular intermetallic Cr → Ln (Ln = Nd, Er, and Yb) ET process via the spin-forbidden Cr ($^2E \leftarrow {}^4A_2$) transition acts as the rate-limiting step in the emission photocycle, as in the polynuclear complex reported in Fig. 3.94 (Aboshyan-Sorgho et al., 2012).

On the other hand, ET through a different lanthanide ion can mediate and remarkably enhance the efficiency of the sensitization process from an organic antenna ligand, as more performing absorption properties of a second lanthanide ion can be coupled with the luminescent properties of the selected emitter. For example, Yb^{III} is an ideal sensitizer for Er^{III} since its upper energy level $^2F_{5/2}$ is perfectly resonant with the $^4I_{11/2}$ of erbium and the absorption cross section for the $^2F_{5/2} \leftarrow {}^2F_{7/2}$ transition is approximately 10 times higher than that associated to the $^4I_{11/2} \leftarrow {}^4I_{15/2}$ transition of Er^{III} (Fig. 3.87). Therefore Yb^{III} can more easily "harvest" the energy absorbed by the antenna chromophore and, at the same time, can transfer energy to the accepting Er^{III} with full efficiency, thanks to the resonance of energy levels. Similar considerations can be made for other suitable pairs of lanthanide ions to achieve different purposes. For instance, Nd^{III}, which has a wealth of absorption lines in the visible spectrum (Fig. 3.87), can provide additional and more convenient excitation channels for the single-wavelength absorbing Yb^{III} (cf. Section 3.3.4.2) and heterolanthanide ET can efficiently occur through Nd^{III} ($^4F_{3/2}$) → Yb^{III} ($^2F_{5/2}$). Such molecular heterolanthanide assemblies can be easily prepared, thanks to the tendency of these ions to form polynuclear architectures, by selecting suitable multidentate or bridging ligands able to ensure a short distance between the two communicating metal ions, as in the examples reported in Fig. 3.95 (André et al., 2002; Artizzu et al., 2013; Chen et al., 2007).

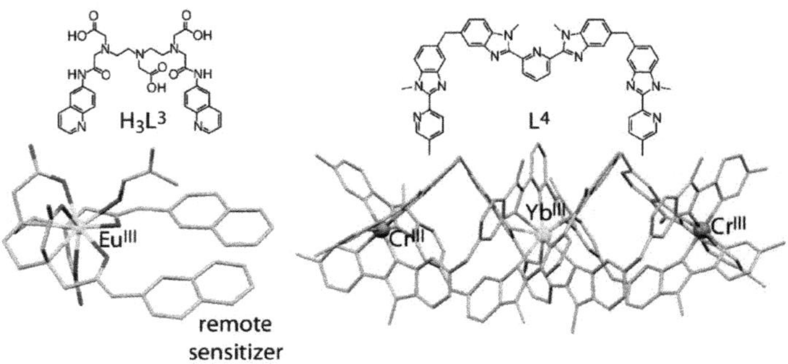

Figure 3.94 Molecular structure of the Eu^{III} complex with the trianion of ligand H_3L^3 [$Eu(L^3)(DMF)$] bearing two quinoline units as remote sensitizers (left) and of the heteropolynuclear molecular helicate complex $[CrYbCr(L^4)_3]^{9+}$ (right). Hydrogen atoms are omitted for clarity.

From this brief overview on this class of luminophores, it becomes evident that the peculiar optical properties of lanthanide ions coupled with the versatility of coordination chemistry can afford an extraordinary variety of optically functional molecular materials that can be suitably tailored for a wide range of applications. However, it must be pointed out that the control of the chemistry and the photophysical properties of lanthanide-based luminophores is far from straightforward and requires a very well thought molecular design encompassing an extensive array of different aspects and variables that are also intimately related to each other. For instance, the chemical design of the antenna ligand in terms of binding ability and/or the absence of quenching groups must be considered in concert with its electronic features (essential for achieving ligand-to-metal ET), keeping in mind that even subtle changes in one aspect can dramatically affect the resulting properties in the other. The best-performing NIR emitter with the longest emission lifetime, as the one reported in Fig. 3.92, obtained by ligand perfluorination, is, however, particularly susceptible to quenching by solvent due to the increased ligand system polarity and the weak shielding effect. Moreover, it shows poor ligand-to-metal sensitization efficiency and no suitability to electrical excitation, as a consequence of the introduction of strongly electron withdrawing CF groups affecting the position of the donor energy levels on the antenna unit.

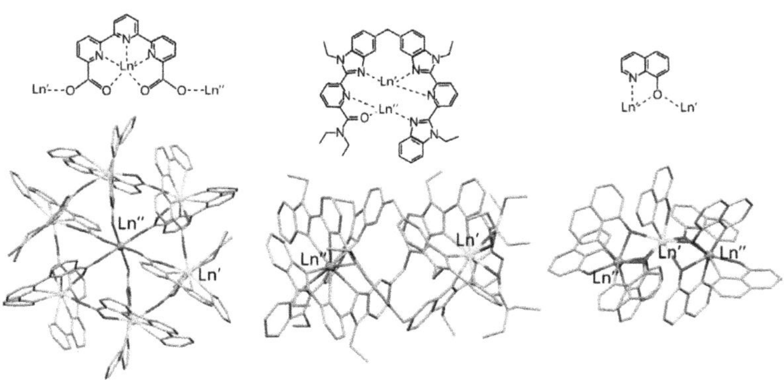

Figure 3.95 Molecular structures and coordination modes of some examples of polynuclear heterolanthanide complexes.

Often, the design of luminescent lanthanide complexes is a matter of finding the right compromises between all the variables that have been herein described, carefully selecting and striving to favor those properties (emission lifetime, overall quantum yield, thermodynamic stability, electrical excitation, etc.) that would best suit the intended application, taking into account that strategies benefitting one aspect are likely to sacrifice another. Moreover, the practical employment of these functional molecular materials often requires additional figures of merit, such as thermal stability, solubility, and chemical purity of the sample. A good amount of research has still to be done in this field, but despite these issues, lanthanide-based molecular optical devices are currently, as a matter of fact, among the most promising functional molecular materials for immediate practical and technological applications, as will be discussed in Chapter 5.

3.3.5 Molecular NLOphores

This class of materials is one of the most extended and composite in view of the several optical phenomena that can be related to the field of nonlinear optics (NLO) and comprises large subclasses of inorganic materials as well as organic molecules and coordination complexes. In this book we will limit the discussion to those materials that most properly can be defined as molecular materials, that is, mostly metal-containing molecules, and hence to the NLO properties that are commonly shown by these compounds besides the most largely studied. In this regard, we will particularly focus on the phenomenon of second harmonic generation (SHG), which is by far the most important in the field of molecular NLO and the most understood in terms of structure/property relationship. Third harmonic generation (THG) and two-photon absorption (TPA)-related properties take on a more limited importance in the framework of molecular materials, being mainly associated with purely inorganic or organic molecules, and the reader should refer to more specialized texts.

Nonetheless, all the above-cited optical phenomena share the same common origin, that is, the simultaneous absorption of more than one photon of incident light. This implies that after light absorption, the energy of the molecule is raised exactly by a multiple of the energy of the incident radiation. The nature of the state reached by the molecular entity after the interaction with light and the subsequent relaxation mode, determines the different types of NLO responses, which basically can be subdivided into nonresonant (SHG and THG) NLO processes whenever the excited state is nonquantized and resonant (TPA) NLO processes whenever the excited state is quantized (cf. Section 2.3.8).

3.3.5.1 SHG NLOphores

In Chapter 2 we have seen that optical nonlinearity in molecules arises from the interaction with an intense light beam that causes a displacement of the electron density, that is, an induced polarization, which does not respond linearly to the electric field of the radiation. This phenomenon gives rise to what is called hyperpolarizability, expressed with the parameter β (first hyperpolarizability), alongside the "regular" linear polarizability (α) (cf. Eq. 2.103). The perturbation of the electron cloud consequent to the interaction with highly intense light brings the molecule to a more energetic state than its initial condition, so it is often said that the molecule reaches a "virtual excited state" (cf. Section 2.3.8 and Fig. 2.59), which is a direct consequence of the absorption of two photons. In Chapter 2 it has also been stated that the rules that must be obeyed to allow for light absorption in the realm of linear optics are essentially two, the Bohr frequency condition and the occurrence of a variation in the dipole moment between the initial and the final state. For electronic transitions, the latter is generally always satisfied, thanks to the different electronic distributions associated with the quantized excited state with respect to the ground configuration. In the field of NLO the Bohr condition obviously loses its validity as multiphoton absorption phenomena and nonquantized excited states are involved. On the contrary, the dipole moment variation requisite still holds, and, similar to first-order processes in the linear regime, the larger the dipole moment variation, the stronger is the light absorption and, consequently, the extent of the NLO response. As the state reached after the nonresonant absorption of two photons is nonquantized, the reason why this second requisite can get somehow fulfilled can be explained by taking this "virtual" excited state as actually arising from the linear combination of the ground and all the excited state wavefunctions of the system ("sum over states" model). Consequently, a change in the charge distribution can be instantaneously achieved upon irradiation with nonresonant light, thanks to this mixing. However, the design of a molecular system that satisfies the dipole moment variation requirement on this general basis would involve a load of theoretical considerations. Therefore, a simplified approach providing a more convenient guideline can be instead opportunely followed. This simplification can be made by considering the first hyperpolarizability as only related to a single excited state (two-state model) and assuming that the NLO process is dominated by only one tensor component of β, which is the case when the first excited state of a molecule implies a significant displacement of electron density along one direction between two separate moieties of the molecular system with respect to the ground state, that is, a CT.

Molecular First Hyperpolarizability in the Two-State Model

The parameter β is a measure of how easily a polarization can be induced in a molecule under the effect of an oscillating electric field. Its value is a function of the frequency ω of the applied field, so it is often indicated as $\beta(-2\omega;\omega,\omega)$, where the first term between the brackets specifies the frequency of the SHG light and the other two the frequencies of the two absorbed photons. This labeling scheme will differ whenever two different light sources are used for inducing other NLO phenomena, such as sum frequency generation and difference frequency generation (cf. footnote 36, Section 2.3.8). The frequency-independent value of the molecular first hyperpolarizability, indicated as β_0, can be also extrapolated for a static field ($\omega = 0$, vide infra).

In the two-state model, the value of β is also dependent on the transition moment μ^r_{if} between the ground and the excited state as well as on the difference between their dipole moments $\Delta\boldsymbol{\mu}$, taking positive values when the excited state $\boldsymbol{\mu}$ is larger that of the ground state and negative values when the excited state $\boldsymbol{\mu}$ is smaller that of the ground state. Because of this dependence, larger SHG responses can be obtained when the frequency double of the incident light is nearly resonant with the absorption band of the molecule. Moreover, the energy E_{if} of the considered electronic transition also dramatically affects the extent of the molecular first hyperpolarizability value through an inverse proportionality, so β is larger for low-energy absorptions. These points can be summarized in the simplified two-state model equation that follows:

$$\beta_0 = \frac{3e^2}{2}\frac{(\mu^r_{\text{if}})^2\Delta\mu}{E^2_{\text{if}}}$$

β values are most often reported in CGS units, ($\text{cm}^4\ \text{statvolt}^{-1}$ = esu), while SI units are $\text{C m}^3\ \text{V}^{-2}$.

From the above concise description of nonresonant NLO response in molecules and recalling that SHG can be only displayed in noncentrosymmetric systems, it becomes immediately evident that the extent of this phenomenon is in the first place dependent on the molecular polarizability, that is, the ability of the electrons to move under the influence of an electric field. Secondly, asymmetry at the molecular level is necessary to provide an asymmetric response of the induced polarization. This means that the primary requirements to observe efficient SHG in molecular materials can be met by resorting to:

- Extensive π conjugation
- Presence of electron donor/acceptor groups
- Systems showing intense CT absorption bands at low energies
- Noncentrosymmetric molecules[38]

For a long time SHG has been investigated only in purely inorganic and organic systems. In particular, organic molecules have paved the way toward a more "chemical" approach toward these kinds of functional materials in view of the low preparation costs and their chemical versatility, which allows fine-tuning of the electronic properties of the material. The first molecules to be studied to this purpose, in view of their structural simplicity and the dipolar character associated with a one-directional CT, were benzene and stilbene derivatives bearing electron-donating (donor, D) and electron-withdrawing (acceptor, A) groups on opposite sides of the molecular backbone, as shown in Fig. 3.96.

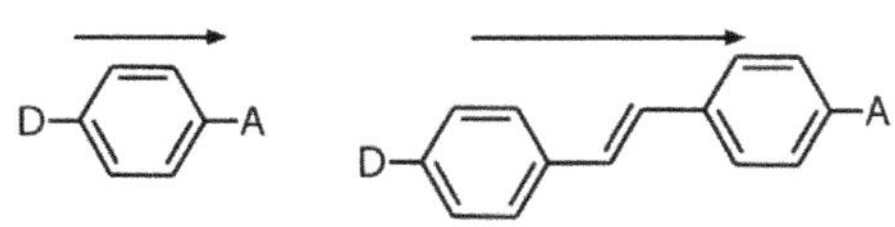

Figure 3.96 Substituted benzene and stilbene structures. The arrows indicate electron density displacement under the influence of the electric field of light.

These molecules are representative examples of the basic structure that can be envisaged for typical SHG molecular materials, which are commonly arranged according to a D-π-A general scheme. These systems are often referred to as "push-pull" after the electron-donating and the electron-withdrawing properties of the D and A groups along a specific direction of the molecular backbone. The D-π-A scheme well evidences that the first hyperpolarizability in

[38]Since the discussion in this book focuses on molecular materials, the requirement of noncentrosymmetry is intended at a molecular level. In the case an ordered/crystalline array of molecules or chemical units is investigated, this requisite applies to the crystal lattice.

molecules depends on the electron donor acceptor properties of the organic substituents as well as on the extent of π-conjugation. Stilbene derivatives usually show larger first hyperpolarizability values than the equally substituted benzene ones due to the more extended π-conjugation path and the resulting higher polarizability. However, extensive studies have taught that the right compromise between the two above-mentioned parameters must be met in order to achieve the largest β values. In fact, the best donor or acceptor moieties do not always prevail over the overall electronic structure of the molecule to afford a high NLO response and, conversely, a large extension of the π-conjugated path does not necessarily coincide with an increased hyperpolarizability due to the decreased communication, in terms of orbital contribution, between the D/A terminal groups. Therefore, it can be concluded that there is an ideal π-bridge length for each pair of donor/acceptor groups that maximizes the value of β. Unfortunately, these are the only two parameters that can be chemically controlled in organic molecules to achieve tunability of the NLO response. These considerations point out that there is actually little room for the adjustment of the electronic properties of organic molecules to afford significant improvement of optical nonlinearity.

On the other hand, organometallic and coordination compounds can widely extend the opportunities to control and tune the SHG properties, thanks to the additional features associated with the presence of the transition metal ion, providing also many benefits toward the achievement of larger β values. These can be summarized as follows:

- Intense CT bands associated with metal-to-ligand, ligand-to-metal, and ligand-to-ligand CT, besides ILCT
- Large dipole moment variation between the ground and excited states due to charge displacement between the metal center and the ligands
- Enhanced donor (or acceptor) properties related to the redox behavior of the metal center, which may be easily oxidized or reduced
- Low-energy absorption bands due to the involvement of metal orbitals in the frontier MOs

All these features offered by metal-based compounds are in addition accompanied by a higher degree of versatility in the chemical properties with respect to organic molecules, such as usual higher solubility, and the opportunity to easily afford charged species where the counterion can bring supplementary features such as chirality, useful to ensure noncentrosymmetry of the system in the solid state.

On the basis of the knowledge gathered from the extensive studies performed on organic molecules, the first metal-containing materials taken into consideration as NLOphores were seen as a modification of the molecular arrangements of the organic backbones that had been established to provide the best results. Therefore, the first archetypal examples of this class of molecular materials can be found in stilbene derivatives where electron-rich-metal-containing units, mostly based on d^6–d^{10} transition metal ions, were incorporated as donor moieties, as in the general examples shown in Fig. 3.97.

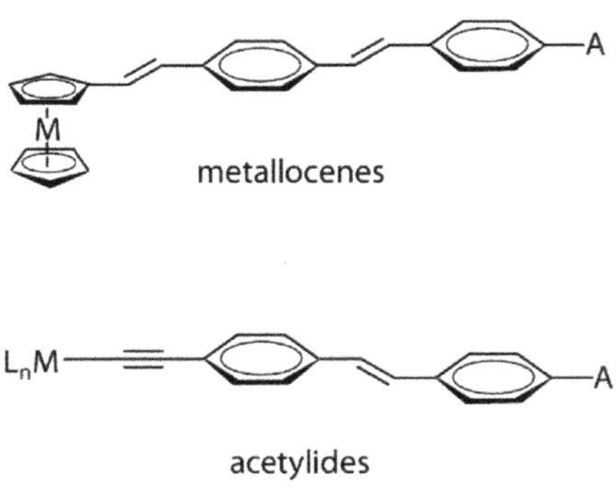

Figure 3.97 Examples of metal-containing SHG NLOphores based on a stilbene-A moiety.

In both examples of Fig. 3.97, the lowest absorption band is related to a metal-to-ligand charge transfer (MLCT) between the low-lying electron-rich orbitals with relevant d-metal contribution and the empty π* orbitals related to the stilbene-A system. The direction of this MLCT transition is practically linear along the molecular plane identified by the stilbene-A moiety in acetylides, whereas for metallocene-containing systems a "verticality" of charge displacement with respect to the molecular plane upon light absorption can be envisaged. For this reason, acetylides usually display much larger β values than analogous metallocene-based molecules due to an increased conjugation between the donor unit's orbitals with those of the stilbene-A moiety. In addition, the electronic properties of acetylide systems can be easily tuned by varying the nature of the L_nM group. The electron-donating properties of the coordinated L ligands and the electronic configuration of the metal in fact remarkably affect the resulting NLO response, although they are not the only factors that count in this regard. The second issue to be considered is the metal oxidability, that is, the ease with which electrons can be removed from the metal center. Systematic experimental studies have shown that Ru^{II}-based systems are among the best donor moieties to be used for the preparation of SHG NLOphores due to the tendency of this metal ion to lose an electron and be oxidized to Ru^{III}. The redox properties of ruthenium also offer an interesting strategy to achieve molecular switches where the NLO response is only detectable when the metal is in its reduced divalent form whereas the corresponding oxidized Ru^{III}-based system would not show optical nonlinearity due to the suppression of the MLCT transition.

This latter observation also introduces the concept that metal-containing moieties actually can be designed to serve either as donor or as acceptor units, depending on:

- The metal electronic configuration
- The metal oxidation state
- The electronic (donor or acceptor) properties of the metal environment

Therefore, as a general rule, electron-rich metal ions in a low oxidation state and coordinated with electron donor ligands (Lewis bases), such as amino complexes, are among the best donor units. On the other hand, electron-poor metal ions coordinated with electron-withdrawing groups can function as good acceptor moieties. In some instances, the metal environment may have intermediate properties, that is, between electron-donating and withdrawing properties. This is, for example, the case of metal carbonyls, where the CO ligand behaves either as a σ acceptor and a π donor. In such systems the electronic properties of the metal-containing moiety are dramatically sensitive to the nature of the coordinated metal ion.

The ample versatility of metal-based units to function as donors or acceptors inspired the investigation of bimetallic systems as potential SHG NLOphores as well. An archetypal simple example of this kind of molecular assembly is represented by the $[(CN)_5Ru^{II}$-CN-$Ru^{III}(NH_3)_5]^-$ bimetallic complex where the two centers Ru^{II} and Ru^{III}, respectively, the donor and the acceptor units, are connected by the short π-bridging CN group. In this case, the transition associated with β can be considered as intravalence CT (IVCT) where the two metal ions exchange one electron.

Besides their versatility in the construction of donor/acceptor units, transition metal ions can be conveniently incorporated into molecular SHG NLOphores to increase the π-conjugation of the system, either by direct π-bridging or by templating a planar molecular arrangement. Some archetypal molecular NLOphores that can be mentioned in this regard are metal complexes with asymmetrically substituted Schiff bases, as the ones reported in Fig. 3.98.

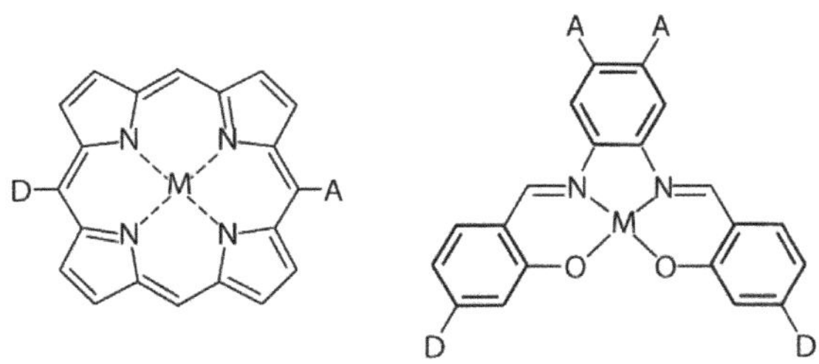

Figure 3.98 Examples of SHG molecular NLOphores based on metal complexes of substituted porphyrines (left) and multidentate Schiff bases (right).

Another interesting class of SHG NLOphores where the metal ion plays a crucial role in determining the electronic features of the molecular system is represented by square-planar dithiolene complexes, similar to the ones already presented in Section 3.2.3 but bearing two different ligands with the aim to deliver asymmetry at the molecular level. Whether the electron-withdrawing (push/pull) properties of the two coordinated ligands are significantly different, the electronic arrangement of the molecular backbone will be best described by one of the two limiting structures depicted in Fig. 3.57

in the case of neutral complexes with divalent d^8 metal ions,[39] where one ligand is in the dianionic dithiolate form and the other one is in the neutral dithione form. The π-localization associated with this arrangement provides relevant asymmetry in the electronic charge distribution that results in frontier molecular orbitals (MOs) where the major contribution to the HOMO is furnished by the C_2S_2 core of the dithiolate ligand, whereas the LUMO mainly involves the C_2S_2 unit of the dithione, as schematically shown in Fig. 3.99.

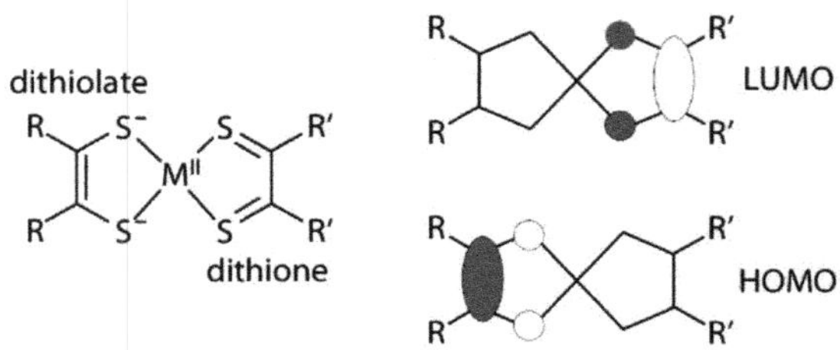

Figure 3.99 Scheme of a typical square planar neutral mixed ligand dithiolene complex with a d^8 metal ion and a π-localized electron arrangement together with a general description of the frontier MOs. The figure omits possible metal contribution to the HOMOs and LUMOs.

Complexes with these features show HOMO-LUMO π–π* transitions associated with the displacement of charge from the dithiolene ligand to the dithione one, giving rise to ligand-to-ligand charge transfer (LLCT) bands usually falling in the visible–NIR spectral region. Besides the electronic properties of the ligands, the separation extent between the HOMOs and LUMOs is also largely determined by the contribution of the metal ion, whose role therefore goes beyond that of a mere π-bridge between the two coordinated ligands.[40] The relevant LLCT character of the transition

[39]Dithiolene ligands are particularly prone to redox activity and may coordinate a metal ion in the dianionic $C_2S_2R_2^{2-}$, the radical monoanionic $C_2S_2R_2^{-}$, or the neutral $C_2S_2R_2$ form, as depicted in Fig. 3.57. Accordingly, bis-dithiolene complexes of divalent d^8 transition metal ions can assume any oxidation state, ranging from the dianionic form to the dicationic form. For simplicity's sake, we will herein restrict the discussion to neutral complexes, taking into account that it applies also to monoanionic and monocationic molecules, whereas the –2 and +2 charged complexes do not display NLO activity (vide infra).

[40]The nature of the metal ion is also crucial for determining the overall oxidation state of the bis-dithiolene complex as the extent of the metal–ligand orbital mixing determines the stabilization of the frontier MOs.

between the ground and the excited state, usually accompanied by a large dipole moment variation, and giving rise to a low-energy absorption band, makes these complexes good NLOphores for SHG,[41] where the nonlinear response can be easily tuned by varying the electronic properties of the system, in particular the nature of ligand substituents.

We have, so far, considered only planar or almost-planar donor acceptor dipolar molecular systems where the variation in hyperpolarizability is associated with a one-directional charge displacement along the molecular backbone. However, it is worth remarking that also 3D molecules, such as the Ru^{II} tris-bypiridyl complex shown in Fig. 3.77, can show relevant SHG activity. Although in the light of the previous discussion this observation may initially sound curious since the octahedral $[Ru^{II}(bipy)_3]^{2+}$ complex has an overall symmetric shape, the explanation for this behavior comes clearly into mind when addressing this system as octupolar. In such a circumstance the overall NLO response can be regarded as the result of the combination of dipolar responses of different directionality associated with the MLCT transitions typical of this complex and its derivatives (cf. Section 3.3.3.2). Following a similar approach, insertion of suitable substituents on the ligand periphery can afford good SHG NLOphores of octahedral or even tetrahedral geometry where the NLO response is relatable to ILCT transitions.

The charge-related features of a system, in terms of dipolar or apolar overall character and oxidation state, are also crucial for selecting the appropriate method for measuring the first hyperpolarizability value, as described in the box titled "Methods for the Measurement of β."

Methods for the Measurement of β

Different techniques can be employed for the measurement of first-order hyperpolarizabilities in molecular materials. Among them, the three most commonly used are described below.

[41]Dianionic and dicationic bis-dithiolene complexes will have a completely filled or empty ground-state configuration for the orbitals depicted in Fig. 3.99. Therefore, these orbitals will no longer correspond to the frontier MOs of the system and no LLCT transition can occur. This is the reason why complexes in the –2 and +2 oxidation states usually do not show SHG NLO activity.

Kurtz powder technique: This is a direct method of measurement based on the detection of the SHG-emitted light from the powdered sample after laser irradiation. The output signal is then compared with that of a reference sample to obtain a measure of the SHG efficiency. Although this method is definitely the simplest one, it is, however, not very accurate as the detected signal is significantly affected by several factors, such as particle size and homogeneity. Moreover, being a solid-state technique, the Kurtz method requires the sample to display bulk second-order susceptibility ($X^{(2)}$), that is, only molecular materials that crystallize in a noncentrosymmetric space group can be investigated. Therefore this technique is not useful for assessing hyperpolarizability at the molecular level.

Electric-field-induced second harmonic generation (EFISH): This technique is one of the most popular for molecular NLOphores since it allows the study of liquid or solid solutions of the investigated compound, thus ruling out the effects associated with the crystal lattice and allowing the correlation of the nonlinear response directly with the molecular properties. In a typical experiment, sample solutions are held into an optical cell embedded between two electrodes and exposed to an external DC electric field. This external field tends to parallel-align the molecules through interaction with the molecular dipole moments, thus destroying the macroscopic centrosymmetry of the system arising from the random thermal orientation of the molecules. The sample solution is then irradiated with laser light of frequency ω, and the second harmonic response of the solution at a frequency 2ω can be detected. Both the sample and the solvent responses are collected simultaneously. Therefore a careful correction of the background signal must be made. The measured value refers to the component of the tensor β along the direction of the molecular dipole moment and is retrieved as the scalar product $\boldsymbol{\mu}\cdot\beta$, where $\boldsymbol{\mu}$ is the dipole moment of the ground state. Therefore, β cannot be determined without knowing the value of $\boldsymbol{\mu}$, which can be nonetheless retrieved through theoretical calculations. Moreover, it has to be taken into account that the value of the first hyperpolarizability is dependent on the wavelength of the light source ("dynamic" hyperpolarizability, β_λ). Therefore, to compare the SHG response of different samples the "static" β_0 values at zero oscillating field should be extrapolated through the following equation:

$$\boldsymbol{\mu}\beta_0 = \boldsymbol{\mu}\beta_\lambda\left[1-\left(\frac{2\lambda_{\max}}{\lambda}\right)^2\right]\left[1-\left(\frac{\lambda_{\max}}{\lambda}\right)^2\right]$$

The EFISH technique is a widely used method for the measurement of SHG in molecular materials, although it is somehow limited in that it is only applicable in the case of dipolar molecules suitable to be easily oriented under a DC electric field. Moreover, molecules should have a well-defined CT direction along the molecular axis identified by the dipole moment. In this circumstance, the dominant component of β would effectively coincide with the measured one and reliable results can be afforded.

Hyper–Rayleigh scattering (HRS): This technique consists of measuring the intensity of the second-harmonic-generated scattered light at a frequency 2ω of a sample dissolved in a solvent and irradiated with laser light of frequency ω. The measurement apparatus is suitably designed to discriminate the doubled frequency light from the fundamental component, and the detected signal output can be correlated to the mean β value for molecules dispersed in isotropic solutions. Appropriate references are used to assess molecular first hyperpolarizability values and subtract solvent effects. This method offers the advantage over the EFISH technique of being suitable for the investigation of apolar molecules or conductive solutions of salts (charged species will migrate to the electrodes in EFISH measurements). Moreover, it provides information on all the components of the first hyperpolarizability without the need for knowing the value of the molecular dipole moment.

3.3.5.2 General considerations

As usually concluded for several types of functional molecules, the introduction of transition metal ions into molecular architectures allows for an upgrading of the control that can be achieved over the tuning of the physical properties of a molecular material. For SHG NLOphores, the metal ion plays a real multitasking role for the design of such materials as it can be envisioned either as one of the three fundamental units typically constituting these molecular devices, that is, a donor, an acceptor, or a π-bridging unit, or can even be used to obtain molecular structures not otherwise affordable in mere organic molecules. With this approach, a large reach of molecular architectures that can be employed as SHG NLOphores becomes available, and even 3D structures such as octahedral complexes can display such functionality. Moreover, the metal ion can significantly enhance the NLO optical response of molecules since it crucially influences the electronic properties of the frontier MOs of the system. In fact it can determine the type of CT transition involved in the process and also somehow allows for a fine control of the parameters that determine the value of β, such as the transition moment and energy of the absorption transition (cf. equation in the box titled "Molecular First Hyperpolarizability in the Two-State Model)."

Although metal-based compounds have not yet bested the results displayed by some organic SHG NLOphores, the extraordinary chemical/physical versatility offered by these materials undoubtedly brings the perspectives of research in this field to a promoted level.

Chapter 4

Multifunctional Molecular Materials

In Chapter 3, we have shown the design principles and physical properties of the most common classes of monofunctional molecular materials belonging to the realms of magnetic, conducting, and optical materials. Now we will show how selected molecular building blocks can be combined to prepare hybrid and multifunctional molecular materials in which different physical properties coexist or are in interplay.

In multifunctional molecular materials, the augmented number of functionalities within the same material is usually associated with a more complex structural arrangement with respect to that of the constituting monofunctional molecular units. However, since in most cases each building block maintains its molecular structure unchanged when incorporated into more sophisticated architectures, either by supramolecular arrangement or through proper chemical bonding, the description of the properties displayed by the assembled material can often be accounted for simply by taking into consideration the sum of the separate contributions of the different components as carriers of a specific physical property. This approach usually affords a reliable interpretation of the observed properties as long as the functionalities of the associated building blocks are superimposed and not mutually interacting. In such circumstances, as we will see in the framework of specific examples, the study of appropriate reference compounds, which most often must be designed and synthesized to this specific

Functional Molecular Materials: An Introductory Textbook
Matteo Atzori and Flavia Artizzu

ISBN 978-981-4774-76-5 (Hardcover), 978-1-351-23366-8 (eBook)
www.panstanford.com

purpose, is necessary to be able to draw up conclusions on the basis of a comparative investigation method.

On the other hand, the assembly of preformed molecular units, although the most immediate, is not the only design strategy to afford multifunctional molecular materials. A well-thought-out chemical engineering approach can be adopted to more subtly blend different chemical/structural/electronic elements, bringing different physical properties into the same material, by appropriately selecting metal centers, ligands, and synthetic methods to achieve specific molecular architectures, without actually resorting to predesigned, stand-alone, molecular units.

The aim of this chapter is to provide an overview of the different approaches toward materials design in this research field through a discussion of the most significant and instructive examples of this class of compounds, taken as archetypes.

4.1 A Family of Paramagnetic Superconductors

The combination of selected molecular building blocks as carriers of specific magnetic and conducting properties can lead to hybrid molecular materials with technologically useful properties. Aside from the interest in deeply understanding the interactions between localized electrons responsible for the magnetic properties and delocalized electrons that are responsible for the charge transport, the preparation of hybrid conducting/magnetic materials represents a challenging task in the field of materials science, which captured the interest of several research groups worldwide. Relevant examples of such materials are the compounds prepared and investigated by Day and coworkers (Kurmoo et al., 1995). By combining the bis(ethylenedithio)-tetrathiafulvalene (BEDT-TTF) organic donor and mononuclear paramagnetic trisoxalatometallate complexes, $[M(C_2O_4)_3]^{3-}$ (M = Fe^{III}, Cr^{III}), Day and coworkers succeeded in the preparation of a family of molecular materials of the general formula $[BEDT\text{-}TTF]_4[AM(C_2O_4)_3]\cdot S$ (A = H_3O^+, K^+, NH_4^+; Fe^{III}, Cr^{III}; S = C_6H_5CN, $C_6H_5NO_2$, C_5H_5N). The crystal structure of such materials consists of alternate layers containing BEDT-TTF donors and $[AM(C_2O_4)_3]^-$ anions. The anionic layers contain alternating A^+ cations and $[M(C_2O_4)_3]^{3-}$ complexes forming an approximately hexagonal network, while the BEDT-TTF molecules are arranged in segregated layers with specific crystal packing arrangements (Fig. 4.1) (vide infra).

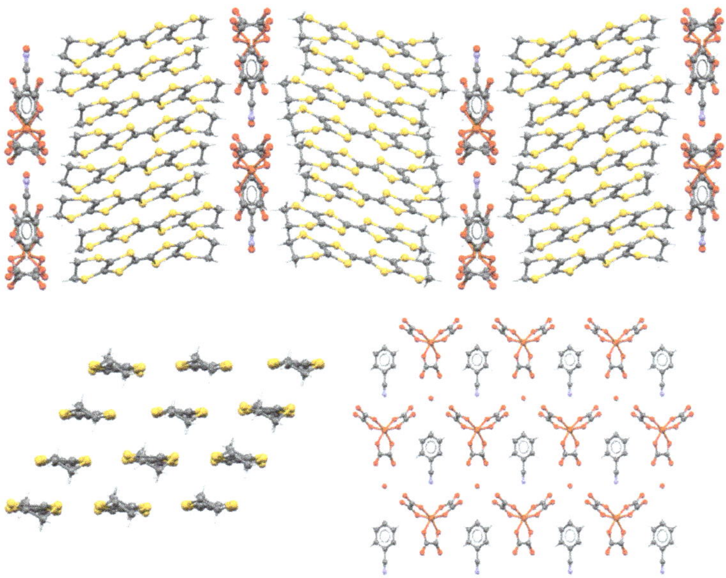

Figure 4.1 Views of the crystal structure of β″-[BEDT-TTF]$_4$[(H_3O)Fe(C_2O_4)$_3$]·C_6H_5CN. Overall structure with alternated segregated organic-inorganic layers (top), detailed view of the β″-phase donor packing (bottom left), and hexagonal arrangement of the inorganic layers (bottom right). Color codes: yellow, sulfur; gray, carbon; white, hydrogen; red, oxygen; cyano, nitrogen; and orange, iron.

From the analysis of the properties of the distinct building blocks, the potential physical properties that such hybrid materials may exhibit can be rationalized as follows:

- The presence of noninteracting mononuclear paramagnetic metal anions will be likely associated with the retention of the intrinsic magnetic properties of the isolated metal ion also in the hybrid system.
- The presence of partially oxidized organic donors will lead to relevant conducting properties (semiconductivity, metal-like conductivity, or superconductivity) whose nature is

not easily predictable a priori as it depends on the type of supramolecular crystal packing, the oxidation degree of each donor molecule, and the mutual interaction between them.

- The interplay between magnetic and conducting properties may be observed. However, since only weak interactions of supramolecular nature are present between the organic and inorganic layers, observation of the interplay between such properties is expected to be unlikely.

While the stoichiometry and molecular packing of the inorganic anionic layers are identical for all components of the family, appreciable differences can be observed as far as the packing of the donor molecules and their oxidation states are concerned. Without going deep into details, it is sufficient to say that the compounds of this family are split into two series: one where the donor molecules adopt a pseudo-κ phase packing and the other in which they arrange in the β″-phase packing (cf. Figs. 4.1 and 3.54). As stated before, the conducting properties are strongly dependent on the molecular crystal packing and, as expected, the two series show different transport properties. Compounds belonging to the pseudo-κ phase class are paramagnetic semiconductors with room-temperature conductivity of the order of 10^{-4} S cm^{-1} and an activation energy of ca. 140–230 eV, while compounds arranged in the β″ phase are paramagnetic metals with a conductivity of ca. 10^{2} S cm^{-1} at 200 K (6 orders of magnitude higher than the semiconducting ones), which increases monotonically down to a temperature of ca. 6–8 K (Fig. 4.2), below which they become paramagnetic superconductors (Table 4.1).

As far as magnetic properties are concerned, all paramagnetic components of the series behave as Curie paramagnets in almost the entire investigated temperature range (2–300 K), and the resulting properties of the material derive from the sum of two paramagnetic contributions, that is, that related to the localized electrons of the paramagnetic metal ions and that associated with the unpaired electrons of the organic radical cations. Only small deviations for temperatures lower than the metal–superconductor transition temperature are observed between semiconducting and superconducting materials.

Table 4.1 Structural details and physical properties of selected multifunctional radical-cation salts of the general formula $[BEDT\text{-}TTF]_4[AM(C_2O_4)_3]$ (Coronado and Day, 2004)

Metal	Cation	Solvent	Type of donor packing	Conducting properties	Magnetic properties
Fe	NH_4^+	C_6H_5CN	pseudo-κ	Semiconductor (E_a = 140 eV)	Paramagnet
Fe	K^+	C_6H_5CN	pseudo-κ	Semiconductor (E_a = 141 eV)	Paramagnet
Cr	H_3O^+	C_6H_5CN	pseudo-κ	Semiconductor (E_a = 153 eV)	Paramagnet
Cr	H_3O^+	C_6H_5CN	β''	Superconductor (T_c = 6 K)	Paramagnet
Fe	H_3O^+	C_6H_5CN	β''	Superconductor (T_c = 8.3 K)	Paramagnet
Cr	H_3O^+	$C_6H_5NO_2$	β''	Superconductor (T_c = 6 K)	Paramagnet
Fe	H_3O^+	$C_6H_5NO_2$	β''	Superconductor (T_c = 4 K)	Paramagnet

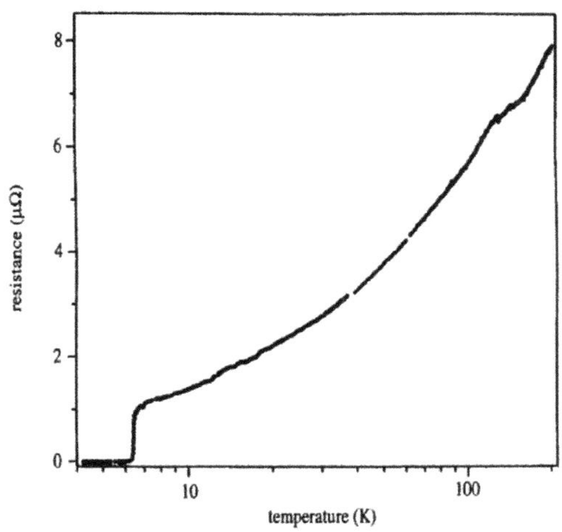

Figure 4.2 Temperature dependence of the resistance of β''-$[BEDT\text{-}TTF]_4[(H_3O)Fe(C_2O_4)_3]\cdot C_6H_5CN$. The resistivity decreases monotonically down to ca. 7 K. Below this temperature the resistivity assumes an almost zero value, typical of a superconductor. Adapted with permission from Kurmoo et al. (1995). Copyright (1995) American Chemical Society.

In conclusion, this example represents a case in which selected conducting and magnetic building blocks are combined together to obtain a family of hybrid conducting/magnetic materials. While the magnetic properties are, as expected, those typical of isolated magnetic complexes plus the contribution of the organic donor, the conducting properties are determined by the packing assumed by the partially oxidized donor molecules and are not easy predictable a priori. As clearly demonstrated with this example, two different crystal phases can be obtained starting from the same building blocks. Indeed, even if the stoichiometry is the same for all components, differences in the packing of the donor molecules lead to definitely different conducting properties. In particular, coexistence of paramagnetism and superconductivity is clearly evidenced.

4.2 Magnetic-Field-Induced Superconductivity in a Paramagnetic Metal

Superconducting states of matter in the presence of an external magnetic field are very rarely observed because, in general, the effect of the magnetic field is to destroy the superconductivity. However, if the material has a peculiar crystal structure, as for example that of 2D superconductors, the effect of the magnetic field in destroying the superconductivity can be reduced if the magnetic field is applied parallel to the conducting layers. We have already seen in Chapter 3 that 2D molecular conductors can be prepared by partial oxidation of organic donors to radical cations. If intrinsic magnetic properties are also desired, a paramagnetic counterion can be introduced to counterbalance the charge, as exemplified in the previous section. By following this design strategy Kobayashi and coworkers (Uji et al., 2001) have prepared a hybrid system in which electrical conductivity is provided by the bis(ethylenedithio)-tetraselenafulvalene (BETS) organic donor and paramagnetism is provided by the magnetic moments of the $FeCl_4^-$ counteranions. The resulting material is the λ-$[BETS]_2[FeCl_4]$, where λ indicates the crystal phase, whose structure is reported in Fig. 4.3.

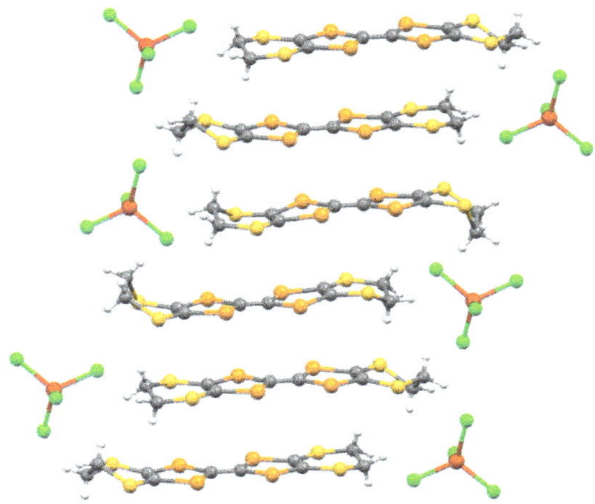

Figure 4.3 View of the crystal structure of λ-[BETS]$_2$[FeCl$_4$]. Color codes: pale yellow, sulfur; deep yellow, selenium; gray, carbon; white, hydrogen; green, chlorine; and orange, iron.

At zero magnetic field, this compound shows metal-like conductivity down to ca. 8 K, at which a metal–insulator transition occurs (cf. Section 2.2.4). The origin of such behavior was attributed to the concomitant antiferromagnetic ordering of the Fe^{III} magnetic moments. This hypothesis is supported by the interesting observation that the isostructural compound λ-[BETS]$_2$[GaCl$_4$], in which the Fe^{III} is replaced by the diamagnetic Ga^{III}, undergoes a metallic-superconducting transition (cf. Section 2.2.4) at ca. 6 K. When a magnetic field of 1 T is applied to the Fe^{III} derivative at a temperature below the T_c, the magnetic moments are canted by the magnetic field but the conductivity remains that of an insulator. If the strength of the magnetic field is increased to above 10 T, the paramagnetic metallic state is recovered (Fig. 4.4). If the strength of the magnetic field is increased further, beyond 17 T, at very low temperatures (<1.0 K) the resistivity suddenly drops to reach an almost zero value, typical of a superconducting state (Fig. 4.4). Thus, Kobayashi et al. (2001) have demonstrated that in molecular materials based on the BETS organic donor, superconductivity can be induced by a strong magnetic field if the magnetic field direction is carefully chosen on a structural basis, and additionally, it can coexist even in the presence of an intrinsic paramagnetic component, such as $FeCl_4^-$ anions.

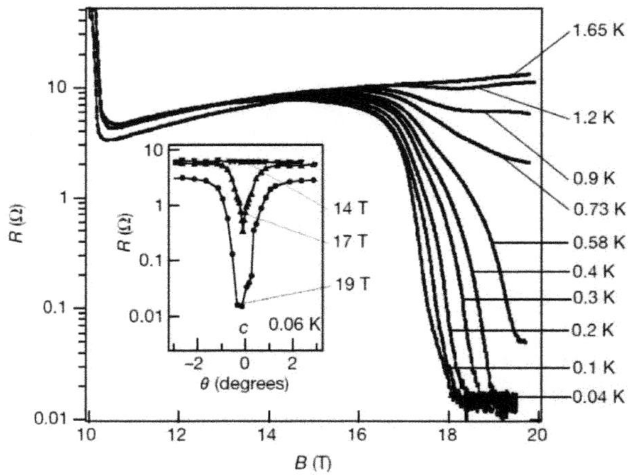

Figure 4.4 Magnetic field dependence of the electrical resistivity of the λ-$[BETS]_2[FeCl_4]$ at different temperatures. It can be clearly seen that the resistivity drops for magnetic field values higher than 10 T and that by further increasing the magnetic field strength, a superconducting state is observed for temperatures lower than 0.4 K. The inset shows the resistivity at 0.06 K at different magnetic field values as a function of the crystal orientation. Reprinted by permission from Macmillan Publishers Ltd: [*Nature*] (Uji et al., 2001), copyright (2001).

In the framework of the purpose of this book, this example helps us in pointing out that additional and unexpected physical properties can be observed when different molecular materials are combined in a multifunctional hybrid system.

4.3 Ferromagnetic Metal-Like Conductors

Coexistence of ferromagnetism and metallic conductivity has been achieved for the first time in the compound [BEDT-TTF]$_3$[MnCr(C_2O_4)$_3$] (Coronado et al., 2000). This system has been obtained by combining two functional molecular building blocks: the oxalate-based heterobimetallic coordination polymers of general formula $[M^{II}M^{III}(C_2O_4)_3]^-$ showing ferromagnetic properties (cf. Section 3.1.2) and the organic donor BEDT-TTF as a conductivity carrier (cf. Section 3.2.1). The two components are organized in alternate layers to give the assembled supramolecular crystal structure reported in Fig. 4.5.

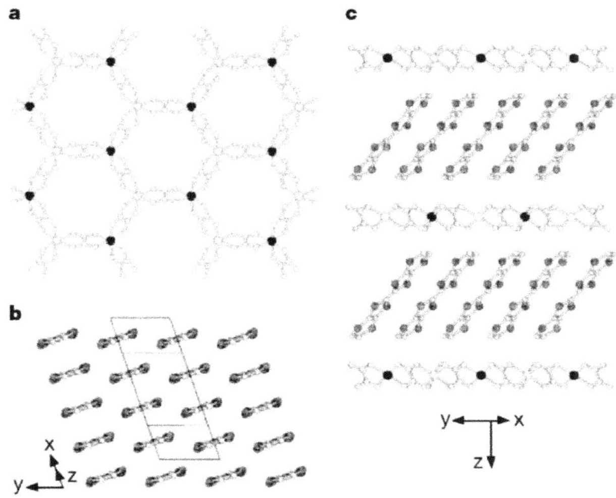

Figure 4.5 View of the inorganic (a) and organic (b) networks of [BEDT-TTF]$_3$[MnCr(C_2O_4)$_3$] hybrid system and a global view of its crystal structure (c). Reprinted by permission from Macmillan Publishers Ltd: [*Nature*] (Coronado et al., 2000), copyright (2000).

From the magnetic point of view, the $[Mn^{II}Cr^{III}(C_2O_4)_3]^-$ building block has been selected from among other compounds having the same molecular formula and different combinations of metal ions since the magnetic exchange interactions between Cr^{III} and Mn^{II} favor, more than other metal ions, the observation of ferromagnetic, instead of ferrimagnetic, ordering. The reason for that is related to the Goodenough–Kanamori–Anderson rules illustrated in Section 2.1.12, stating that orthogonal interactions between magnetic orbitals favor ferromagnetic interactions. In this example, in fact, the Cr^{III} ion has threefold-occupied t_{2g} orbitals that are symmetrically orthogonal to the twofold-occupied e_g orbitals of Mn^{II}.

The selection of a magnetic building block with a 2D structural arrangement and an anionic nature is of crucial importance for the realization of a material coupling ferromagnetic ordering and conductivity for several reasons. First, the 2D structural arrangement guarantees the propagation of the ferromagnetic exchange interaction over the entire layer. Previously undertaken studies in analogous compounds with nonfunctional counterions (e.g., $[(n\text{-Bu})_4N]^+$) have evidenced that this structural arrangement is very likely to induce ferromagnetic ordering with a T_c of ca. 4–5 K, regardless of the nature of the counterion. Second, the low structural hindrance provided by the planar arrangement of the magnetic molecular component allows a layered ordering of the conductivity molecular carrier that is required to afford good transport properties (cf. Section 3.2). Finally, the global anionic charge is required to sustain the overall structure through electrostatic interactions with the positively charged functional counterion.

With these clues in mind, it can be reasonably understood that when the organic donor BEDT-TTF is slowly oxidized to a radical cation in the presence of $[Cr^{III}(C_2O_4)_3]^{3-}$ and Mn^{II}, the building blocks self-assemble together to form layers of $[M^{II}M^{III}(C_2O_4)_3]^-$ coordination polymers and layers of partially oxidized BEDT-TTF radical cations.

While the magnetic properties of the magnetic carrier can be easily predicted, thanks to the "rigid" structural features of the $[M^{II}M^{III}(C_2O_4)_3]^-$ polymer, the conducting properties of radical cation salts cannot be foreseen even in this case. The reason why the BEDT-TTF has been selected as a conductivity carrier can be explained on the basis of the discussion made in Chapter 3. This molecule, in fact, shows a planar structure that yields good transport properties as a result of the efficient intermolecular interactions between highly polarizable atoms at the complex periphery. The good conducting properties of this building block have been evidenced in most of the multifunctional molecular materials containing this component, so they can be reasonably expected also when the BEDT-TTF is associated with the coordination polymer mentioned herein.

The hybrid molecular material $[\text{BEDT-TTF}]_3[MnCr(C_2O_4)_3]$ shows the structural features predicted on the basis of the above-mentioned considerations and the following physical properties:

- Ferromagnetic ordering below a Curie temperature T_c of 5.5 K. This is evidenced by the presence of a magnetization curve with a magnetization saturation value of ca. 8 μ_B (as expected for the ferromagnetic interaction between Cr^{III}, 3 μ_B, and Mn^{II}, 5 μ_B) (Fig. 4.6).
- High conductivity at room temperature (ca. 250 S cm^{-1}) with a temperature dependence of the resistivity that shows a metal-like behavior down to 2 K (Fig. 4.6).

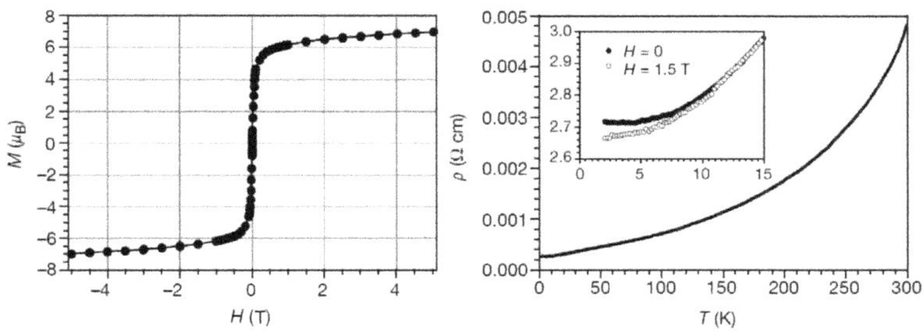

Figure 4.6 Magnetization versus magnetic field plot (left) and thermal dependence of the resistivity (right) for the [BEDT-TTF]$_3$[MnCr(C_2O_4)$_3$] hybrid system. The inset shows the low-temperature conductivity with and without an applied external static magnetic field. Reprinted by permission from Macmillan Publishers Ltd: [*Nature*] (Coronado et al., 2000), copyright (2000).

If the conductivity measurements are performed in the presence of a static external magnetic field perpendicular to the layers, the resistivity shows a small decrease in proximity of the Curie temperature. This phenomenon is called *negative magnetoresistance* and, in this specific case, it can be ascribed to the presence of additional internal magnetic fields generated by the ferromagnetic layers. No effect on the conductivity is observed if the magnetic field is applied parallel to the layer. The influence of the magnetic properties on the conductivity can be regarded as an interplay between different properties occurring at low temperatures, while no alterations of the magnetic behavior is detected. This indicates that, overall, these physical properties predominantly coexist rather than interact.

In conclusion, the [BEDT-TTF]$_3$[MnCr(C_2O_4)$_3$] system represents an archetype example of multifunctional material showing coexistence between ferromagnetic ordering and metal-like conductivity, obtained through a judicious selection of molecular building blocks.

After the discovery of the first multifunctional molecular material where ferromagnetism and metal-like conductivity coexist, synthetic efforts have been devoted to the introduction of a third property in such molecular architectures. In particular chirality was seen as a chemical feature that might be able to induce novel properties, such as the electrical magnetochiral anisotropy presented in Section 3.2.5.2, or improve the conductivity (cf. Section 3.2.5.1). With this view, a multifunctional molecular material consisting of an enantiopure tetramethylated derivative of the BEDT-TTF, the (*S,S,S,S*)-tetramethyl-BEDT-TTF (TM-BEDT-TTF), assembled with the Cr-Mn oxalate-based building block already presented, was investigated. The combination of these building blocks resulted in a material of general formula [(*S,S,S,S*)-TM-BEDT-TTF]$_x$[$MnCr(C_2O_4)_3$]·CH_2Cl_2 (x = ca. 3), which represents the first chiral molecular material where coexistence of ferromagnetism and metallic conductivity has been observed (Galán-Mascarós et al., 2010).

Although the crystal structure of this compound has not been solved by single-crystal X-ray diffraction experiments, the combination of different characterization techniques and the similitude of the building blocks employed for its preparation with respect to the [BEDT-TTF]$_3$[$MnCr(C_2O_4)_3$] compound allow the description of the structural arrangement for this material. It reasonably consists of alternate layers of inorganic and organic building blocks, similar to that shown in Fig. 4.5c, with the only exception that the BEDT-TTF is replaced by the TM-BEDT-TTF donor, which assumes a β-type structural packing.

As far as the physical properties are concerned, [(*S,S,S,S*)-TM-BEDT-TTF]$_x$[$MnCr(C_2O_4)_3$]·CH_2Cl_2 shows the expected magnetic properties of the [$MnCr(C_2O_4)_3$] coordination polymer, with ferromagnetic ordering below 5.5 K, a saturation magnetization of ca. 8 μ_B, and a low coercive field, while the transport measurements show a room-temperature conductivity of ca. 65 S cm^{-1}, with a temperature dependence typical of a metal-like behavior down to ca. 190 K. When the temperature is decreased further, the conductivity decreases. This behavior was attributed to a charge localization phenomenon, which limits the conductivity and can be attributed to the Peierls distortion described in Section 2.2.7. At ca. 5 K, the ordering temperature of the ferromagnetic layers, the conductivity shows a second minimum and starts to increase rapidly with temperature (Fig. 4.7).

Although there is no direct evidence of any influence of the chirality on the transport properties of this material, the [(*S,S,S,S*)-

TM-BEDT-TTF]$_x$[$MnCr(C_2O_4)_3$] system nonetheless represents an archetype example of a multifunctional material showing coexistence of ferromagnetic ordering, metal-like conductivity, and chirality.

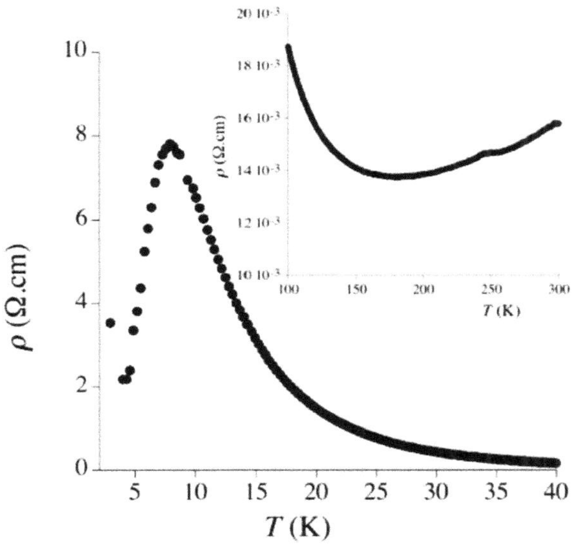

Figure 4.7 Thermal dependence (2–40 K) of the resistivity for the [(*S,S,S,S*)-BEDT-TTF]$_x$[$MnCr(C_2O_4)_3$] hybrid system. The inset shows the conductivity at high temperature (100–300 K). Adapted with permission from Galán-Mascarós et al. (2010). Copyright (2010) American Chemical Society.

4.4 Coexistence of Electrical Conductivity and Slow Magnetic Relaxation in a Hybrid Material

Hybrid molecular materials showing coexistence of electrical conductivity and magnetic properties have been extensively studied in the past years due to their potential technological applications. These materials also attract great interest from the point of view of the advancement of fundamental science, specifically in regard to the investigation of the interaction between localized electrons of the magnetic building blocks and delocalized electrons of the conducting counterpart. So far, we have described materials in which electrical conductivity is combined with magnetic carriers showing paramagnetism and ferromagnetism, and we have seen that in certain materials the two properties can also interact with each other.

Another class of magnetic building blocks that offer a stimulating playground to investigate peculiar interactions between magnetic and conducting building blocks is represented by metal complexes showing slow relaxation of the magnetization, thus behaving as single-molecule magnets (SMMs). Motivated by the interest in observing synergism between slow magnetic relaxation and electrical conductivity, Yamashita and coworkers (Hiraga et al., 2007) have combined a Mn^{II}/Mn^{III} tetranuclear complex ($[Mn^{II}_2Mn^{III}_2(hmp)_6(MeCN)_2(H_2O)_4]^{4+}$ (hmp^- = 2-hydroxymethylpyridinate), hereafter abbreviated as $[Mn_4]^{4+}$), which behaves as a SMM, and $[Pt(mnt)_2]^-$ (mnt^{2-} = maleonitriledithiolate) as a conducting building block. Their combination led to a hybrid material of formula $[Mn^{II}_2Mn^{III}_2(hmp)_6(MeCN)_2\{Pt(mnt)_2\}_4][Pt(mnt)_2]_2$ (Hiraga et al., 2007). Although at a first view the molecular formula of this material seems very complicated, it consists in the association of the $[Mn_4]^{4+}$ complex where the four coordinating molecules of water are replaced by four $[Pt(mnt)_2]^{n-}$ units and two uncoordinated $[Pt(mnt)_2]^{n-}$ molecular anions that serve to counterbalance the residual positive charge of the cation (Fig. 4.8).

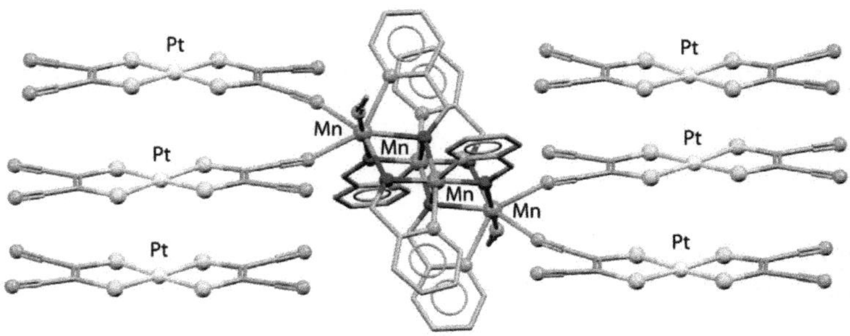

Figure 4.8 View of the molecular structure of the magnetic and conducting building blocks composing the hybrid material $[Mn^{II}_2Mn^{III}_2(hmp)_6(MeCN)_2\{Pt(mnt)_2\}_4][Pt(mnt)_2]_2$. Coordinated and uncoordinated $[Pt(mnt)_2]^{n-}$ units are clearly highlighted.

This arrangement leads to 1D stacks of $[Pt(mnt)_2]^{n-}$ units between layers of cations (Fig. 4.9), with the peculiar feature that, contrary to the materials described in the other sections, covalent coordinating bonds between the conducting and magnetic layers are established.

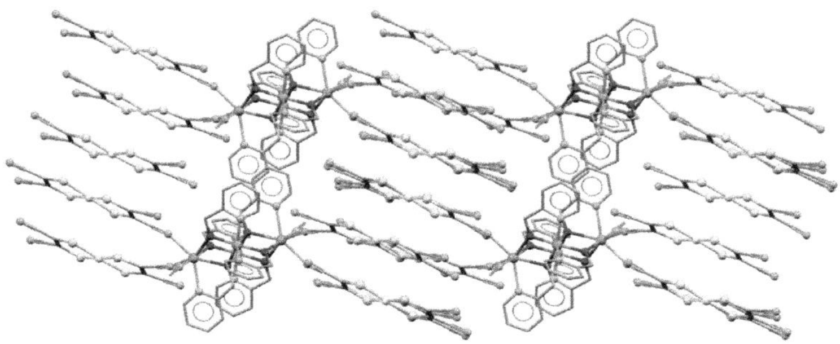

Figure 4.9 View of the crystal structure of the $[Mn^{II}_2Mn^{III}_2(hmp)_6(MeCN)_2\{Pt(mnt)_2\}_4]$ $[Pt(mnt)_2]_2$ hybrid material where the alternation of segregated stacks of conducting and magnetic building blocks is highlighted.

The $[Pt(mnt)_2]^-$ starting material is partially oxidized so that each unit bears a fractional negative charge *n* of 0.66, which is equally distributed between the coordinated and uncoordinated $[Pt(mnt)_2]^{n-}$ units.

From the analysis of the structural features the following consideration regarding the expected physical properties for this material can be highlighted:

- The partial oxidation of the $[Pt(mnt)_2]^{n-}$ molecules leads to a mixed-valence system in which the charge is delocalized between units.
- The formation of segregated 1D stacks of the conducting building blocks, combined with the features in the point above, is expected to provide appreciable transport properties.
- The $[Mn_4]^{4+}$ cationic complex retains the overall molecular structure, with the only exception that peripheral coordinating water molecules are replaced by terminal nitrogen atoms of the mnt ligand. In this circumstance, the SMM properties of the cationic metal complex can be preserved.
- Interactions between magnetic and conducting properties may, in principle, be observed.

Investigation of the physical properties of the $[Mn^{II}_2Mn^{III}_2(hmp)_6(MeCN)_2\ \{Pt(mnt)_2\}_4][Pt(mnt)_2]_2$ hybrid system revealed a room-temperature electrical conductivity of 0.22 S cm^{-1}, which gradually decreases as the temperature is lowered, thus indicating semiconductivity (the activation energy is 136 meV). Unfortunately, as it can be evidenced by inspection of Fig. 4.10, below 110 K the material no longer shows an appreciable conductivity and becomes an insulator.

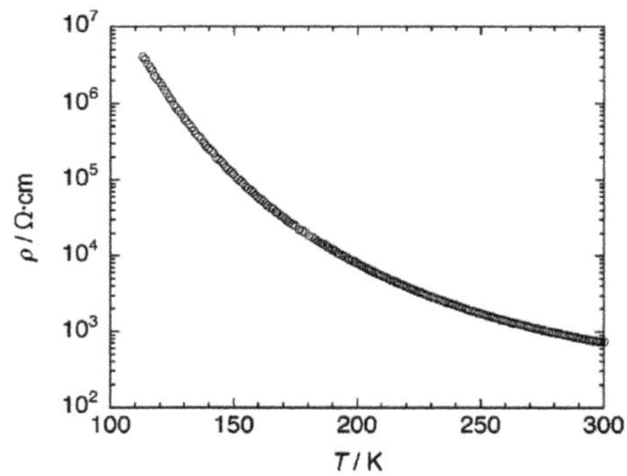

Figure 4.10 Temperature dependence of the resistivity of $Mn^{II}_2Mn^{III}_2(hmp)_6(MeCN)_2\{Pt(mnt)_2\}_4][Pt(mnt)_2]_2$. Adapted with permission from Hiraga et al. (2007). Copyright (2007) American Chemical Society.

The analysis of the static magnetic properties through susceptibility measurements reveals that the ferromagnetic interactions between Mn^{II} and Mn^{III} of the $[Mn_4]^{4+}$ tetranuclear metal complex are preserved in the hybrid material (Fig. 4.11), while an investigation of the magnetization dynamics as a function of temperature and applied static magnetic field shows slow relaxation of the magnetization below 3 K and an open hysteresis cycle (Fig. 4.11).

In conclusion, the investigation of the physical properties of the $[Mn^{II}_2Mn^{III}_2(hmp)_6(MeCN)_2\{Pt(mnt)_2\}_4][Pt(mnt)_2]_2$ hybrid system allows one to identify this system as an archetypal example of SMM/semiconducting multifunctional molecular material. Both hypothesized conducting and magnetic properties are observed, although, unfortunately, the semiconducting transport properties do not allow coexistence between conductivity and magnetism, since they are featured in quite different temperature ranges (conductivity > 110 K and slow relaxation of the magnetization < 3 K). Consequently, evident synergism of these two properties is still to be found. Nevertheless, such material is of utmost interest at a fundamental level as it demonstrates the potential preparation of highly conductive hybrid materials showing slow magnetic relaxation. To date, only a few examples of such materials have been reported, and this example really brings to attention that a judicious selection of building blocks can lead to the preparation of new functional materials where unexpected physical properties can be observed.

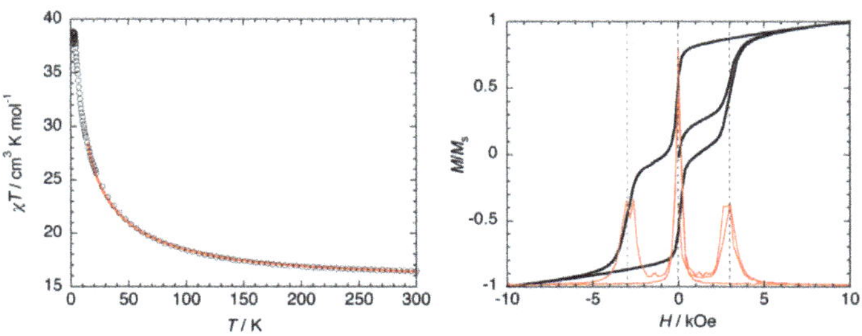

Figure 4.11 Temperature dependence of the molar magnetic susceptibility times the temperature (left) and field dependence of the magnetization at $T = 0.470$ K (right) for $[Mn^{II}_2Mn^{III}_2(hmp)_6(MeCN)_2\{Pt(mnt)_2\}_4][Pt(mnt)_2]_2$. The red lines represent the fit of the static magnetic susceptibility data (graph on the left) and the derivative of the magnetization curve (graph on the right). The cusps on the latter indicate the magnetic field values at which fast relaxation of magnetization occurs through quantum tunneling. Adapted with permission from Hiraga et al. (2007). Copyright (2007) American Chemical Society.

4.5 Synergism between Electrical Conductivity and Spin-Crossover Behavior

In the previous sections, we have described the design principles and the resulting physical properties of multifunctional molecular materials exhibiting coexistence or synergism between conductivity and magnetism. In particular, the discovery of field-induced superconducting transition in λ-$[BETS]_2FeCl_4$ provides an impressive example of the synergistic action achievable between conducting electrons and localized magnetic moments in hybrid systems. We have also explained that the conducting properties of molecular materials are very sensitive to small structural modifications, implying that they can be controlled by introducing functional molecules whose structure can be modified by applying an external stimulus. With these ideas in mind, Sato and coworkers (2006) designed a multifunctional molecular material in which the electrical conductivity, provided by a well-known conductivity carrier, the monoanionic dithiolene complex $[Ni(dmit)_2]^-$, is modulated by the structural contraction/expansion that accompanies the low-spin (*ls*)/high-spin (*hs*) transition of the paramagnetic $[Fe^{III}(qsal)_2]^+$ spin-crossover complex (Takahashi et al., 2006). The combination of these two building blocks provides the hybrid material reported in Fig. 4.12.

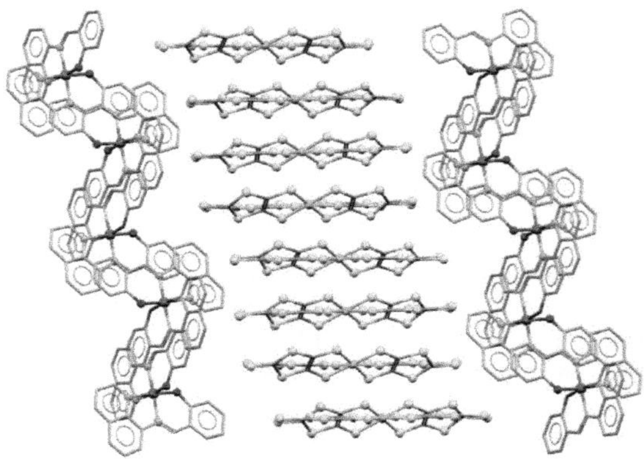

Figure 4.12 View of the crystal structure of the $[Fe(qsal)_2][Ni(dmit)_2]_3{\cdot}CH_3CN{\cdot}H_2O$ hybrid system. Crystallization molecules are omitted for graphical clarity.

The $[Fe(qsal)_2][Ni(dmit)_2]_3{\cdot}CH_3CN{\cdot}H_2O$ compound shown in Fig. 4.12 consists of alternate layers of $[Ni(dmit)_2]^-$ radical anions, which are responsible for the conducting properties, and layers of the $[Fe^{III}(qsal)_2]^+$ cations displaying a spin-crossover behavior. As a result of the combined properties of its constituting building blocks, such a material displays the following physical features:

- A gradual spin-crossover transition as a function of the temperature; it goes from an *ls* state below 70 K to an *hs* state above 250 K.
- A relatively high room-temperature conductivity of ca. 2.0 S cm^{-1}.

- Thermal dependence of the electrical resistivity, which increases on lowering the temperature with the typical trend of a semiconductor.
- A change in the slope of the resistivity curve as a function of the temperature in correspondence to the spin-transition temperature (Fig. 4.13).

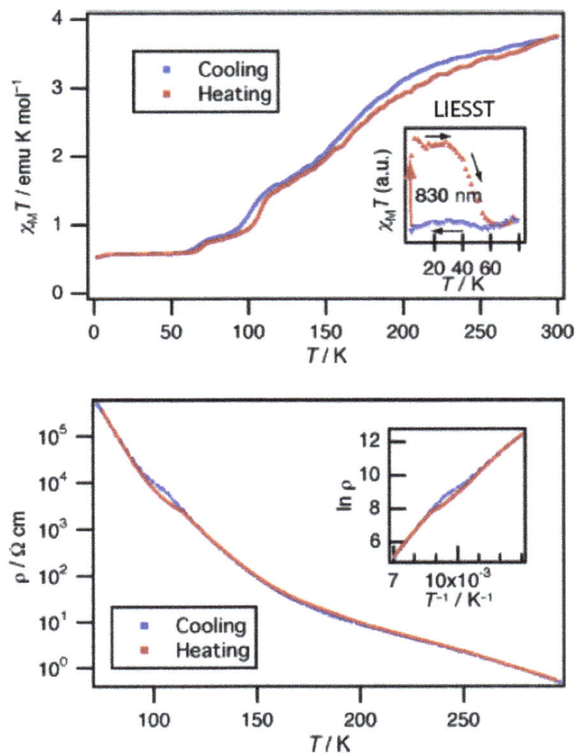

Figure 4.13 Temperature dependence of the molar magnetic susceptibility times the temperature (above) and the electrical resistivity (below) for the $[Fe(qsal)_2][Ni(dmit)_2]_3{\cdot}CH_3CN{\cdot}H_2O$ hybrid system. The gradual spin-crossover transition centered at ca. 150 K and the spin transition induced by light irradiation at low temperature, due to the LIESST effect (inset), can be appreciated in the χT versus T plot above. The resistivity plot below shows a change in the slope at ca. 150 K, indicating that the spin-crossover complex plays a role in modulating the conducting properties. The inset shows the Arrhenius plot used to extrapolate the activation energy (E_a = 0.17 eV for T < 150 K and 0.25 eV for T > 150 K). Adapted with permission from Takahashi et al. (2008). Copyright (2008) American Chemical Society.

Analogous but even more striking results of the effect of the structural contraction/expansion of a building block on the conductivity properties were reported by the same team of scientists with the compound $[Fe^{III}(qnal)_2][Pd(dmit)_2]_5{\cdot}CH_3COCH_3$, where the $[Fe^{III}(qnal)_2]^+$ cation acts again as a spin-crossover complex and the $[Pd(dmit)_2]$ building block is the conductivity carrier (Fig. 4.14) (Takahashi et al., 2008).

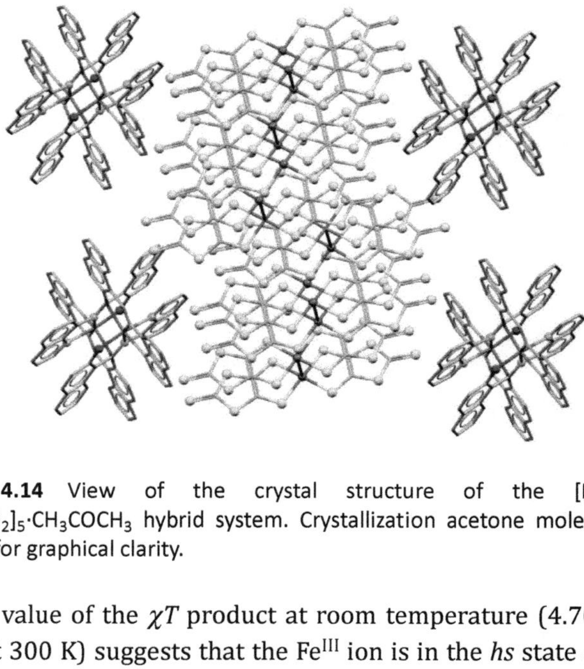

Figure 4.14 View of the crystal structure of the $[Fe(qnal)_2][Pd(dmit)_2]_5{\cdot}CH_3COCH_3$ hybrid system. Crystallization acetone molecules are omitted for graphical clarity.

The value of the χT product at room temperature (4.70 cm^3 K mol^{-1} at 300 K) suggests that the Fe^{III} ion is in the *hs* state and one out of five $[Pd(dmit)_2]$ molecules has one unpaired electron. On lowering the temperature, the χT value gradually starts decreasing at around 220 K and a plateau appears below 150 K, indicating a partial conversion (93%) of the Fe^{III} ions to the *ls* state, whose contribution is summed up to that of the unpaired electrons of the conducting layer. The temperature dependence of the electrical resistivity is that of a semiconductor but displays an anomalous behavior occurring in the same temperature range where the spin transition takes place (Fig. 4.15).

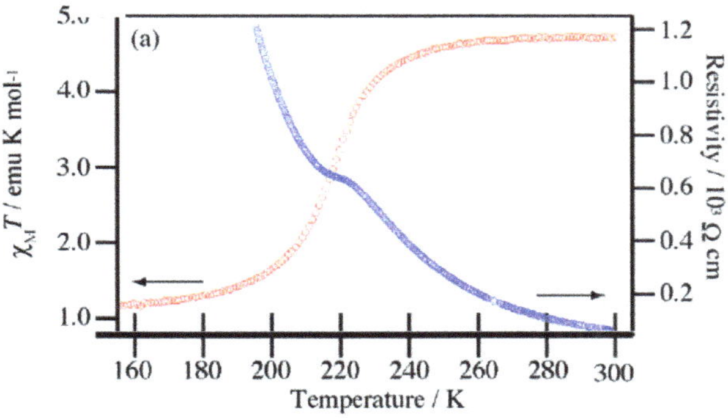

Figure 4.15 Temperature dependence of the molar magnetic susceptibility times the temperature (red circles) and the electrical resistivity (blue circles) for the $[Fe(qnal)_2][Pd(dmit)_2]_5{\cdot}CH_3COCH_3$ hybrid system. The anomalous behavior of the electrical conductivity in correspondence with the spin-crossover transition (ca. 220 K) is evident. Adapted with permission from Takahashi et al. (2008). Copyright (2008) American Chemical Society.

In conclusion, both $[Fe(qsal)_2][Ni(dmit)_2]_3{\cdot}CH_3CN{\cdot}H_2O$ and the $[Fe(qnal)_2][Pd(dmit)_2]_5{\cdot}CH_3COCH_3$ systems are multifunctional molecular materials in which a lattice deformation induced by a cocrystallized building block affects the physical properties of the adjacent component. In particular, the magnetic spin transition of the spin-crossover complex modifies the overall supramolecular interactions of the crystal structure and has a crucial effect on the electrical conductivity of the material. Remarkably, these interactions can be controlled by acting on an external stimulus, that is, the temperature, thanks to the relevant thermal response of the magnetic building block.

4.6 A Molecular Ferromagnet with an Increased Coercive Field

The development of permanent magnets for practical applications, for example, as components of electric motors and generators, is oriented toward the maximization of the product between the saturation magnetization and the coercive field. These parameters determine the maximum energy density that a magnet can store. In classical inorganic materials such as metallic alloys, optimal performances have been achieved by combining the high saturation magnetizations of 3d ferromagnetic metals (Fe or Co) with 4f ions (Nd or Sm), which contribute to the enhancement of the magnetocrystalline anisotropy of the material and to the resulting magnetic coercivity. Inspired by these materials, Luis and coworkers used a molecular approach to combine in the same material high saturation magnetization and coercive fields (Clemente-León et al., 2014). To achieve this goal a strategy based on the combination of charge-compensating complementary building blocks has been undertaken, similarly to what is explained above for the combination of conducting and magnetic properties. In this framework, two building blocks have been specifically selected: (i) the polymeric oxalate-based anionic network $[M^{II}M^{III}(C_2O_4)_3]^-$ (cf. Section 3.1.2), which exhibits ferromagnetic properties with a high saturation magnetization if proper coordinated metal ions are selected, and (ii) a dimeric cationic SMM of formula $[Mn^{III}(salen)(H_2O)]_2^{2+}$ (salen^{2-} = *N,N'*-ethylenebis-(salicylideneiminate)), which shows large uniaxial magnetic anisotropy accompanied by a slow relaxation of the magnetization in the presence of bulky counterions (cf. Section 3.1.7).

The cocrystallization of these two building blocks leads to the $[Mn^{III}(salen)(H_2O)]_2[Mn^{II}Cr^{III}(C_2O_4)_3]_2\cdot CH_3OH\cdot CH_3CN$ hybrid material, whose portion of the crystal structure is reported in Fig. 4.16.

The structure is formed by an anionic 3D polymeric oxalate-bridged bimetallic network whose cavities are occupied by cationic $[Mn^{III}(salen)(H_2O)]_2^{2+}$ complexes and by disordered methanol and acetonitrile solvent molecules.

As far as the magnetic properties of such material are concerned, we should in principle expect that the total magnetic response will be given by the sum of the contribution of the diverse magnetic ions composing the crystal structure Mn^{II} (**S** = 5/2) and Cr^{III} (**S** = 3/2) in the anionic network and two coupled Mn^{III} (**S** = 4) of the cationic complex.

However, to achieve a reliable picture of the resulting properties of an assembled multifunctional material, the independent magnetic response of each sublattice should be studied in advance. This issue was elegantly addressed for the $[Mn^{III}(salen)(H_2O)]_2[Mn^{II}Cr^{III}(C_2O_4)_3]_2\cdot CH_3OH\cdot CH_3CN$ compound by investigating the magnetic properties of two analogous materials where (i) the Mn^{III}-based cation is inserted into a paramagnetic network that cannot undergo long-range magnetic ordering and (ii) a diamagnetic cationic complex is introduced into the $Mn^{II}Cr^{III}$ network. Following this approach, Luis, Clemente-León, and coworkers (2014) were able to disclose the physical properties of the assembled hybrid system on the basis of those observed for the two separated building blocks inserted into the same structural motif.

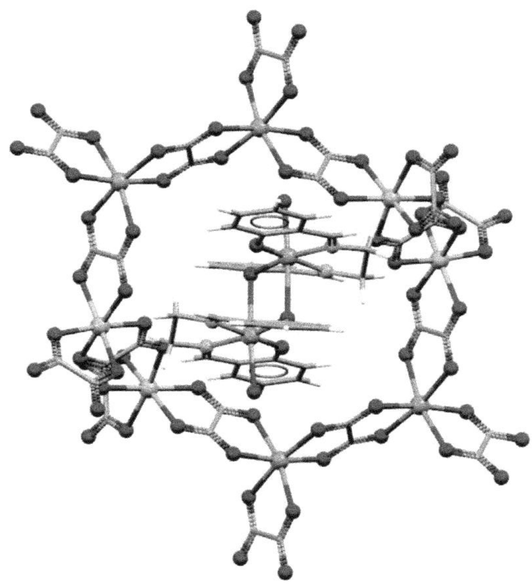

Figure 4.16 View of a portion of the crystal structure of $[Mn^{III}(salen)(H_2O)]_2[Mn^{II}Cr^{III}(C_2O_4)_3]_2{\cdot}CH_3OH{\cdot}CH_3CN$ where only one dimeric cation is depicted in one cavity of the extended molecular network. Solvent molecules are omitted for graphical clarity.

Without going into details, we can simply summarize that the Mn^{III}-based cationic complex behaves as an SMM, although with a low blocking temperature and quite short relaxation times, while the $Mn^{II}Cr^{III}$ network shows the onset of ferromagnetism at temperatures below the ordering temperature of 5 K.

The hybrid system $[Mn^{III}(salen)(H_2O)]_2[Mn^{II}Cr^{III}(C_2O_4)_3]_2{\cdot}CH_3OH{\cdot}CH_3CN$ reveals a magnetic behavior that cannot be described as the simple sum of two separated contributions, thus indicating interplay between the cationic and anionic networks. This is particularly evidenced by the magnetization curves at $T = 0.43$ K, a temperature at which both ferromagnetic ordering and slow relaxation of the magnetization occur (Fig. 4.17).

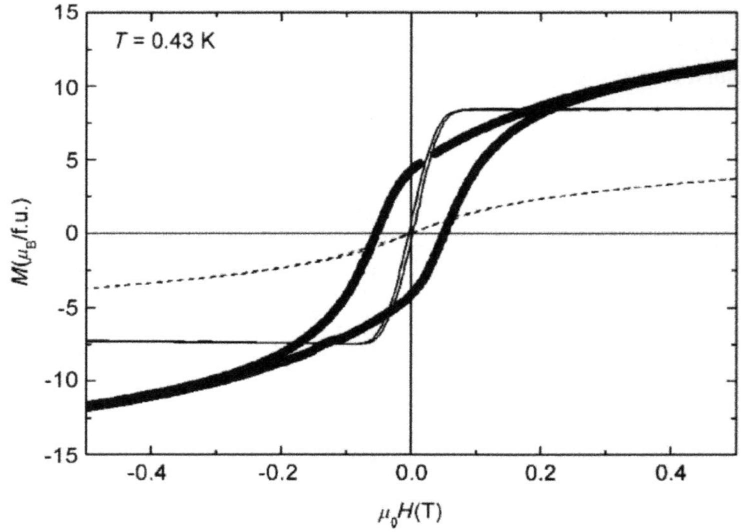

Figure 4.17 Magnetization hysteresis loops measured at $T = 0.43$ K for the multifunctional hybrid material (solid dots), the monofunctional SMM material (dashed line), and the monofunctional ferromagnetic material (solid line) discussed in the text. Adapted with permission from Clemente-León et al. (2014). Copyright 2014 Wiley-VCH.

Indeed, the multifunctional hybrid material shows a large magnetization hysteresis below 1 K, which strongly contrasts with the nearly reversible behavior of both the ferromagnet and the SMM (Fig. 4.17). By investigating the magnetization dynamics, a slowdown of magnetization relaxation for both the SMM and the ferromagnetic network is also evidenced, whereas the ferromagnetic ordering temperature seems not strongly influenced by the presence of the SMM cation. The overall magnetic behavior was rationalized by assuming that an antiferromagnetic coupling mediated by H-bonds occurs between the ferromagnetic network and the SMM cation. Such mutual magnetic interactions were taken as responsible for both the slowed magnetization of the two networks as well as the increased coercive field and remnant magnetization.

This example clearly demonstrates the reciprocal influence that two different building blocks may have once associated in a hybrid system, evidencing in particular the interplay between ferromagnetic ordering and strong magnetic anisotropy. The current challenge in this field is now to achieve such magnetic behaviors and resulting properties at technologically useful temperatures.

4.7 Optical Control in a Chiral Photomagnet

The control of the physical properties and functionalities of materials through optical stimulation is attractive from the point of view of technological applications. An example of such optical control has been presented and explained in Chapter 3 for the class of molecular materials called spin-crossover complexes. Briefly, irradiation with a light of a certain wavelength switches the magnetic state of such systems. Furthermore, the introduction of chirality in the structure of such materials can allow the observation of peculiar physical effects, such as magnetization-induced second harmonic generation (MSHG) (vide infra) and magnetochiral dichroism, due to interaction between the electromagnetic radiation and magnetic and electric fields. Ohkoshi and coworkers have prepared a peculiar molecular material constituted by a 3D chiral cyanobridged bimetallic assembly with spin-crossover sites that exhibits spin-crossover-induced second harmonic generation (SHG), light-reversible long-range magnetic ordering, and photoreversible switching of the MSHG effect (Ohkoshi et al., 2014). Such material is a coordination polymer formulated as (±)-$[Fe(4\text{-bromopyridine})_4]_2[Nb(CN)_8]\cdot 2H_2O$ and constituted by alternated $[Fe(4\text{-bromopyridine})_4]^{2+}$ cations and $[Nb(CN)_8]^{4-}$ anions self-assembled in a neutral 3D structure (Fig. 4.18).

An analysis of the crystal structure revealed that the self-assembling process leads to the formation of two enantiomers featuring a chiral helical structure.

Before entering into the description of the physical properties of this fascinating material, the reader should notice a fundamental structural difference with respect to the materials described above.

Here the coexistence of different functionalities is not obtained as a result of a judicious combination of different charge-compensating building blocks with specific physical properties in a hybrid crystal structure. On the contrary, the combination of functional building blocks templates the formation of a continuous 3D structure, whose chiral nature is fundamental in providing additional properties (vide infra).

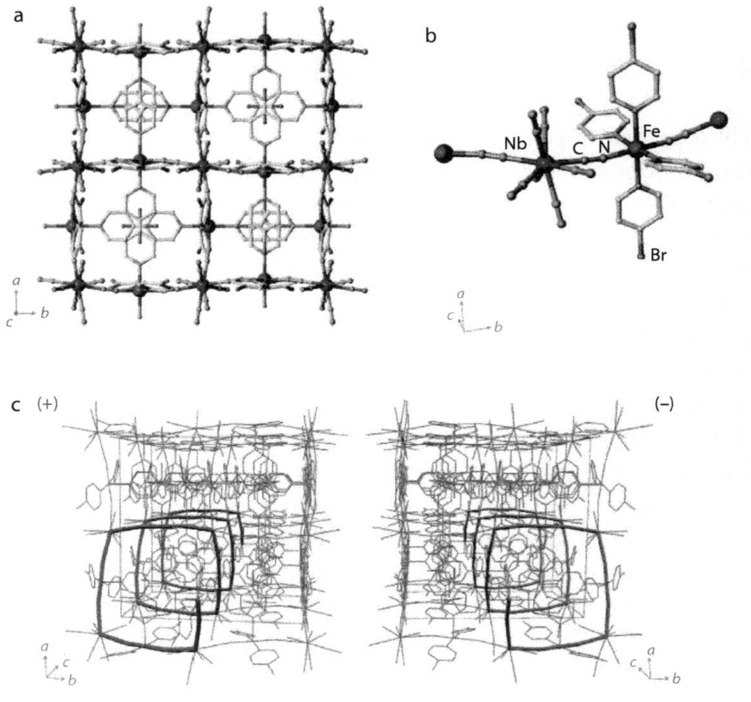

Figure 4.18 View of the crystal structure of (±)-[Fe(4-bromopyridine)$_4$]$_2$[Nb(CN)$_8$]·2H_2O along the *c* axis (a), coordination environments around the Fe^{II} and Nb^{IV} ions (b), and helical structures along the fourfold screw axis (*c* axis) of the (+)- and (–)-enantiomers (c). Reprinted by permission from Macmillan Publishers Ltd: [*Nature Photonics*] (Ohkoshi et al., 2014), copyright (20134).

This material shows a spin-crossover behavior related to the Fe^{II} sites, which adopt an *hs* configuration at room temperature down to ca. 120 K, when the *hs* to *ls* conversion begins and completes at a slightly lower temperature. The χT versus T curve reported in Fig. 4.19 shows a hysteretic behavior of the magnetic susceptibility during cooling/heating cycles and visually identifies the two temperature ranges where the Fe^{II} ions are entirely in the *hs* and *ls* states.

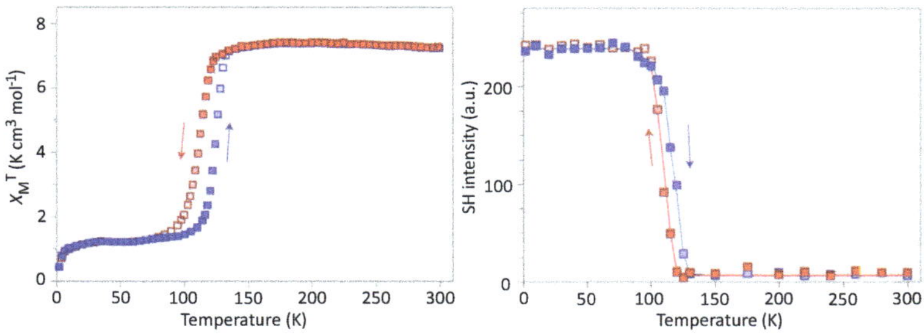

Figure 4.19 Temperature dependence of χT product (left) and SH intensity (right) for (±)-[Fe(4-bromopyridine)$_4$]$_2$[Nb(CN)$_8$]·2H_2O in both cooling (red symbols) and heating (blue symbols) modes (left). Reprinted by permission from Macmillan Publishers Ltd: [*Nature Photonics*] (Ohkoshi et al., 2014), copyright (2014).

The change in the spin state causes important modifications in the structural and electronic properties of the material, which in turn result in significant changes of its physical properties besides the magnetic behavior. One of these remarkable changes is the variation of the nonlinear optical (NLO) response, which shows SHG of light when irradiated by a laser beam of $\lambda = 1064$ nm. Interestingly, the SH intensity is much higher in the temperature region associated with the *ls* state than at higher temperatures corresponding to the *hs* state (Fig. 4.19). The dependence of the NLO response on the spin state of the material is confirmed by the thermal hysteresis observed during cooling/heating cycles that well agrees with that obtained through magnetic susceptibility measurements. The reasons for this behavior are related to the change in space group, accompanied by a variation of the polarizability tensor (cf. Section 2.3.8), as well as to the change of the absorption properties when passing from the *ls* to the *hs* state. Therefore, a remarkable switch of the optical properties (both linear and nonlinear) of the material can be obtained by means of an external stimulus, such as the temperature. However, temperature is not the only parameter that can be acted upon to attain a significant variation of the physical properties of this compound. Light can be a convenient tool to achieve control of the magnetic properties. In fact, when irradiating the material at low temperatures with a blue light beam ($h\nu_1$), exciting the lowest energy $^1T_2 \leftarrow {}^1A_1$ transition of *ls* Fe^{II}, a singlet-quintet transition occurs as a consequence of the successive population of the lower-lying $^5T_2$ metastable *hs* state (cf. Fig. 2.55). This light-induced excited-state spin-trapping (LIESST) effect (cf. Section 3.1.5) is accompanied by a photoinduced magnetization upon application of an external magnetic field as a result of the antiferromagnetic interaction between the photo-generated *hs* Fe^{II} centers with Nd^{IV} ions through the cyano bridging ligands. Such noncompensated antiferromagnetism (*hs* Fe^{II} has a ground spin state $\mathbf{S}$ = 2, while for Nd^{IV} $\mathbf{S}$ = 1/2) results in a spontaneous and photoinduced ferrimagnetic ordering. As the material is further irradiated in the absorption region corresponding to the $^5E \leftarrow {}^5T_2$ transition ($h\nu_2$), the *hs* Fe^{II} state is depopulated in favor of the *ls* ground state. This results in a lower number of paramagnetic Fe^{II} ions that can interact with the Nd^{IV} ions, with a concomitant reduction of the magnetization saturation (Fig. 4.20).

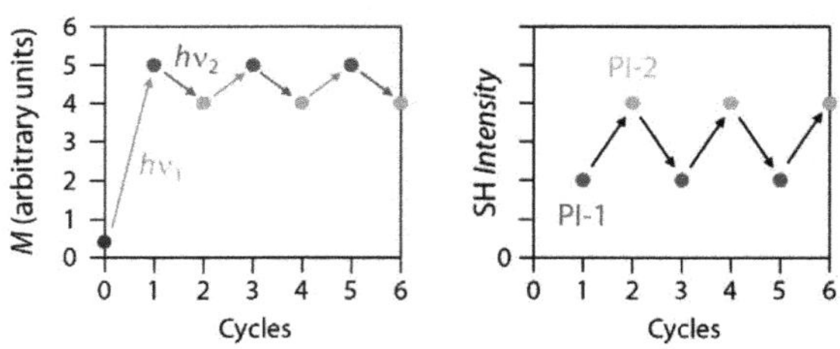

Figure 4.20 Variation of the magnetization saturation (left) and the SH intensity (right) after alternate irradiation with $h\nu_1$ and $h\nu_2$ light.

From these observations, it can be concluded that the material can adopt two photoinduced phases, PI-1 and PI-2, after successive irradiation of the *ls* Fe^{II} metal centers and then of the as-generated *hs* Fe^{II} ions, with light sources of appropriate frequency ($h\nu_1$ and $h\nu_2$, respectively). Crystallographic studies revealed that a change in the structural phase between the first photoinduced (PI-1) and the second photoinduced (PI-2) phases occurs, as expected in accordance with the observed spin-crossover behavior. Even more interestingly, alternate irradiation with $h\nu_1$ and $h\nu_2$ light causes reversible photoswitching between PI–1 and PI–2 and both magnetic and NLO responses of the material vary accordingly (Fig. 4.20) (Ohkoshi et al., 2014).

Moreover, the rotation angle θ of the polarized SHG signal with respect to the plane of the incident laser light displays a crucial dependence on the phase of the material. In fact, the angle of maximum SH intensity, found at $\theta = 90°$ before irradiation, switches at ca. 0° for the PI-1 phase and then goes back to values close to 90° when the material adopts the PI-2 phase. If an external magnetic field $\mathbf{H}_0$ is applied after illumination, a simultaneous switching between paramagnetism and ferrimagnetism can be achieved.[1] These phenomena are schematically depicted in Fig. 4.21.

Thus, through LIESST and reverse-LIESST effects, light irradiation controls the magnetic ordering of the material as well as the intensity and polarization plane of the output SHG light. This latter property, which can be fruitfully employed for magneto-optic applications, is achieved by the introduction of chirality in the structure and thanks to the light-induced magnetization.

This compound represents a remarkable example of a multifunctional molecular material in which different chemical units, as carriers of specific properties, are assembled in a chiral 3D structure. As a result of simultaneous and cooperative processes, this material exhibits three main peculiar physical phenomena: spin-crossover-induced SHG variation, light-controlled and reversible long-range magnetic ordering, and photoswitching of MSHG with a change of the polarization plane of the output SH light. This example is of particular interest since it clearly demonstrates that optical control of the magnetic properties can be achieved.

4.8 Multiemissive Molecular Materials

Systems able to generate emission at multiple wavelengths are at the basis of several applications in optical device technologies such as lighting and advanced telecommunication systems. Molecular

[1]The rotation angle θ can assume positive or negative values according to the direction of the applied external magnetic field $\mathbf{H}_0$.

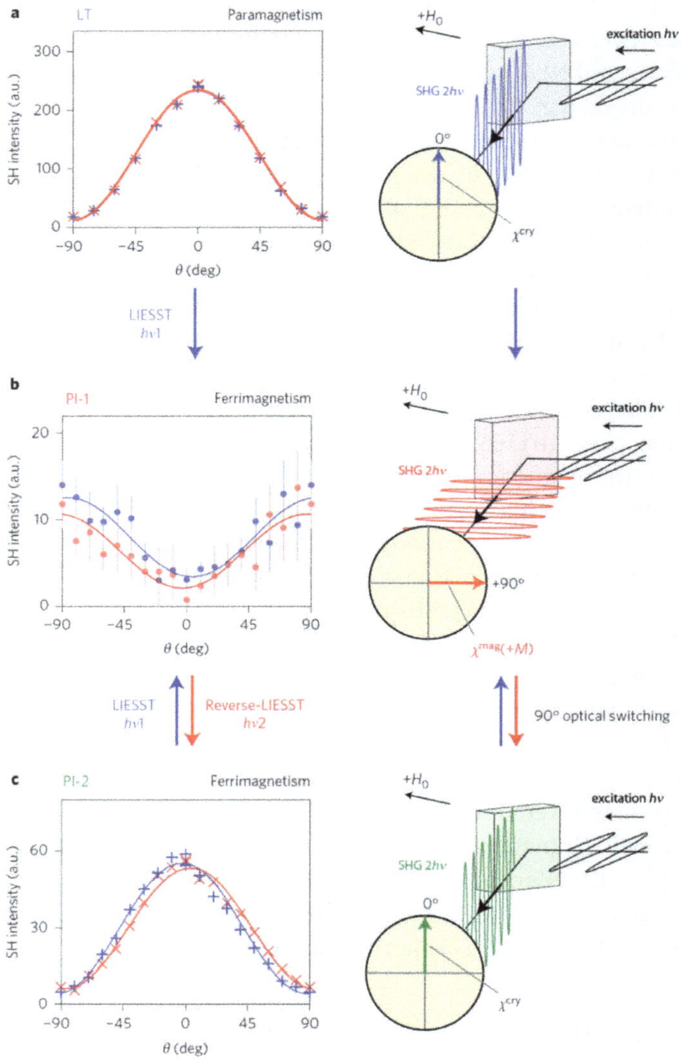

Figure 4.21 Optical switching of the SH intensity and angular dependence (θ) for (+)- or (−)-[Fe(4-bromopyridine)$_4$]$_2$[Nb(CN)$_8$]·2H_2O in the paramagnetic *ls* state (a), the first photoinduced ferrimagnetic phase PI-1 (b), and the second photoinduced ferrimagnetic phase PI-2 (c), with magnetic field parallel (red symbols) or antiparallel (blue symbols) aligned with respect to the incident light. Schematic representations of polarized incident and output lights planes are reported on the right side. Reprinted by permission from Macmillan Publishers Ltd: [*Nature Photonics*] (Ohkoshi et al., 2014), copyright (2014).

materials offer the opportunity to combine different emission colors, generated from specifically tailored chemical objects, into the same molecular architecture. This molecular approach toward multiemissive materials is greatly advantageous in terms of device fabrication as it allows the processing of highly homogeneous materials constituted of a single molecular entity, which often can be carried out through "soft routes," in contrast to the assembly of diverse and heterogeneous units.

Depending on the nature of the emitting units and the molecular structural arrangement, the final optical output of the assembled molecular material can result from the simple superposition or from the cooperation of the properties of the constituting building blocks. This additional opportunity leaves further room for the tailoring of specific molecular emitters for target use and applications, provided a careful chemical design is addressed.

4.8.1 Superimposed Optical Properties

One representative example of this class of multipurpose luminescent molecular materials is represented by the polynuclear complex reported in Fig. 4.22 and developed by De Cola and coworkers (Coppo et al., 2005) through an elegant strategy. This molecular architecture is constituted by two luminescent building blocks: a blue emitter based on an Ir^{III} cyclometallated complex and a red emitter constituted by a Eu^{III}-coordinated Schiff base. These two units are joined together by means of covalent bonding through the peripheral groups of the ligands, giving rise to a single discrete molecular entity where the color combination of the individual emissions gives rise to almost-white light.

In this molecular system, the Ir^{III} molecular unit, having triplet energy states high enough in energy to be suitable to transfer energy to the Eu^{III} upper levels (cf. Section 3.3.4.1), serves as an antenna for the sensitization of lanthanide emission as well as an individual light emitter. The double function of the Ir^{III}-based moiety is in this case allowed owing to the fact that the efficiency of the energy transfer to Eu^{III} is lower than 50%. Therefore, the blue emission from the Ir^{III} complex does not get completely quenched, and at the same time, the Eu^{III} ion can be adequately sensitized to be able to display red emission. As a consequence, a suitable balance between red and blue luminescence intensity is achieved, providing white light as the resulting emission color.

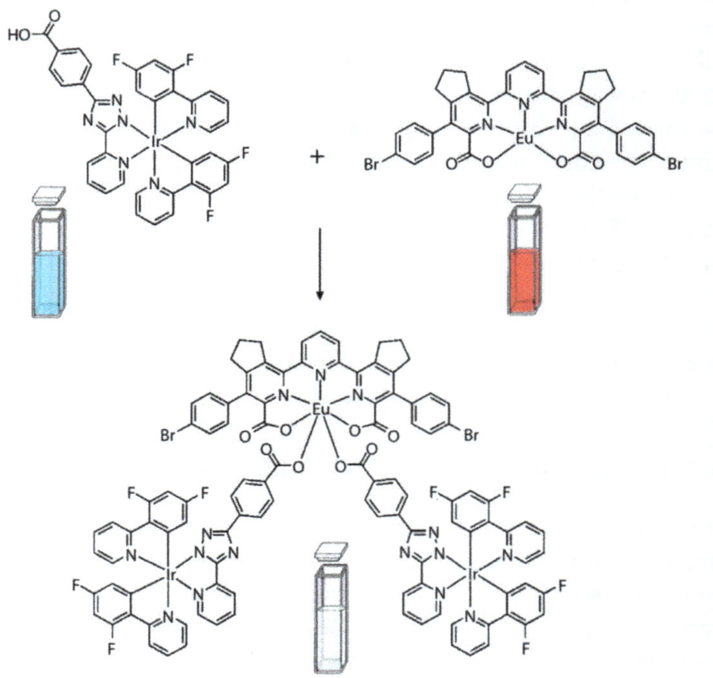

Figure 4.22 Molecular building blocks and assembled polynuclear structure of an Ir^{III}- and Eu^{III}-based molecular multiluminescent compound which gives rise to white light emission. The color of the solution in the cuvettes refers to the emitted light.

Another remarkable benefit of this excitation process is that simultaneous emission from the two individual molecular units can be obtained after single-wavelength irradiation. This point is a considerable advantage for applications in photoluminescent devices as it conveniently allows the one-step conversion of light from a single excitation source into a multicolored emission. The photoluminescence spectrum of this Ir^{III}/Eu^{III} assembly, obtained after irradiation at 400 nm, consists principally of a broad band related to the ^{3}MLCT (metal–ligand charge transfer) state of the Ir^{III} complex and the narrow emission lines originated from intra-atomic transitions of Eu^{III}, as reported in Fig. 4.23.

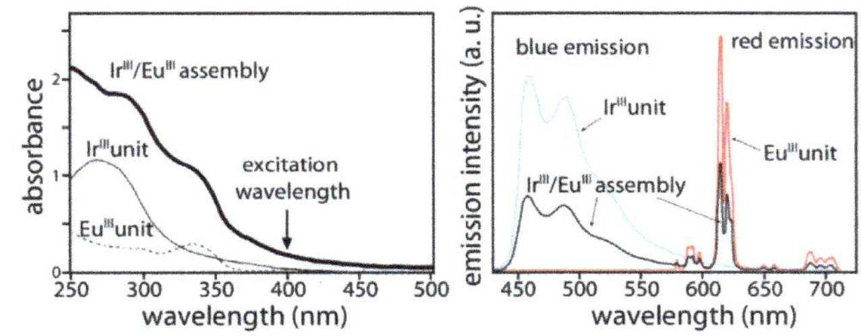

Figure 4.23 Absorption (left) and emission (right) spectra of the polynuclear Ir^{III}/Eu^{III} assembly depicted in Fig. 4.22. Redrawn with permission from Coppo et al. (2005). Copyright 2005 Wiley-VCH.

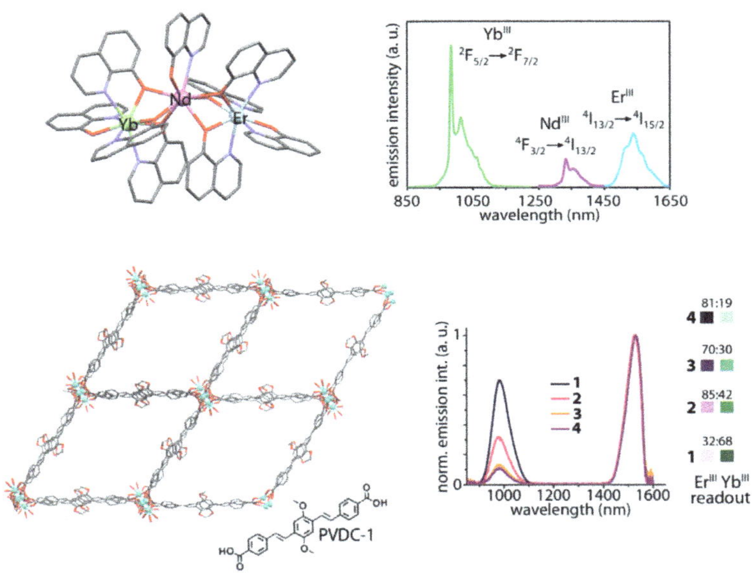

Figure 4.24 (Top) Structure and emission properties of two multiemissive molecular materials based on luminescent lanthanide ions: trinuclear discrete polynuclear structure of NdErYbQ$_9$ (Q = 8-quinolinolate) and emission spectrum upon ligand excitation at 400 nm. Redrawn from Artizzu et al. (2015). Copyright 2015 Wiley-VCH. (Bottom) Mixed Er-Yb coordination polymer with the PVDC-1 ligand and emission output for different percentages of Er^{III} and Yb^{III} normalized on Yb^{III} emission. Hydrogen atoms are omitted for clarity. Adapted from White et al. (2009). Copyright (2009) American Chemical Society.

The drawback of the discussed example arises from the poor color purity associated with the very broad emission related to the Ir^{III}-based complex. On the other hand, lanthanide ions, whose luminescence properties originate from intrashell f–f transitions, bring the advantage of displaying almost monochromatic emission that is solely correlated to the nature of the metal. Therefore, from a certain point of view, lanthanide ions can be seen as "atomic building blocks" for the achievement of multiemissive molecular materials. This allows for a wide variability of chemical architectures, ranging from 3D infinite networks to discrete polymetallic assemblies, as the two examples reported in Fig. 4.24. In such heterolanthanide assemblies the emission output of the material, resulting from the superposition of the different luminescence properties of the embedded lanthanides, can be conveniently controlled simply by adjusting the ratio of the metal emitters. This is possible thanks to the very similar chemical behavior of the lanthanide ions with respect to each other that allows them to be often interchangeable within a certain molecular architecture.

4.8.2 Cooperative Optical Properties

When dealing with multiemissive compounds as the ones described in the previous paragraph, it is important to take into account that the resulting optical behavior of the obtained material may deviate from what is expected on the basis of a simple superposition of the properties of its constituting optically active units. For instance, a relevant interplay of the optical properties of lanthanide emitters can result for specific pairs of these ions having energy levels close enough to allow for intermetallic communication, such as the Er^{III}/Yb^{III}, Tb^{III}/Eu^{III}, and Nd^{III}/Yb^{III} pairs. In this circumstance, the overall luminescence of the material will be crucially influenced by these interactions, new properties can be generated, and the resulting observed behavior can be difficult to predict and interpret. Suitable pairs of lanthanide ions can, in fact, efficiently exchange energy by means of Förster's mechanism (Förster's resonance energy transfer, FRET) between energy-alike levels. This can allow a specific lanthanide ion to act as a very efficient sensitizer toward another emitting ion, a phenomenon that can be particularly useful to enhance light harvesting for weakly absorbing metal emitters, as already described in Section 3.3.4.3. The other side of the coin is, however, that the emission from the sensitizing ion will get quenched, so the multicolor aspect of emitted light is significantly reduced depending on the efficacy of this process.

Additionally, other more sophisticated phenomena, such as *up-conversion*, where two photons absorbed by two ions combine to give rise to light of frequency equal to the sum of the two originating photons (up-converted light), may also occur. These phenomena are schematically represented in Fig. 4.25 for the pair Yb^{III}/Er^{III}.

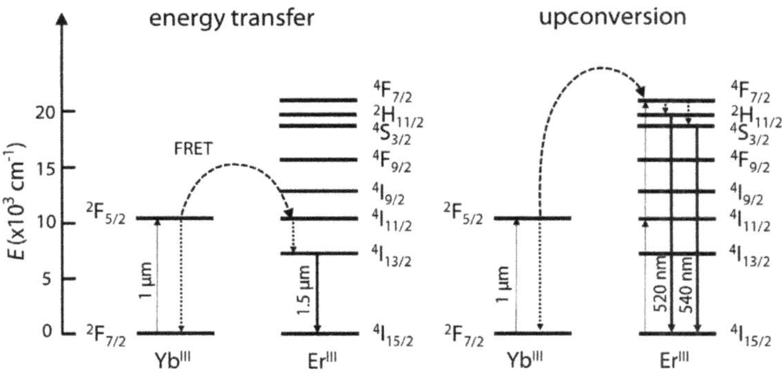

Figure 4.25 Simplified schemes for the metal-to-metal energy transfer and up-conversion mechanisms for the Yb^{III}/Er^{III} pair. In the first case NIR emission at 1.5 μm from Er^{III} is obtained, whereas in the second instance, green light emission is observed. Thin arrows represent absorption, and thick arrows indicate emission. Dashed arrows represent nonradiative processes.

Up-conversion arises from a two-photon absorption (TPA) process and can be considered a nonlinear optical phenomenon (cf. Section 2.3.8) in view of the quadratic dependency of the intensity of the emitted light on the intensity of the incident radiation. Actually, its occurrence is rather unlikely to be observed in coordination complexes of lanthanides since relevant quenching processes, usually taking place for these emitters when immersed in an organic environment as in coordination complexes, prevent the excited levels from living long enough to allow for sequential TPA. A remarkable exception to this empirical rule is represented by the polynuclear d-f Cr^{III}-Er^{III} system depicted in Fig. 4.26, which, in a way similar to the analogous example already reported in Section 3.3.4.3 and Fig. 3.94, is well suited to allow for efficient $Cr^{III} \rightarrow Ln^{III}$ intermetallic communication (Aboshyan-Sorgho et al., 2011). In this compound, green-light emission from Er^{III} $^4S_{3/2} \rightarrow {}^4I_{15/2}$ can be observed after irradiation with near-infrared (NIR) light at 750 nm that excites the ^2E level of Cr^{III}. Up-conversion arises from a sequential energy transfer from two long-lived[2] excited Cr^{III} ions to Er^{III}, leading to the population of the upper $^4S_{3/2}$ level of the latter, as shown in Fig. 4.26. This process is therefore related to the specific configuration of the molecular arrangement, having an Er^{III} ion "sandwiched" between two Cr^{III} ions that sit at a short distance (<9Å).

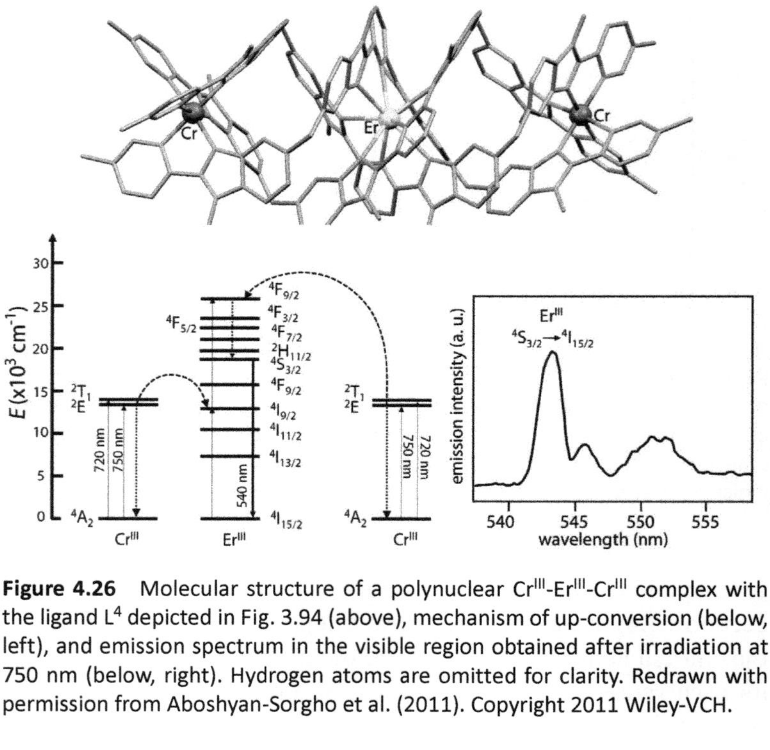

Figure 4.26 Molecular structure of a polynuclear Cr^{III}-Er^{III}-Cr^{III} complex with the ligand L^4 depicted in Fig. 3.94 (above), mechanism of up-conversion (below, left), and emission spectrum in the visible region obtained after irradiation at 750 nm (below, right). Hydrogen atoms are omitted for clarity. Redrawn with permission from Aboshyan-Sorgho et al. (2011). Copyright 2011 Wiley-VCH.

Although up-converted light can be only observed at very low temperatures that would lower vibrational quenching phenomena that can deactivate Er^{III} excitation prior to the absorption of a second photon, this example nonetheless demonstrates that light up-conversion can be even achieved in a molecular material.

As a final consideration, it is important to point out that the above-mentioned cooperative phenomena can only occur for specific

[2]Recall that the ^2E-excited level of Cr^{III} is long lived in view of the spin-forbidden nature of the $^2E \rightarrow {}^4A_2$ transition.

pairs of metal ions where the involved optically active units are close enough in space to exchange energy. Therefore, a careful chemical design is required if a simple superposition, and not a cooperation, of the optical properties of different emitters is sought.

4.9 Conducting Luminescent Materials

The ability of molecular materials to couple emissive properties to good charge transport attracts particular interest in view of their uses in optical devices such as organic light-emitting diodes (OLEDs, cf. Section 5.1). The most archetypal example in this class of compounds is undoubtedly the tris-8-quinolinolato Al^{III} complex (AlQ_3) already described in Section 3.3.2, which is the first example of an electrically excited green-emitting molecular material (Tang and Van Slyke, 1987).

In such compounds, an electric impulse applied to the sample in the solid state and embedded between two electrodes injects electrons into the lowest unoccupied molecular orbitals (LUMOs) and holes in the highest occupied molecular orbital (HOMOs) of molecules near the cathode and the anode, respectively. The resulting charged species are stable owing to the high aromaticity of the Q ligand, and charge mobility is then realized through electron- and hole-hopping mechanisms between adjacent molecules that share extensive intermolecular π–π interactions. As charges recombine at a specific site, an *exciton* is formed and the resulting electrically excited species is then able to emit light after electron decay to the ground state (electroluminescence). The crystal structure of AlQ_3 and a schematic representation of the charge mobility mechanism and exciton formation are reported in Fig. 4.27.

As readily understandable, also in the light of the discussion made in Section 3.2, the charge transport properties of discrete molecular entities such as AlQ_3 crucially depend on the supramolecular arrangement/crystal packing. This feature can be somehow chemically controlled by adjusting the synthetic parameters (e.g., solvent, temperature, and pressure), although for nonplanar molecules such as AlQ_3, additional factors such as coordination chemistry isomerism must be taken into consideration. The extent of intramolecular interactions, determined by the crystal phase and the type of isomers present, not only affects the effective conductivity of the material but also its emission properties (cf. Section 3.3.2.3). Extensive studies on AlQ_3 (Brinkmann et al., 2000) have in fact demonstrated that a significant red shift of the luminescence with respect to the free molecule in solution is observed in crystal phases where relevant π-stacking is established.

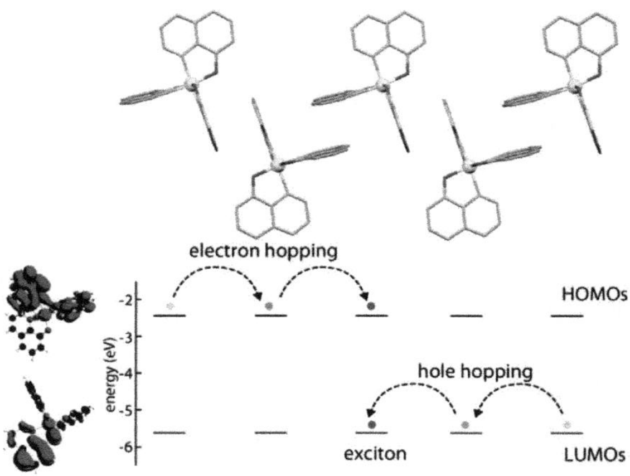

Figure 4.27 Crystal structure and packing of the β-phase of AlQ_3 characterized by extensive π-stacking between adjacent molecules (above); calculated HOMO-LUMO and charge hopping and recombination mechanism (below). Molecular orbitals reprinted from Kwiatkowski et al. (2008) with permission of The Royal Society of Chemistry.

Nonetheless, AlQ_3 retains both conductivity and electroluminescent properties when processed (usually by vacuum deposition or spin coating) as a thin film in a disordered, amorphous phase, as typically done for the fabrication of electro-optical devices. According to Kwiatkowski and coworkers (Kwiatkowski et al., 2008), in this case conductivity is allowed along preferential pathways involving a restricted number of molecules. The probability for the establishment of these pathways is higher for films of lower thickness so that the conductivity of the material is highly dependent on the depth of the deposited layer. Interestingly, the relative orientation of the molecules (both in crystalline and amorphous phases) affects more the hole-transport properties than the electron mobility, due to the more localized nature of the HOMOs with respect to the LUMOs (cf. Fig. 4.27). For this reason, AlQ_3 is usually mainly considered as an electron transport material in multilayered semiconducting electroluminescent diodes such as OLEDs.

As far as the emissive properties of an electrically excited molecular emitter are concerned, one fundamental aspect to consider is that the combination of uncorrelated spins associated with the electron-hole pair trapping (i.e., the exciton formation) may result in a singlet state exciton or a triplet state exciton, with total spin $\mathbf{S} = 0$ or $\mathbf{S} = 1$, respectively. As a consequence, either fluorescence or phosphorescence can be obtained. Statistically, the probability of the formation of a singlet excited state S* is threefold lower than the excited triplet T* (25% S*, 75% T*). Therefore phosphorescent emitters are in principle more efficiently excited upon electric impulse than fluorophores[3]. The downside is that the spin-forbidden phosphorescent emission may occur over a timescale much longer than the exciton formation, leading to saturation of the excited states. This existing upper limit in the emission lifetime values is useful for efficient electrophosphorescence claims for specific molecular designs.

[3] AlQ_3 is actually a fluorescent emitter (cf. Section 3.3.2.1), so a low emission quantum yield is obtained for electrical excitation as a consequence of the low percentage of generated excited singlets.

Chapter 5

Turning Molecular Materials into Devices

In this final chapter we aim to provide the reader a brief overview of how molecular materials can be fruitfully employed for the preparation of technological devices with advanced functional properties. As optical materials are less affected by the temperature limitations that magnetic and conducting materials are typically subjected to, optical devices currently represent the most developed technological application of molecular materials. Among such devices of paramount importance are organic light-emitting diodes (OLEDs), which are commonly employed to build flexible screens and for low-cost lightening, optical fibers and amplifiers for telecommunication technology, and dye-sensitized solar cells (DSSCs) for solar energy conversion. Besides these examples, for which commercial use is a reality, some of the classes of molecular materials presented in the previous chapters, although not yet suitable to be employed for practical technological applications, will nonetheless open revolutionary technological perspectives should their actual limitations be overcome. For this reason, one of the main aims of this chapter is also to highlight some relevant challenges that scientists are undertaking to fill the gap between fundamental science and technological applications.

Functional Molecular Materials: An Introductory Textbook
Matteo Atzori and Flavia Artizzu

ISBN 978-981-4774-76-5 (Hardcover), 978-1-351-23366-8 (eBook)
www.panstanford.com

5.1 Organic Light-Emitting Diodes

Organic light-emitting diodes (OLEDs) are optical devices extensively employed for electroluminescence (EL) applications, owing to their low fabrication and operational costs coupled with unbeaten performances. In the so-called low-molecular-weight-molecule OLEDs, direct conversion of electricity into light is obtained thanks to the excitation of a discrete molecular emitter, that is, a small organic dye or a metal complex, by an electrical pulse (cf. Section 4.9).[1]

In Fig. 5.1 a sketch of a typical multilayer OLED device is depicted. In the simplest configuration, this device consists of a series of three organic semiconducting layers embedded between two electrodes, which inject charge carriers into the diode when a small voltage is applied. The diode is deposited over a thin, flexible glass support, transparent to the light output.

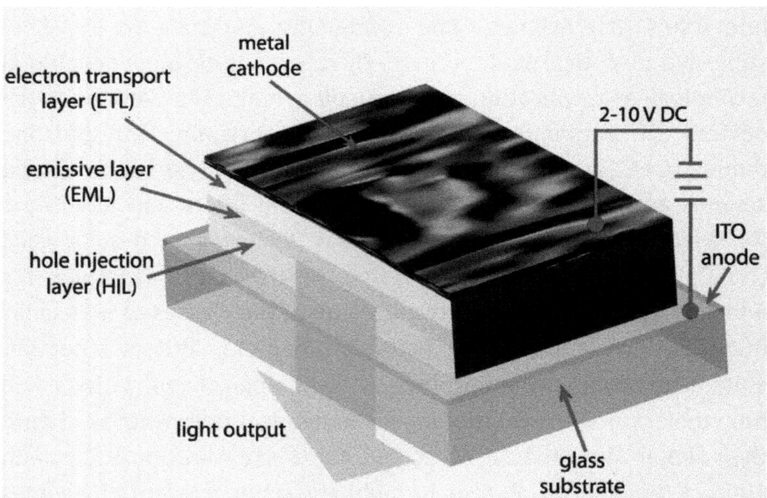

Figure 5.1 Drawing of a typical multilayer OLED device.

As can be seen from the schematic draw in Fig. 5.1, the emissive layer (EML), constituted of the molecular emitters, is embedded between a hole injection layer (HIL) and an electron transport layer

[1]Electroluminescent devices based on polymers, that is, polymer light-emitting diodes (PLEDs), are also widely investigated and employed but will not be discussed herein.

(ETL). The materials constituting the different layers are properly chosen to display suitable conducting properties. The HIL usually consists of aromatic amines of low ionization potential, whereas the ETL is generally made of highly aromatic molecules or polymers characterized by good electron-conduction ability. Not rarely, devices are fabricated in a two-layer structure where the EML serves also as an electron-transport material since electroluminescent molecules often display this property (cf. Section 4.9). Holes are injected into the device by the anode, typically made of indium tin oxide (ITO), a transparent material with a high work function ϕ_1, while electrons come from the cathode, usually consisting of a metal plate of a low work function ϕ_2 (Al or a Mg:Ag 10:1 alloy).[2]

The application of an electric potential at the electrodes (forward bias) induces injection of holes into the HIL and electron injection into the ETL. Under this potential difference, charges start to migrate across the layers toward the opposite electrode, with hole transport occurring through the highest occupied molecular orbitals (HOMOs) (the valence band, VB) and electron transport through the lowest unoccupied molecular orbitals (LUMOs) (the conduction band, CB) of the molecules constituting the semiconducting layers (cf. Section 2.2.5). Depending on the rate of the process and on the charge mobility, the recombination of holes and electrons at the EML generates an excited state (exciton, cf. Section 4.9), suitable to stimulate light emission by direct radiative relaxation or by the subsequent sensitization of a guest emitting molecule through energy transfer (vide infra).

In the "trapping" model, usually taken into consideration to account for this mechanism of electroexcitation, each molecular entity constituting the semiconducting layers represents a potential trap that can host the exciton. However, for efficient emission stimulation, the exciton should be confined solely to the EL. For this reason, the exciton migration across the diode should be blocked by selecting proper hole-transporting materials with a higher-energy CB than that of the EML. At the same time, the HOMOs of the molecular material constituting the HIL must lie just slightly above those of the EML to ensure easy hole injection into the EL.

[2]The work function associated to a solid surface represents the minimum energy (thermodynamic work) necessary to extract an electron to the "vacuum level" immediately outside the surface.

The EML should be thin enough to allow for easy exciton formation close to the interface with the two outer layers (heterojunctions). A scheme of the energy levels of a multilayer diode is reported in Fig. 5.2.

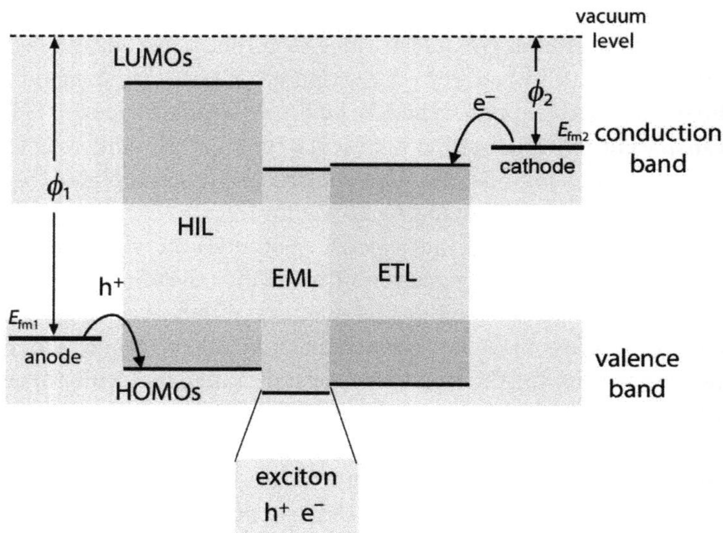

Figure 5.2 Energy level diagram of a multilayer diode.

In the light of this brief description of OLED devices, it is clear that the materials used for the fabrication of these devices must fulfill specific requirements to guarantee efficient luminance performances, low power consumption, and long operational lifetimes. Without going into details perhaps more pertaining to the engineering design of the multilayer diode, we can summarize the key features that a molecular material suitable to be used as the electroluminescent active center in such devices should display:

- High thermodynamic stability
- Reversible redox properties
- Good charge (electron) transport properties
- Suitability to be processed as a thin film
- High luminescence efficiency (fluorescence or phosphorescence)

As already mentioned in Sections 3.3.2 and 4.9, the first example of a luminescent metal complex used as an electroluminescent emitter in OLED devices is the tris-8-quinolinolato Al^{III} complex (AlQ_3), which displays efficient green emission upon electroexcitation. In this material, the role of the metal is crucial for enhancing some desirable characteristics for electroluminescent devices with respect to the performances of free organic luminophores. In fact, the metal center increases the thermal stability and photostability of the overall molecule as well as the rigidity of the system, leading to a more efficient radiative quantum yield (cf. Section 2.3.4).

In Section 4.9 it was briefly explained that in EL experiments the relative percentage of excited singlet (S*) states with respect to excited triplet (T*) states formed upon voltage application is 25% to 75%. Hence, phosphorescent emitters are, in principle, better suited for OLED devices, provided that the emission lifetime, which is related to a spin-forbidden transition, is significantly shorter than the exciton formation to avoid saturation. At the same time, the emission should be long-lived enough to ensure high luminance (ideally it should be between 2 and 50 μs). However, for optimal device efficiency, both singlet and triplet excitons should be effectively harvested and converted into light. A smart strategy to fulfill this purpose is to employ, as EML, host-guest systems composed of specifically designed singlet and triplet emitters. Most often, a hybrid material such as this is made up of a thin film of a fluorescent conductive molecule such as AlQ_3, doped with a suitable phosphorescent luminophore. One significant example of such devices is represented by the white-light emitting OLEDs (WOLEDs) developed by Anzenbacher and coworkers (2008) and Pérez-Bolívar et al. (2011), in which the color of the output radiation results from the combination of blue-green emission, from a complex derivative of AlQ_3 with a substituted 8-quinolinolate ligand, and orange-red phosphorescence of an Ir^{III} complex (the bis-2-phenyl-1-quinolinolate acetylacetonato iridium(III), $Ir(pq)_2(acac)$). In the typical configuration, schematically depicted in Fig. 5.3, the EML consists of two adjacent sublayers: one made of the fluorescent and electron-transporting pure Al^{III} complex and the other of the phosphorescent Ir^{III} complex in the pure phase or as dopant of the former material. While the Al^{III} derivative emits blue-green light after singlet exciton trapping, the phosphorescent emitter is able to harvest the highly diffusive triplet excitons thanks to Förster's energy transfer from excited Al^{III}-based molecules.

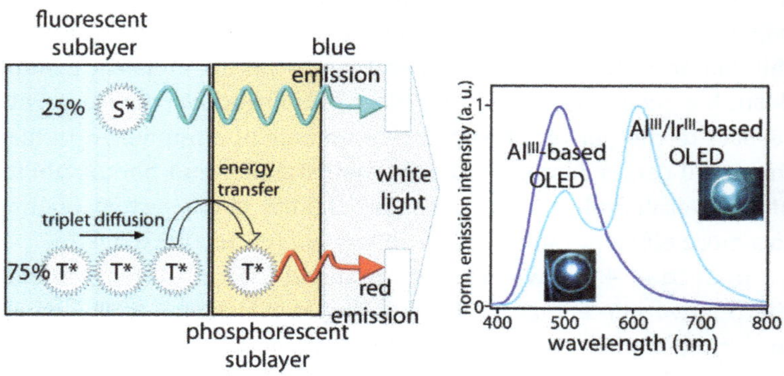

Figure 5.3 Scheme of the EML of a white-light emissive OLED (WOLED) device made of a fluorescent sublayer based on a Al^{III}-based blue emitter and a phosphorescent sublayer containing a red-emitting Ir^{III}-based complex (see text) (left). S* and T* represent singlet and triplet excitons, respectively. Emission spectrum of the WOLED device depicted on the left (light blue) compared with an undoped OLED with the same Al^{III}-based blue fluorophore (blue) (right). Pictures of the two devices are also included as insets. Redrawn with permission from Pérez-Bolívar et al. (2011). Copyright 2011 Wiley-VCH.

In the late 1990s, lanthanide complexes were also proposed as luminescent materials for EL devices and, following the model green-emitting device based on AlQ_3, the Er^{III} quinolinolato complex (Er_3Q_9, formerly wrongly formulated as ErQ_3) was used for the fabrication of the first lanthanide-based near-infrared (NIR)-emitting OLED (Gillin & Curry, 1999). Lanthanide complexes have the advantage to provide high color purity of the luminance output as a consequence of their characteristic sharp emission lines in comparison to the broad bands of common organic luminophores and d- or p-metal complexes. Lanthanide-centered luminescence can be stimulated by energy transfer from the ligand-excited triplet state, directly generated by charge recombination into the diode. Moreover, since the efficiency of intersystem crossing is typically almost unitary in these complexes, singlet excitons can be effectively converted into light, leading to overall device efficiencies in principle close to 100%. As an additional benefit, the use of lanthanide complexes as emitting centers in OLED devices allows one to very simply fine-tune the

EL colors by selecting the proper metal ion, without affecting the stability and physical properties of the emitting material.

OLED devices are currently finding many applications both as signal amplifiers in telecommunication technology and as autoluminescent components for ultrathin and flexible displays (flexible organic light-emitting diodes, FOLEDs). One of the main advantages of this technology, other than the high optical quality achievable, is undoubtedly the low operating cost, which requires a very low voltage with almost no heat dissipation.

5.2 Optical Fibers and Amplifiers for Telecommunication

Optical telecommunication systems are based on the ability of the light to transport information over considerable distances. As the propagation of light is much less subjected to attenuation than electrical signal transmission through wires or traditional silicon devices, technologies based on photonics represent the new frontier for highly performing global telecommunication.

The control and manipulation of the optical signal necessary for data storage and transmission are realized through special component devices such as beam splitters, amplifiers, and waveguides. All these components can be easily and conveniently integrated on planar substrates for the fabrication of photonic integrated circuits similar to the traditional electronic integrated circuits.

One of the most important components of photonic telecommunication technology are optical waveguides, processed as fibers, that have the function to contain and guide the transmission of light over a distance at higher data rates than wired technologies. Optical networks based on silica fibers[3] are nowadays widely used as waveguides for telecommunication purposes in view of the low fabrication costs, performances, and accessibility. NIR radiation (in the 1–1.6 μm range) is typically used to transport the signal since silica displays high transparency in this spectral region.

[3]Optical fibers can be made either of silica glass or of plastic materials (polymers), the latter having some advantages such as mechanical resistance and processing ease. However, glass fibers are preferentially used for long-haul telecommunications in view of their lower optical attenuation.

A typical optical fiber is a cylindrical dielectric waveguide that transmits light along its axis by *total internal reflection*. The fiber consists of a core of a high-refractive-index material surrounded by a cladding layer of a lower refractive index. The difference of the refractive indexes between the internal waveguide and the external coating ensures the confinement of the light beam to the core, and optical signal transmission occurs as a result of a constant bouncing of the light at the interface between the two layers (Fig. 5.4).

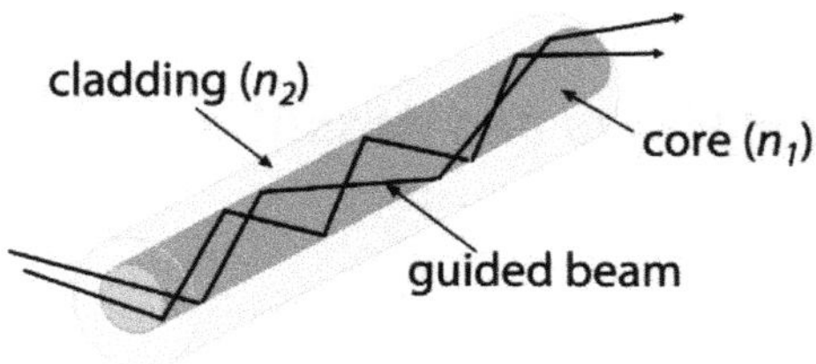

Figure 5.4 Sketch of a typical optical fiber. The core and the cladding materials have different refractive indexes n, where $n_1 > n_2$.

Optical fibers are usually coated with a tough resin *buffer* layer, which may be further surrounded by a *jacket* layer, typically plastic. These layers add strength to the fiber but do not contribute to its optical waveguiding properties. A light-absorbing ("dark") glass between the fibers prevents light that leaks out of one fiber from entering another (crosstalk) and reduces interferences and noise.

Although glass optical fibers are characterized by a very low optical attenuation compared to other optical waveguides, they nonetheless still suffer from optical losses arising from scattering or absorption phenomena mostly due to impurities (e.g., the presence of O–H groups) present in the materials constituting the fiber. To compensate for this drawback, signal amplification is necessary to ensure highly performing data transmission over long-haul fibers. This can be achieved by incorporating into the fibers optically active entities able to strengthen (amplify) the signal. These dopant amplifiers have the function of emitting light at a suitable wavelength when optically pumped by an external source to allow for optical gain. Some trivalent NIR-emitting lanthanide ions (Er^{III}, Yb^{III}, Nd^{III}) can be conveniently employed for this purpose. In particular, Er-doped materials (Er-doped fiber amplifiers, EDFAs) are nowadays the most diffused amplification systems used in the long-haul telecommunication technology operating in the 1.5 μm spectral window.

However, this technology still presents some shortcomings that result in the fabrication of long and expensive devices. The first difficulty arises from the low solubility of lanthanide inorganic salts and oxides in all the matrices optical fibers are usually made of. This hampers the optimal loading and homogenous dispersion of the optically active dopant into the fiber, leading to poor optical gain, often further attenuated by concentration quenching effects related to cluster formation. In addition, bare lanthanide ions have very low absorption and emission cross sections, which make optical pumping, required to obtain laser population inversion and stimulated emission, very inefficient, difficult to achieve, and costly (cf. Section 3.3.4).

These drawbacks can be overcome by resorting to complexes with suitable organic ligands that can play the double role of improving the solubility of the metal emitters into silica or polymeric matrixes and acting as efficient light harvesters ("antennae") enabling efficient indirect optical pumping (cf. Section 3.3.4). In this way, optical gain can be easily attained by using a readily available and rather inexpensive laser pump.

To be suitable for incorporation into inorganic and polymeric matrices, lanthanide complexes must be chemically stable and should not undergo photobleaching under operating conditions. Fortunately, the doping of these compounds into optical waveguides can be easily accessed by using "soft" chemistry routes, most often based on solution methods and mild conditions (low temperatures and pressures). Moreover, thanks to the high solubility of these complexes, which can be also adjustable by chemical modifications, high dopant concentrations and homogeneous dopant dispersion can be achieved. One of the most common methods for dispersing these metal complexes into a solid matrix, commonly made of silica, is the sol-gel method described in the box titled "Doped Silica Sol-Gel Glasses."

Doped Silica Sol-Gel Glasses

Silica glasses doped with optically active entities are convenient materials to be used for the fabrication of several optical devices, such as lasers and optical fibers, since they are highly transparent and homogeneous and feature good mechanical properties. From the point of view of optical properties, the silica glass matrix is an ideal host material since it hardly influences the spectral features of the guest entity (dopant) and displays negligible absorption coefficients at the excitation or emission wavelengths normally used in photonic devices. As additional benefit, the incorporation of luminescent complexes into a rigid host material, may increase their emission efficiency as the number of vibrational degrees of freedom is significantly lowered than in a liquid environment (solution) or even with respect to samples in the pure crystalline form.

Doped silica glasses can be prepared through the convenient *sol-gel method* in solution, which allows opportune and controllable loading and ensures optimal dispersion of the dopant. More importantly, the most valuable advantage of this method is that it can be carried out under mild conditions, especially in regard to temperature, so that thermal dissociation or degradation of the dopant complexes is avoided.

The sol-gel method consists of different sequential steps, namely the preparation of the sol, its gelation, and subsequent treatment of the gel to obtain the desired product. The process is usually carried out in an alcoholic or hydroalcoholic solution, starting from alkoxysilane precursors of general formula $Si(OR)_4$. Among them, the most commonly used are TEOS (tetraethylalkoxysilane) and TMOS (tetramethylalkoxysilane), but a variety of alkoxydic or alkyl chains can be introduced (organically modified silicon alkoxides, ORMOSILs) to tune the chemical/physical properties of the material. The reactions involved in this process are hydrolysis and condensation of the precursors to form a tridimensional silica network, as shown in the scheme below.

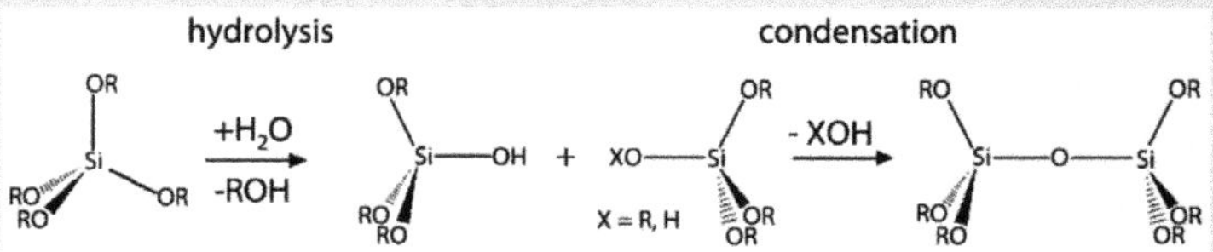

These nucleophilic substitution reactions involve, in the first step, the replacement of alkoxide groups, –OR, by hydroxyl groups, –OH, with concomitant alcohol release (*hydrolysis reaction*). Condensation reactions then follow, leading to the formation of Si–O–Si bridges with water release (*water condensation*) or alcohol release (*alcohol condensation*). Generally, hydrolysis can be catalyzed by acids whereas basic conditions speed up the condensation process. It must be remarked that the described reactions, once started, do not occur in sequence but involve simultaneous equilibria.

The schematic depiction of the doped sol-gel glass formation is reported in the following figure:

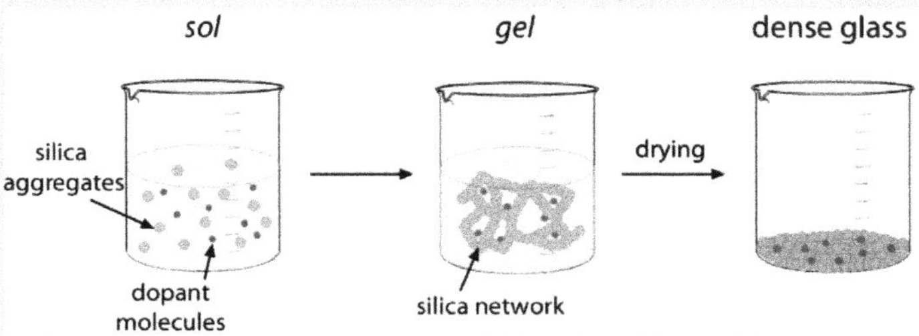

The colloidal solution obtained from the initial formation of the siloxane Si–O–Si bridges as a consequence of the condensation of the monomers is defined as a *sol*. As the reactions go on, a continuous silica-network begins to form until the *gelation point* is reached. The gelation point corresponds to the reaction time in which the sol-to-gel transition occurs. A *gel* is defined as a colloidal system of solid character in which the dispersed substance forms a continuous, coherent framework that is interpenetrated by a liquid system consisting of units smaller than colloidal entities (in the nanometer scale). If the dopant complex is dissolved in the starting sol solution, molecules of this material will be entrapped inside the pores created by the silica network. When the sol becomes a gel, the 3D silica network (*wet gel*), retains most of the by-products of the sol-gel process, namely H_2O and ROH, which have to be removed to obtain a dry gel. Solvent removal by slow evaporation at ambient pressure induces a collapse of the silica porous structure and leads to a dense ceramic material, giving rise to the desired doped sol-gel glass.

An example of a (thick) sol-gel glass doped with a generic green-emitting molecular material is shown below. It is possible to notice the homogeneity of the dopant dispersion by the uniform yellow color of the material under ambient lighting (left) and the waveguiding effect at the glass edge for the molecular luminescence under ultraviolet (UV) illumination (right).

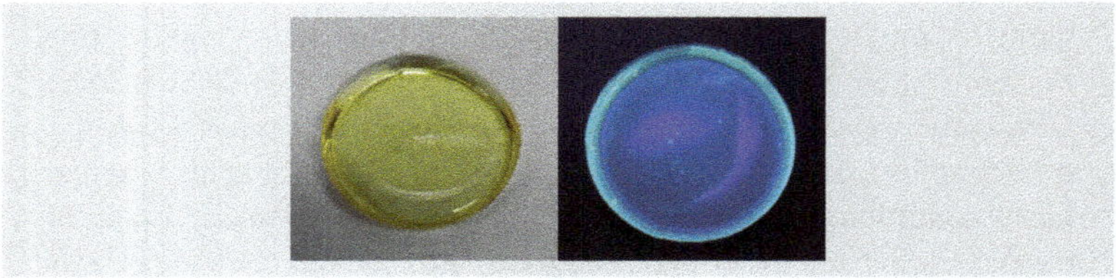

Although lanthanide complexes suffer from quenching phenomena related to the presence of CH or OH groups in the immediate surroundings of the metal emitter (cf. section 3.3.4.2), they offer remarkable benefits for applications as signal amplifiers in optical fibers, especially in regard to pumping efficiency. Sloof and coworkers (2002) have demonstrated that higher optical gain at 1.5 μm with lower pump power can be achieved in polymeric optical waveguides doped with Er^{III} polydentate hemispherand organic cage complexes when compared to fibers doped with Er^{III} inorganic colloids, despite the much lower luminescence lifetimes ($\tau < 1$ μs versus $\tau \approx 17$ ms, respectively) (Fig. 5.5).

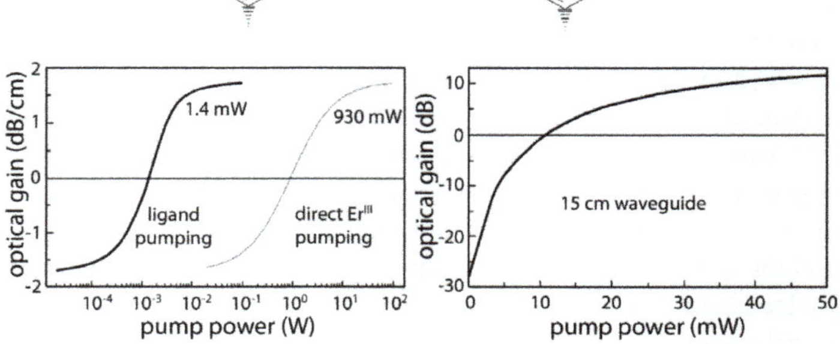

Figure 5.5 (Top) Structures of two Er^{III} polydentate hemispherand organic cage complexes used as optically active dopants in polymeric optical fiber amplifiers. (Below) Plots of the optical gain at 1.5 μm as a function of the pump laser power for polymeric optical waveguides doped with the Er^{III} complex (left) upon indirect ligand excitation (thick line) and direct metal ion pumping (thin line) and with Er^{III} inorganic silica colloids (right). Adapted from Sloof et al. (2002), with the permission of AIP Publishing.

5.3 Dye-Sensitized Solar Cells

Unlike OLED devices, which convert electricity into emitted light, solar cells transform absorbed light into electrical current. To overcome the drawbacks of traditional silicon-based solar cells[4] in regard to high fabrication costs and inefficient light-harvesting properties, O'Regan and Grätzel (1991) developed quite a revolutionary concept for a new generation of these devices, that is, the DSSCs[5]. The main difference between conventional solar cells and DSSCs is that the operational principle of the latter is based on the separation of the two main tasks: light absorption and transport of electrons. A typical scheme of a DSSC is depicted in Fig. 5.6.

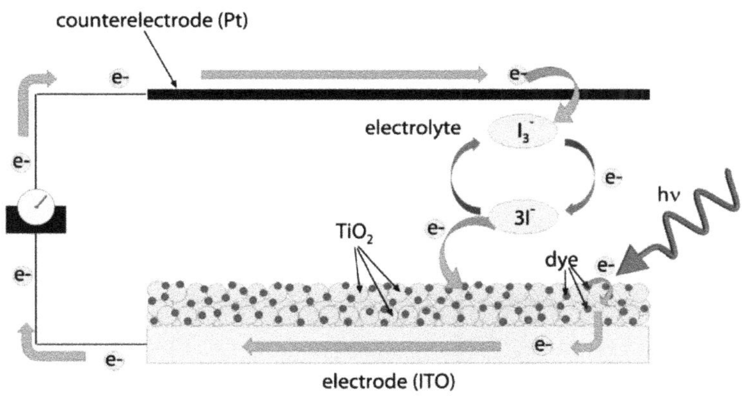

Figure 5.6 Simplified scheme of a dye-sensitized solar cell with electron flow highlighted.

A DSSC consists of a layer of semiconductor, typically titania (TiO_2) nanoparticles, coated with a molecular dye, submerged into a solution of an electrolyte, most often I_3^-/I^-, embedded between a transparent ITO anode and a metallic (Pt) counterelectrode.

[4]Silicon solar cells are simple devices whose operating principle is based on the promotion of one electron from the valence band (VB) to the conduction band (CB) of a layer of semiconductor after light irradiation. Second-generation solar cells are distinct from first-generation devices, made of a single layer of pure silicon, in that they consist of a doped multilayered p-n heterojunction, allowing a directional photocurrent.

[5]These devices are also often called *Grätzel cells* in honor of the scientist who first designed and developed them.

An external circuit connects the two electrodes and allows the photocurrent to circulate. In these devices, the electron flow is mediated by a molecular dye and an electrolyte and occurs through a series of steps briefly summarized as follows:

1. The incident light is absorbed by the dye (D), which is promoted to an excited state D*: $D + h\nu \rightarrow D^*$.
2. The excited D* species undergoes oxidation by transferring one electron to the CB of the semiconducting TiO_2 layer: $D^* \rightarrow D^+ + e^-$.
3. The extra electron in the TiO_2 layer is then captured by the ITO anode and flows in the external circuit until it reaches the Pt counterelectrode.
4. The oxidized form of the electrolyte I_3^- is reduced to I^- by the cathodic current: $I_3^- + 2e^- \rightarrow 3I^-$.
5. The as reduced form of the electrolyte (I^-) transfers one electron to the D^+ species, thus regenerating the initial dye D: $2D^+ + 3I^- \rightarrow 2D + I_3^-$.

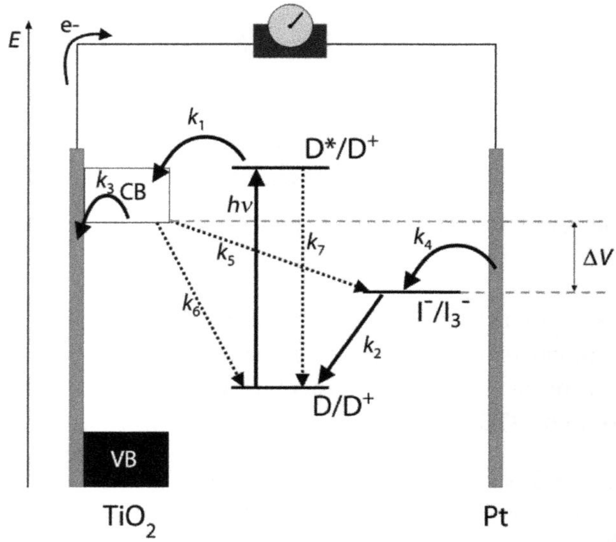

Figure 5.7 Energy diagram and sequential electron flow in a DSSC. The main electron transfer processes are indicated with thick solid arrows, while dotted arrows refer to competing charge recombination mechanisms. Kinetic rate constants are also indicated. CB, conduction band; VB, valence band.

The above described cycle delivers a net production of electricity and, at the same time, ensures the full regeneration of the system. The *cell voltage* is equal to the potential difference ΔV between the bottom edge level of the CB of the semiconductor and the electrochemical potential of the electrolyte, as visualized in the diagram of Fig. 5.7, where all the involved sequential processes of the electron flow are schematically depicted.

The efficiency of such a device has to be intended as a measure of how effectively light is converted into electricity. Most often, it is expressed in terms of *incident photon-to-current efficiency (IPCE)* as explained in the box titled "Solar Cell Efficiency."

Solar Cell Efficiency

There are various approaches to quantify the performance of a solar cell, as many parameters concur in its determination, such as thermodynamic, conduction, charge separation, and light-harvesting efficiencies. All these parameters sum to more technical aspects related to the fabrication of the device, for example, light reflectance, diodes losses, and the electrical resistance of the external circuit. Moreover, the outcome of solar energy conversion may be affected over time by other external factors, not always predictable, such as device degradation and accumulation of dust over the surface.

However, since the above factors generally cannot be measured directly, only two fundamental parameters are taken into consideration to provide a numerical quantification of the efficiency of a solar cell, that is, the *quantum efficiency* (or IPCE) and the cell voltage ΔV (cf. text and Fig. 5.7). In particular, IPCE is often used as an indication of the spectral efficiency of the cell, that is, it provides a measure of the amount of electricity produced by the cell when being irradiated at a given wavelength. There are two types of quantum efficiency:

- The external quantum efficiency (EQE) is the ratio between the number of generated current electrons and the number of incident photons on the device surface at a given wavelength.
- The internal quantum efficiency (IQE) is the ratio between the number of generated current electrons and the number of photons absorbed by the molecular dye at a given wavelength.

The main difference between EQE and IQE is that whereas the former considers the total incident light, the latter only takes into account effectively absorbed photons. Then, it is clear that the IQE parameter is, in general, always higher than the EQE, as it excludes all the (mostly unavoidable) light losses (particularly reflection phenomena).

EQE is indeed the most accessible value as it does not require the discrimination and exclusion of the light reflected and/or absorbed by the device components other than the dye and it is simply directly measured by performing a scan of the photocurrent generated in the circuit over the whole solar spectrum. This type of measurement delivers a curve of quantum efficiency (IPCE percentage) as a function of the illumination wavelength (cf. Fig. 5.10). The integration of this curve over the entire solar spectrum yields a reasonable parameter that accounts for the efficiency of the solar cell. The solar power distribution in watts per square meter per nanometer of bandwidth as a function of the wavelength is reported in the figure below as a reference.

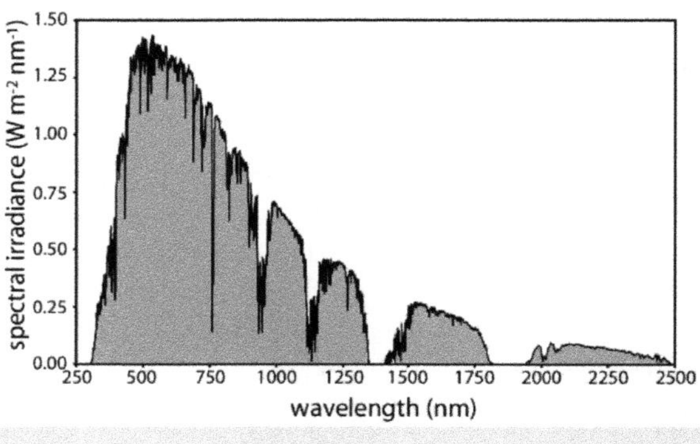

As evidenced by the diagram in Fig. 5.7, to achieve high operational efficiency, the sequential steps constituting the cyclic electron flow must be kinetically favored with respect to competing mechanisms of charge recombination not leading to the generation of photocurrent in the external circuit. The role of the electrolyte is particularly meaningful in this regard since it is employed precisely with the purpose of speeding up the process of dye regeneration (usually occurring within a few μs), otherwise very slow (a few ms), for electron transfer by the cathode. To ensure the most favorable kinetics of the photovoltaic cycle, the three main components of the solar cell, that is, the dye, the electrolyte, and the semiconductor, must be appropriately designed as each one has crucial effects on the efficiency and fabrication costs of the device. Since the first report by O'Regan and Grätzel in 1991, scientists have striven for the improvement of the performances of DSSCs by exploring a wide range of different materials as active components in these devices.

The actual innovative idea behind the development of the first reasonably performing (7.1%) DSSC by O'Regan and Grätzel lies in the use of intimately connected dye and semiconducting materials to guarantee effective electron injection from the former to the latter. This was achieved by chemically binding the molecular dye to TiO_2 nanoparticles deposited as a film with a high total surface area. The archetype dye used in DSSCs is the $[Ru^{II}(SCN)_2(L)_2]$ complex, where L is the 2,2' bypiridil-4,4'-dicarboxylic acid, often labeled *N3 dye*. This complex is capable of anchoring to the surface of TiO_2 in its dianionic bis-deprotonated form, whose tetrabutylammonium salt takes the name of N719, or "red" dye, through covalent ester bonding with the peripheral carboxylate groups, as shown in Fig. 5.8.

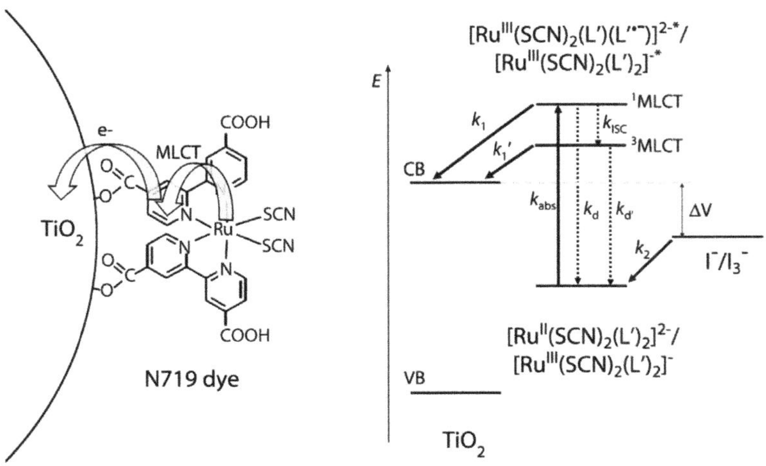

Figure 5.8 Structure of the $[Ru^{II}(SCN)_2(L')_2]^{2-}$ complex (N719, or red dye), where L' is the anionic form of ligand L (2,2' bypiridil-4,4'-dicarboxylic acid) anchored to the surface of a TiO_2 nanoparticle (left), and the partial-energy-level diagram scheme for the dye, the semiconductor, and the electrolyte in a typical DSSC (right). Solid arrows indicate electron transfer, and dotted arrows refer to intersystem crossing (k_{ISC}) and decay mechanism of the dye from excited singlet (k_d) or triplet ($k_{d'}$) states.

The Ru^{II}-based complex depicted in Fig. 5.8 (N719) fully satisfies the main requirements for a suitable DSSC dye:

- A LUMO high enough in energy to lie above the CB of TiO_2 to allow for energetically favored electron transfer
- A HOMO energy lower than that of the reductant electrolyte, facilitating the regeneration of the initial form by reduction
- Reversible redox properties
- High molar extinction coefficients
- Presence of anchoring groups on the ligand periphery
- Charge separation after illumination as a consequence of a metal-to-ligand charge transfer (MLCT) (cf. Section 3.3.3.2).

The last point is of particular relevance since it is the actual key for highly efficient electron injection into the semiconductor as a result of the close contact and chemical bonding of the transiently reduced ligand to the surface of titania. The electron transfer can occur through either the singlet or the triplet excited states of the dye (cf. Fig. 5.8), with a significant difference in the kinetics of the two pathways, falling in the fs timescale for the former and in the ps range for the latter. Actually, the rate of this process is of utmost importance since it competes with the internal decay processes of the excited dye as well as with the back charge recombination from the semiconductor to the metal-centered HOMO (rate constant k_6, Fig. 5.7). This latter process, which usually occurs in the μs–ms timescale, crucially depends on the distance between the charges, up to the point that a 4 Å increase in the electron-hole separation slows it down by about 3 orders of magnitude. With this view, alternatives to the archetypal N3 (or N719) dye were investigated, such as dyes bearing spatially protracted anchoring pendant arms as spacers between the metal and the surface of the semiconductor, as shown for a generic example in Fig. 5.9.

Nevertheless, it has to be taken into account that the introduction of any organic group as a substituent on the ligand, either as a spacer or a binder, may significantly affect the absorption properties of the molecule.[6] This is indeed one of the biggest concerns in finding the most performing dye in terms of solar-light-harvesting efficacy for application in DSSCs. In this regard, the main goal is to extend the range of absorption wavelengths of the dye molecule as far as possible in order to achieve the best match with the spectrum

[6]For example, phosphonic $-PO_3H_2$ groups are very efficient binders to the titania surface and are less hydrolysable than carboxylate groups. However, they usually shift the absorption of the complex toward the blue region.

of the solar radiation, which spreads from the UV up to the NIR. To this purpose, several Ru^{II} complexes have been explored over the years, principally thanks to the work of Nazeeruddin and coworkers, who developed the so-called black dye (N749), bearing a terpyridil derivative ligand (Nazeeruddin et al., 1997) shown in Fig. 5.10. Following the first studies on Ru^{II} complexes, analogous Os^{II} compounds were tested as active dyes in DSSCs, since they typically show broader absorption bands thanks to the increased spin-orbit coupling, which allows a more likely singlet-triplet conversion and the direct excitation of the ^{3}MLCT state, as in the example shown in Fig. 5.10 (Altobello et al., 2005).

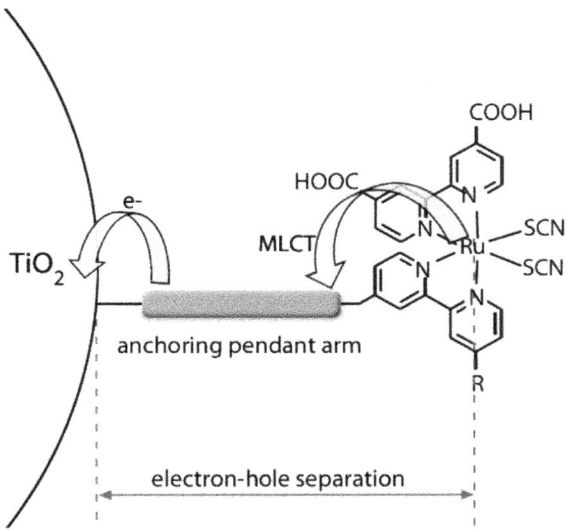

Figure 5.9 Generic example of a dye anchored to the surface of TiO_2 through an extended peripheral pendant arm to increase electron-hole separation between the semiconductor and the transiently oxidized metal center.

Also square planar Pt^{II} dithiolenes have been studied for these applications in view of the significant degree of charge transfer accompanying the LLCT absorption transition and the opportunity to modulate the HOMOs and LUMOs independently (cf. Section 3.3.5.1), despite the typically rather low extinction coefficients. Nonetheless, Ru^{II} complexes have represented for several years the most performing dyes for DSSCs, providing solar energy conversion efficiencies up to 11%, so far unbeaten by Os^{II}- and Pt^{II}-based devices. This observation points out that, once again, finding the best molecular material for a specific application is very often a matter of establishing the right compromise between different features, that is, light absorption, ability of charge injection, and facility to undergo oxidation/reduction, that are not actually independent from each other.

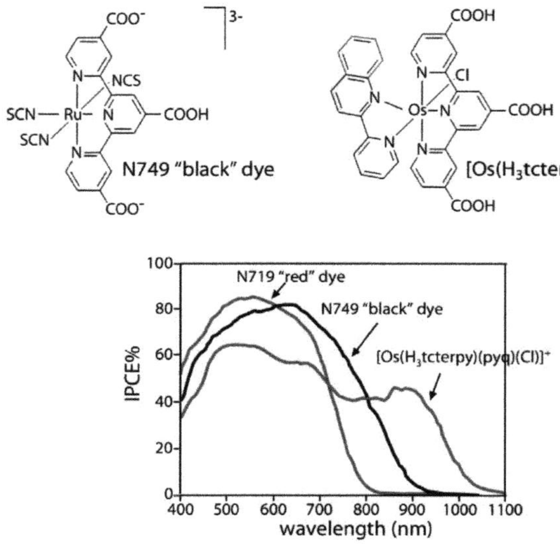

Figure 5.10 Structures of the N749 black and of the $[Os(H_3tcterpy)(pyq)(Cl)]^+$ ($H_3tcterpy$ = 4,4′,4″-tricarboxy-2,2′:6′,2″-terpyridine; pyq = 2-(2′-pyridylquinoline)) complexes (above) and corresponding IPCE% spectra compared to that of the N719 red dye (cf. Fig. 5.8) (below). Partially redrawn from Altobello et al. (2005). Copyright (2005) American Chemical Society.

However, the record for the best-performing DSSC up to now was reported in 2014 for a solar cell based on a Zn^{II}-modified porphyrin dye (SM315) showing a remarkable efficiency of 13% (Mathew et al., 2014). Such a compound, reported in Fig. 5.11, displays a significant charge transfer character of the HOMO-LUMO transition associated with a broad absorption spectrum and a high oscillator strength, favoring both electron injection as well as light-harvesting properties. In addition, another key feature behind the excellent performances of this device relies on the replacement of the traditional I_3^-/I^- electrolyte with the couple $[Co^{III}(bipy)_3]^{3+}/$

$[Co^{II}(bipy)_3]^{2+}$, allowing an increase of the overall cell voltage, as shown in Fig. 5.11.

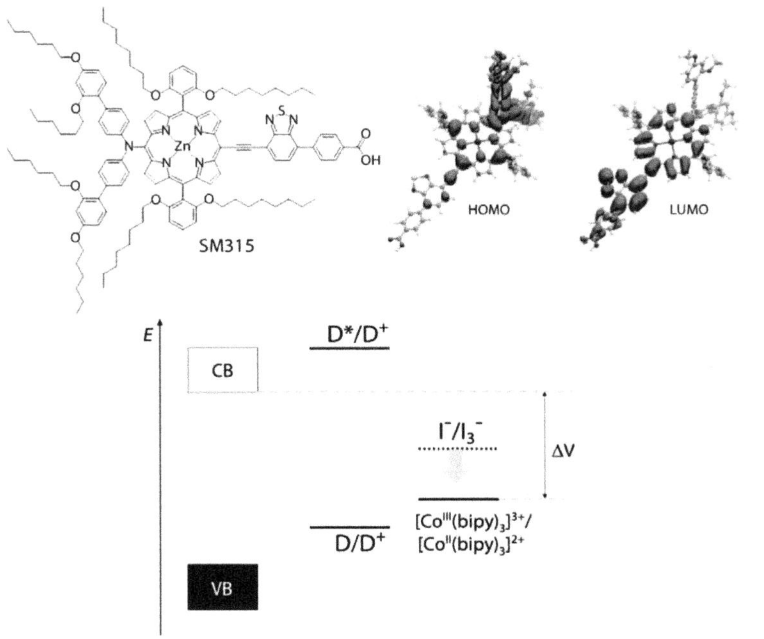

Figure 5.11 Structure and frontier molecular orbitals of the SM315 dye (above) and the partial-energy-level diagram of a solar cell, including the $[Co^{III}(bipy)_3]^{3+}$/ $[Co^{II}(bipy)_3]^{2+}$ electrolyte (below). Partially reprinted by permission from Macmillan Publishers Ltd: [*Nature Chemistry*] (Mathew et al. 2014), copyright (2014).

Although the efficiencies of DSSCs are currently still quite far from those attainable (up to 46%) in more conventional photovoltaic devices based on purely inorganic materials such as Si or GaAs, the low costs of fabrication and the large room for improvements that can be foreseen, thanks to the extensive opportunities of chemical modulation, still make this type of solar cells absolutely promising. It must be, however, remarked that nowadays, a new generation of photovoltaic devices based on perovskites[7] is gaining

[7]The name *perovskite* refers to the crystal structure of materials having an ABX_3 general formulation. The most studied perovskite dye is methylammonium lead trihalide ($CH_3NH_3PbX_3$, where X = Cl, Br, or I), which can be easily obtained through solution methods under mild conditions.

increasing popularity due to their excellent efficiencies (up to 22%), accessibility, and ease of the manufacture process.

5.4 Single-Molecule Magnets on the Surface

Molecular spintronics aims to take advantage of the quantum properties of electrons in molecular materials, in particular the high degree of freedom of the spin, to realize devices for electronics. For example, the single-molecule magnets (SMMs) described in Chapter 3 are molecular objects that can be exploited to store information on the basis of their intrinsic magnetic remanence, due to the high ground spin state and axial anisotropy. Thus, SMMs are seen as fundamental units for the fabrication of spintronic devices such as single-molecule-based memories. For such applications, these molecular materials cannot be employed in the usual crystalline bulk form. Indeed, their deposition and organization on the surface of a conducting material are seen as fundamental steps toward their integration into a device. Anchoring SMMs on conducting surfaces is thus required to make each molecule individually addressable by scanning probes. Although challenging, this is not the only requisite they should fulfill. More importantly, they must retain magnetic remanence once deposited on the surface, a condition that is not easy to meet due to the noninnocent behavior of surfaces.

After some preliminary investigation on the archetypal family of SMMs based on Mn_{12} (cf. Section 3.1.7.1), which was found to be unstable on a gold surface, with a concomitant loss of relevant magnetic properties (Mannini et al., 2008), Sessoli and coworkers succeeded in observing magnetic hysteresis in a tailor-made Fe_4 SMM (cf. Section 3.1.7.1) grafted onto a conducting surface (Mannini, Danieli et al., 2009). This result was accomplished by taking advantage of preliminary promising results on the retention of both static and dynamic properties of Fe_4 SMMs deposited as thin films (a few hundred nanometers) on a gold substrate (Mannini, Joly et al., 2009). Thus, the family of Fe_4 SMMs was found to show the appropriate redox stability and structural robustness to conduct experiments on a self-assembled monolayer (SAM) of these molecules on gold.

A typical Fe_4 SMM is constituted of three outer Fe^{III} ions having a ground spin state $\mathbf{S} = 5/2$, which antiferromagnetically interact with the fourth (inner) one to give a global ground spin state **S** = 5. This magnetic system is characterized by an anisotropy energy barrier of ca. 16 K that has to be overcome to allow for the reversal of the magnetization. The ligands lying above and below the plane of the iron ions are derivatives of 2-hydroxymethyl-1,3-propanediol, which can be functionalized to promote interactions with different substrates. Sessoli and coworkers (Mannini, Danieli et al., 2009) have designed and prepared a sulfur-functionalized derivative, $[Fe_4(L)_2(dpm)_6]$, where H_3L = 11-(acetylthio)-2,2-bis(hydroxymethyl)undecan-1-ol and dpm = dipivaloylmethanate, which has an optimal anchoring group for deposition on a gold surface (Fig. 5.12a).

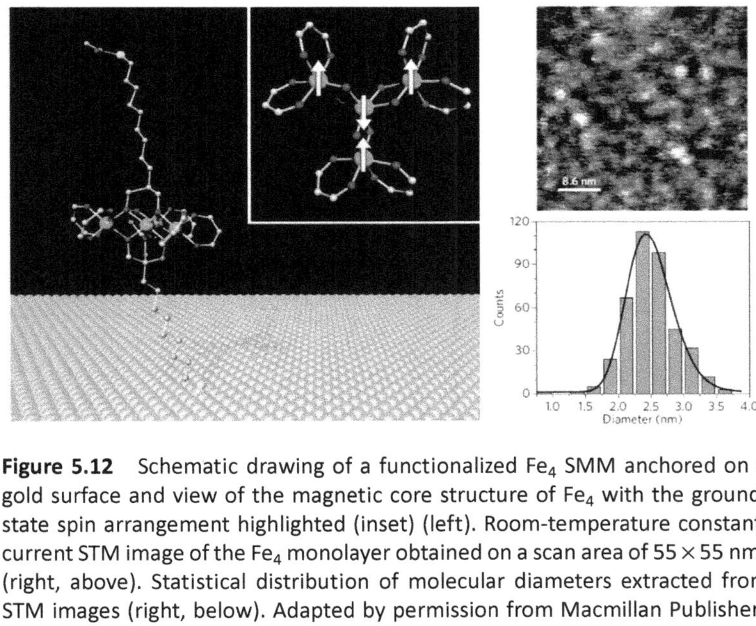

Figure 5.12 Schematic drawing of a functionalized Fe_4 SMM anchored on a gold surface and view of the magnetic core structure of Fe_4 with the ground-state spin arrangement highlighted (inset) (left). Room-temperature constant-current STM image of the Fe_4 monolayer obtained on a scan area of 55 × 55 nm^2 (right, above). Statistical distribution of molecular diameters extracted from STM images (right, below). Adapted by permission from Macmillan Publishers Ltd: [*Nature Materials*] (Mannini, Danieli et al. (2009), copyright (2009).

A SAM was obtained by self-assembly from a solution of Fe_4 molecules in contact with a Au(111) surface. The homogeneous coverage consisting of a layer of objects of the size of the molecules (2.5 ± 0.4 nm, Fig. 5.12c) was confirmed by scanning tunneling microscopy (STM) (Fig. 5.12b).

However, an investigation of the magnetic response of such nanometric scale layers is particularly impervious since standard magnetometry measurements do not allow for the required sensitivity. Therefore, alternative routes must be employed to retrieve such information. This can be accomplished by means of advanced spectroscopic techniques, such as X-ray absorption spectroscopy (XAS) and X-ray magnetocircular dichroism (XMCD) with synchrotron light sources. Without entering into details, it should be herein sufficient to say that the XMCD signal, obtained as the difference between the XAS spectra of a magnetic atom (by probing the 2p → 3d inner electrons transition, $L_{2,3}$ edge, of Fe ions) measured by applying right-hand and left-hand circularly polarized X-rays and a magnetic field, is proportional to the magnetic response of the system. Figure 5.13 shows the field dependence of the XMCD signal recorded at two different temperatures.

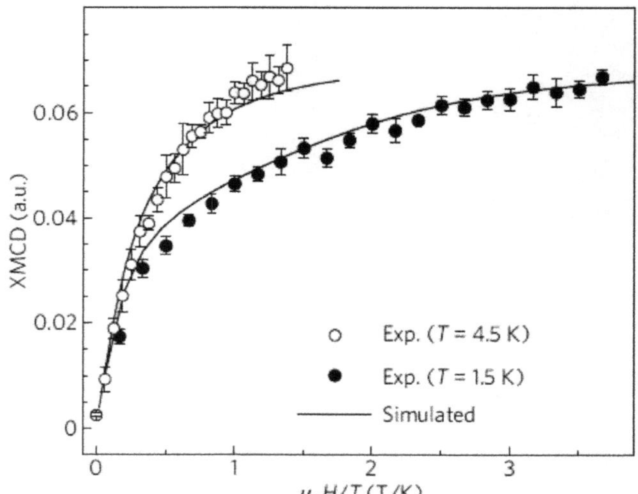

Figure 5.13 Field dependence of the XMCD signal at 709.2 eV (taken as positive) recorded on the Fe_4 monolayer on scanning the field up to 5.5 T at two different temperatures. Adapted by permission from Macmillan Publishers Ltd: [*Nature Materials*] (Mannini, Danieli et al. (2009), copyright (2009).

As indicated in Fig. 5.13, the experimental data, which should be seen as magnetization curves obtained through spectroscopic experiments, are well reproduced by the model described by the spin Hamiltonian $H = D\mathbf{S}z^2 + \mathbf{g}\mu_B\mu_0\ \mathbf{H}\cdot\mathbf{S}$, where D is the axial anisotropy parameter, $\mathbf{g}$ is the Landé factor, μ_B is the Bohr magneton, and $\mathbf{H}$ is the applied magnetic field. This is the sum of the field dependence of the magnetization for a SMM with an integer **S** ground spin state and the field dependence of the magnetization for an isolated paramagnet with a ground spin state **S**. The fit provides values of D and **g** of −0.62 K and 2.00, respectively, which are in agreement with those found for the bulk material. This indicates that the SAM of Fe_4 molecules maintains unchanged its static magnetic properties once grafted on gold.

To check whether also the dynamic properties, that is, the presence of an opened hysteresis cycle and a net remanence magnetization, are retained on surface, similar XAS/XMCD experiments were recorded by scanning the magnetic field in the ±1.5 T range at three different temperatures, 1.0 K, 0.70 K, and 0.50 K (Fig. 5.14).

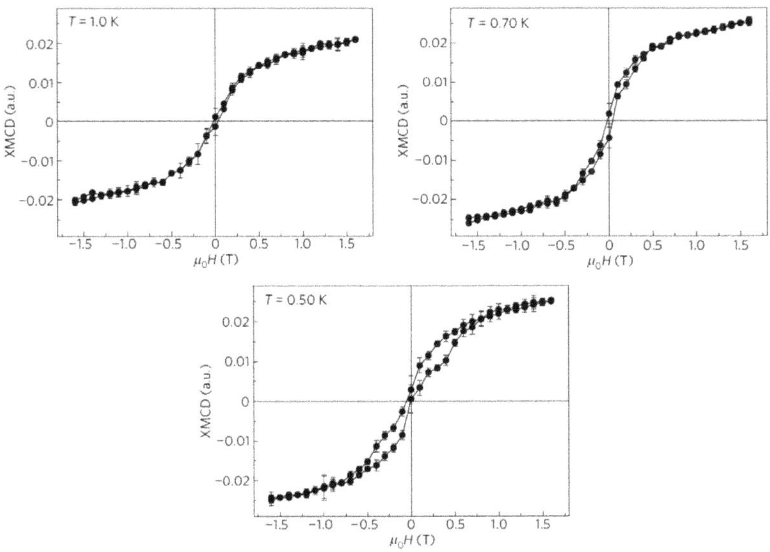

Figure 5.14 Magnetic hysteresis for the Fe_4 monolayer at different temperatures. Adapted by permission from Macmillan Publishers Ltd: [*Nature Materials*] (Mannini, Danieli et al. (2009), copyright (2009).

Whereas a simple paramagnetic behavior was found at 1.0 K, a gradual opening of the hysteresis loop was detected at subkelvin temperatures. At $T = 0.50 \pm 0.05$ K, the loop displays a butterfly shape, resulting from the fast relaxation of the magnetization through quantum tunneling at zero field, and unambiguously demonstrates that the SMM behavior is observable on gold-wired molecules.

Despite the very low blocking temperature of the investigated SMM and its low remanent magnetization due to quantum tunneling in zero field, these results gave a positive answer to a key question, that is, there are no fundamental reasons that preclude the observation of magnetic hysteresis when SMMs are wired to a conducting substrate. Although extra efforts to increase the blocking temperature are required before SMMs can be realistically incorporated into storage devices, this study by Sessoli and coworkers (2009) paves the way for fundamental developments in the field of molecular spintronics. Among them, the possibility to control the reversal of the magnetization through conduction electrons in a metal-molecule-metal nanojunction can be particularly valuable for the development of molecule-based memory storage applications.

5.5 Light-Induced Spin-Crossover Dynamics on the Surface

Molecules showing spin-crossover (SCO) behavior are potential molecular units for the realization of spintronic devices since they possess two available electronic spin states that are controllable through external stimuli. We have already explained in Chapter 3 that the occurrence of the spin transition in such systems is accompanied by a structural compression/expansion of the molecular structure. Depending on the structural features of the metal complex, and thus the entity and magnitude of supramolecular interactions between neighboring molecules in the crystal structure, this process can have a *domino effect*, leading to a cooperative behavior that governs the efficacy and the dynamics of the spin transition. While this process is well understood in bulk materials, to date little information is instead reported on the spin transition dynamics of SCO molecules deposited on a surface. Analogously to what is discussed above for SMMs, the understanding of the dynamics of such molecular materials on a surface is fundamental to developing systems able to retain the properties of the bulk at the nanometric scale.

Recent studies on ultrathin layers (from nanometric thickness to monolayers or even submonolayers) of molecules deposited on the surfaces of different substrates clearly indicate that a part of the molecules that are in closer contact with the surface are structurally and electronically modified by the established interactions. This leads to a partial loss of the magnetic properties at the nanoscale, that is, only a fraction of the molecules can be converted between the two states, while a part of them are trapped in one of the two available electronic states. This observation points out that specific challenging experiments are required to understand the dynamics of such molecules on a surface.

As mentioned above for the case of SMMs, standard magnetic measurements do not have the appropriate sensitivity to allow for detecting the magnetic response of such a restricted amount of molecules, and alternative techniques, such as XAS, have to be used to provide valuable information. However, although XAS provides clear evidence of the degree of spin state switching of SCO molecules on a surface (Warner et al., 2013), it furnishes information related to the overall response of the layer and not to that of each single deposited molecule.

Very recently, Bellac and coworkers (Bairagi et al., 2016) have provided a detailed description of the molecular-scale dynamics of light-induced SCO molecules deposited on a gold surface through scanning tunneling spectroscopy (STS) techniques. This has been achieved by performing single-molecule spectroscopic measurements on a submonolayer (less than a monolayer) deposit of the $[Fe^{II}((3,5\text{-}(CH_3)_2Pz)_3BH)_2]$ (Pz = pyrazolyl) complex on a Au(111) surface. STM images of the molecules on the surface show, at a positive bias, the *hs*-state molecules appearing as bright spots, while the low-spin (*ls*)-state molecules are dark (Fig. 5.15), as confirmed through the support of theoretical calculations.

Thus, the two available spin states of the same magnetic molecule provide different STM responses, allowing the direct observation of the surface organization of the molecules. In this specific case, at low temperatures, the molecules self-organize in a long-range ordered phase, with a domain size up to 200 × 200 nm, characterized by a superstructure where molecules in the *hs* state are separated by two *ls*-state molecules along the $a + b$ direction (Fig. 5.15), with one out of three *hs* molecules within a supercell ($S_{1/3}$ superstructure).

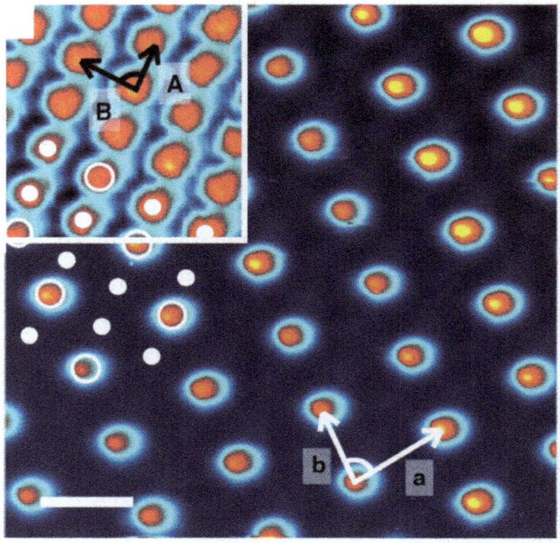

Figure 5.15 STM image of $[Fe^{II}((3,5\text{-}(CH_3)_2Pz)_3BH)_2]$ deposit on Au(111) showing the mixed *hs* and *ls* states in the $S_{1/3}$ superstructure (see text). Bright spheres represent *hs*-state molecules, while the white spots on the left side indicate the position of the *ls*-state molecules (dark background). The inset shows a portion of the same structure where all molecules are in the *hs* state. The scale bar corresponds to 2 nm (Bairagi et al., 2016).

To gain information on the dynamics of the spin transition of SCO molecules on a surface, Bellac and coworkers (Bairagi et al., 2016) have irradiated the molecules with a laser beam of appropriate frequency to induce a *ls*-*hs* transition through a light-induced excited-state spin-trapping (LIESST) effect (cf. 3.1.5) and then analyzed the response of the molecules once relaxed back after a timeframe equivalent to that of light exposure. Before irradiation, the molecules adopt an almost perfect $S_{1/3}$ superstructure (Fig. 5.16a). Upon illumination, part of the *ls*-state molecules is converted to the *hs* state, as can be clearly seen in Fig. 5.16b, to give, on average, another superstructure with two out of three *hs*-state molecules ($S_{2/3}$ superstructure) within a supercell. After irradiation, the system is allowed to relax and the $S_{1/3}$ superstructure is restored, although not completely (Fig. 5.16c).

Figure 5.16 STM images at 4.6 K of the same area in its initial state (left) where the $S_{1/3}$ superstructure is evident, under blue light illumination after exposure of 9 h 45 min (middle), and in its relaxed state 9 h 45 min after stopping the illumination (right). The scale bars correspond to 10 nm (Bairagi et al., 2016).

Interestingly, the cooperative interaction between molecules is clearly evidenced in the relaxed phase by the presence of large domains in the $S_{1/3}$ superstructure separated by small domains of the less thermodynamically stable $S_{2/3}$ superstructure. Complete recovery of the defect-free $S_{1/3}$ superstructure can be achieved by heating the sample up to room temperature.

The analysis of the time evolution of the peak intensity for the *hs*-state molecules yielded a characteristic time constant (131 ± 5 min) for the thermal relaxation of the photoinduced phase (Fig. 5.17).

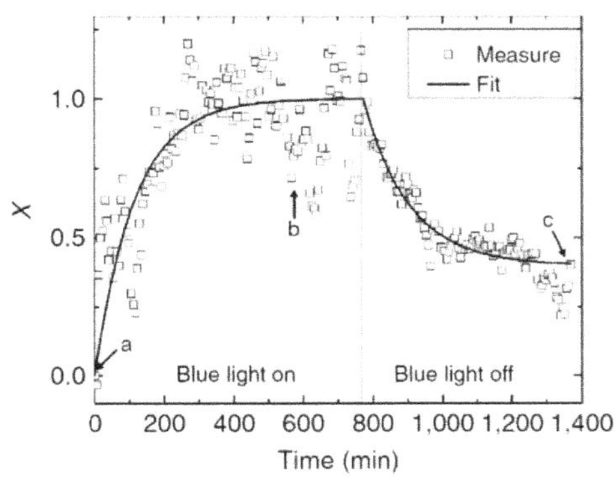

Figure 5.17 Time evolution of the normalized STS peak intensities during light irradiation and during the thermal relaxation. a, b, and c indicate the time at which the STM images of Fig. 5.16 are recorded (Bairagi et al., 2016).

The observed time constant of ca. 130 min is very long when compared to that observed in the bulk phase (a few minutes). For the purpose of this book, this example helps us to highlight how striking the changes of the physical properties of molecular materials can be when such molecular systems are placed in contact with a surface and organized in an ordered phase at the nanometric scale.

Bellac and coworkers (Bairagi et al., 2016) also succeeded in the investigation of the temporal relaxation dynamics of each single molecule under light irradiation, determining an averaged time constant for the *hs-ls* transition. This characteristic time constant, of ca. 10 ± 1 min for both $ls \rightarrow hs$ and $hs \rightarrow ls$ conversions, delivered for the first time accurate and clear information of the single-molecule dynamics of SCO molecules on a surface. This example clearly demonstrates that control of the bistability of SCO molecules on a surface can be achieved.

Bibliography

Aboshyan-Sorgho, L., Nozary, H., Aebischer, A., Bünzli, J.-C. G., Morgantini, P.-Y., Kittilstved, K. R., Hauser, A., Eliseeva, S. V., Petoud, S., Piguet, C. (2012). Optimizing millisecond time scale near-infrared emission in polynuclear chrome(III)–lanthanide(III) complexes, *J. Am. Chem. Soc.,* **134**, pp. 12675–12684.

Aboshyan-Sorgho, L., Besnard, C., Pattison, P., Kittilstved, K. R., Aebischer, A., Bünzli, J.-C. G., Hauser, A., Piguet, C. (2011). Near-infrared → visible light upconversion in a molecular trinuclear d–f–d complex, *Angew. Chem. Int. Ed.,* **50**, pp. 4108–4112.

Aguilà, D., Barrios, L. A., Velasco, V., Roubeau, O., Repollés, A., Alonso, P. J., Sesé, J., Teat, S. J., Luis, F., Aromí, G. (2014). Heterodimetallic [LnLn′] lanthanide complexes: toward a chemical design of two-qubit molecular spin quantum gates, *J. Am. Chem. Soc.,* **136**, pp. 14215–14222.

Al-Rasbi, N. K., Sabatini, C., Barigelletti, F., Ward, M. D. (2006). Red-shifted luminescence from naphthalene-containing ligands due to π-stacking in self-assembled coordination cages, *Dalton Trans.,* **40**, pp. 4769–4772.

Altobello, S., Argazzi, R., Caramori, S., Contado, C. Da Fré, S., Rubino, P., Choné, C., Larramona, G., Bignozzi, C. A. (2005). Sensitization of nanocrystalline TiO_2 with black absorbers based on Os and Ru polypyridine complexes, *J. Am. Chem. Soc.,* **127**, pp. 15342–15343.

Anderson, P. W. (1963). *Magnetism* (Academic Press, New York).

André, N., Scopelliti, R., Hopfgartner, G., Piguet, C., Bünzli, J.-C. (2002). Discriminating between lanthanide ions: self-assembly of heterobimetallic triple-stranded helicates, *Chem. Commun.,* **48**, pp. 214–215.

Anzenbacher, P, Jr., Montes, V. A., Takizawa, S. (2008). High-purity white light from a simple single dopant host-guest white organic light-emitting diode architecture, *App. Phys. Lett.,* **93**, pp. 163302-1-3.

Armelao, L., Quici, S., Barigelletti, F., Accorsi, G., Bottaro, G., Cavazzini, M., Tondello, E. (2010). Design of luminescent lanthanide complexes: from molecules to highly efficient photo-emitting materials, *Coord. Chem. Rev.,* **254**, pp. 487–505.

Artizzu, F., Deplano, P., Marchiò, L., Mercuri, M. L., Pilia, L., Serpe, A., Quochi, F., Orrù, R., Cordella, F., Meinardi, F., Tubino, R., Mura, A., Bongiovanni, G. (2005). Structure and emission properties of Er_3Q_9 (Q = 8-quinolinolate), *Inorg. Chem.,* **44**, pp. 840–842.

Artizzu, F., Deplano, P., Marchiò, L., Mercuri, M. L., Pilia, L., Serpe, A., Quochi, F., Orrù, R., Cordella, F., Saba, M., Mura, A., Bongiovanni, G. (2007). New insights on near-infrared emitters based on Er-quinolinolate complexes: synthesis, characterisation, structural and photophysical properties, *Adv. Funct. Mater.,* **17**, pp. 2365–2376.

Artizzu, F., Bernot, K., Caneschi, A., Coronado, E., Clemente-Juan, J. M., Marchiò, L., Mercuri, M. L., Pilia, L., Serpe, A., Deplano, P. (2008). Synthesis, structure, spectroscopic studies and magnetic properties of the tetrakis(5,7-dichloro-8-quinolinolato)gadolinium(III) complex, *Eur. J. Inorg. Chem.,* **24**, pp. 3820–3826.

Artizzu, F., Quochi, F., Marchiò, L., Sessini, E., Saba, M., Serpe, A., Mura, A., Mercuri, M. L., Bongiovanni, G., Deplano, P. (2013). Fully efficient direct Yb-to-Er energy transfer at molecular level in a near-infrared emitting heterometallic trinuclear quinolinolato complex, *J. Phys. Chem. Lett.,* **4**, pp. 3062–3066.

Artizzu, F., Quochi, F., Marchiò, L., Fonseca Correia, R., Saba, M., Serpe, A., Mura, A., Mercuri, M. L., Bongiovanni, G., Deplano, P. (2015). Ln_3Q_9 as a molecular framework for ion-size-driven assembly of heterolanthanide (Nd, Er, Yb) multiple near-infrared emitters, *Chem. Eur. J.,* **21**, pp. 3882–3885.

Atzori, M., Benmansour, S., Mínguez Espallargas, G., Clemente-León, M., Abhervé, A., Gómez-Claramunt, P., Coronado, E., Artizzu, F., Sessini, E., Deplano, P., Serpe, A., Mercuri, M. L., Gómez García, C. J. (2013). A family of layered chiral porous magnets exhibiting tunable ordering temperatures, *Inorg. Chem.,* **52**, pp. 10031–10040.

Atzori, M., Artizzu, F., Marchiò, L., Loche, D., Caneschi, A., Serpe, A., Deplano, P., Avarvari, N., Mercuri, M. L. (2015), Switching-on luminescence in anilate-based molecular materials, *Dalton Trans.,* **44**, pp. 15786–15802.

Atzori, M., Tesi, L., Morra, E., Chiesa, M., Sorace, L., Sessoli, R. (2016). Room-temperature quantum coherence and rabi oscillations in vanadyl phthalocyanine: toward multifunctional molecular spin qubits, *J. Am. Chem. Soc.,* **138**, pp. 2154–2157.

Atzori, M., Morra, E., Tesi, L., Albino, A. Chiesa, M., Sorace, L., Sessoli, R. (2016). Quantum coherence times enhancement in vanadium(IV)-based potential molecular qubits: the key role of the vanadyl moiety, *J. Am. Chem. Soc.,* **138**, pp. 11234–11244.

Avarvari, N., Wallis, J. D. (2009). Strategies towards chiral molecular conductors, *J. Mater. Chem.,* **19**, pp. 4061–4076.

Bader, K., Dengler, D., Lenz, S., Endeward, B., Jiang, S.-D., Neugebauer, P., van Slageren, J. (2014). Room temperature quantum coherence in a potential molecular qubit, *Nat. Commun.,* **5**, p. 5304.

Bain, G. A., Berry, J. F. J. (2008). Diamagnetic corrections and Pascal's constants, *Chem. Educ.,* **85**, pp. 532–536.

Bairagi, K., Iasco, O., Bellec, A., Kartsev, A., Li, D., Lagoute, J., Chacon, C., Girard, Y., Rousset, S., Miserque, F., Dappe, Y. J., Smogunov, A., Barreteau, C., Boillot, M.-L., Mallah, T., Repain, V. (2016). Molecular-scale dynamics of light-induced spin cross-over in a two-dimensional layer, *Nat. Commun.,* **7**, p. 12212.

Ballardini, R., Varani, G., Indelli, M. T., Scandola, F. (1986). Phosphorescent 8-quinolinol metal chelates: excited-state properties and redox behavior, *Inorg. Chem.,* **25**, pp. 3858–3865.

Barra, A. L., Caneschi, A., Cornia, A., Fabrizi de Biani, F., Gatteschi, D., Sangregorio, C., Sessoli, R., Sorace, L. (1999). Single-molecule magnet behavior of a tetranuclear iron(III) complex. The origin of slow magnetic relaxation in iron(III) clusters, *J. Am. Chem. Soc.,* **121**, pp. 5302–5310.

Beni, E., Dei, A., Laschi, S., Rizzitano, M., Sorace, L. (2008). Tuning the charge distribution and photoswitchable properties of cobalt–dioxolene complexes by using molecular techniques, *Chem. Eur. J.,* **14**, pp. 1804–1813.

Bleaney, B., Bowers, K. D. (1952). Anomalous paramagnetism of copper acetate, *Proc. Roy. Soc.,* **214**, pp. 451–465.

Brinkmann, M., Gadret, G., Muccini, M., Taliani, C., Masciocchi, N., Sironi, A. (2000). Correlation between molecular packing and optical properties in different crystalline polymorphs and amorphous thin films of *mer*-tris(8-hydroxyquinoline)aluminum(III), *J. Am. Chem. Soc.,* **122**, pp. 5147–5157.

Brossard, L., Ribault, M., M., Valade, L., Cassoux, P. (1986). The first 3-D molecular superconductor under pressure: $TTF[Ni(dmit)_2]_2$, *Physica B,* **143**, pp. 378–380.

Bünzli, J.-C. G. (2015). On the design of highly luminescent lanthanide complexes, *Coord. Chem. Rev.,* **293–294**, pp. 19–47.

Canadell, E., Doublet, M.-L., Iung, C. (2012). *Orbital Approach to the Electronic Structure of Solids* (Oxford University Press, Oxford).

Candela, G. A., Swartzendruber, L. J., Miller, J. S., Rice, M. J. (1979). Metamagnetic properties of one-dimensional decamethylferrocenium 7,7,8,8-tetracyano-p-quinodimethanide (1:1), *J. Am. Chem. Soc.,* **101**, pp. 2755–2756.

Caneschi, A., Gatteschi, D., Lalioti, N., Sangregorio, C., Sessoli, R., Venturi, G., Vindigni, A., Rettori, A., Pini, M. G., Novak, M. A. (2001). Cobalt(II)-nitronyl nitroxide chains as molecular magnetic nanowires, *Angew. Chem. Int. Ed.,* **40**, pp. 1760–1763.

Carter, R. L. (1998). *Molecular Symmetry and Group Theory* (Wiley, NY).

Chen, X.-Y., Bretonnière, Y., Pécaut, J., Imbert, D., Bünzli, J.-C., Mazzanti M. (2007). Selective self-assembly of hexameric homo- and heteropolymetallic lanthanide wheels: synthesis, structure, and photophysical studies, *Inorg. Chem.,* **46**, pp. 625–637.

Clemente-León, M., Coronado, E., Gómez-García, C. J., López-Jordà, M., Camón, A., Repollés, A., Luis, F. (2014). Insertion of a single-molecule magnet inside a ferromagnetic lattice based on a 3D bimetallic oxalate network: towards molecular analogues of permanent magnets, *Chem. Eur. J.,* **20**, pp. 1669–1676.

Coppo, P., Duati, M., Kozhevnikov, V. N., Hofstraat, J. W., De Cola, L. (2005). White-light emission from an assembly comprising luminescent iridium and europium complexes, *Angew. Chem. Int. Ed.,* **44**, pp. 1806–1810.

Coronado, E., Galán-Mascarós, J. R., Gómez-García, C. J., Laukhin, V. (2000). Coexistence of ferromagnetism and metallic conductivity in a molecule-based layered compound, *Nature,* **408**, pp. 447–449.

Coronado, E., Galán-Mascarós, J. R., Gómez-García, C. J., Martínez-Agudo, J. M. (2001). Molecule-based magnets formed by bimetallic three-dimensional oxalate networks and chiral tris(bipyridyl) complex cations. The series $[Z^{II}(bpy)_3][ClO_4][M^{II}Cr^{III}(ox)_3]$ (Z^{II} = Ru, Fe, Co, Ni; M^{II} = Mn, Fe, Co, Ni, Cu, Zn; ox = oxalate dianion, *Inorg. Chem.,* **40**, pp. 113–120.

Coronado, E., Day, P. (2004). Magnetic molecular conductors, *Chem. Rev.,* **104**, pp. 5419–5448.

Cotton, F. A. (1990). *Chemical Applications of Group Theory*, Chapter 7 (Wiley, New York).

de Sá, G. F., Malta, O. L., de Mello Donegá, C., Simas, A. M., Longo, R. L., Santa-Cruz, P. A., da Silva, E. F. (2000). Spectroscopic properties and design of highly luminescent lanthanide coordination complexes, *Coord. Chem. Rev.*, **196**, pp. 165–195.

Christou, G., Gatteschi, D., Hendrickson, D. N., Sessoli, R. (2000). Single-molecule magnets, *MRS Bull.*, **25**, pp. 66–71.

Dexter, D. L. (1953). A theory of sensitized luminescence in solids, *J. Chem. Phys.*, **21**, pp. 836–850.

DiVincenzo, D. P. (1995). Quantum computation, *Science*, **270**, pp. 255–271.

Ferbinteanu, M., Miyasaka, H., Wernsdorfer, W., Nakata, K., Sugiura, K., Yamashita, M., Coulon, C., Clérac, R. (2005). Single-chain magnet $(NEt_4)[Mn_2(5\text{-MeOsalen})_2Fe(CN)_6]$ made of Mn^{III}–Fe^{III}–Mn^{III} trinuclear single-molecule magnet with an $S_T = 9/2$ spin ground state, *J. Am. Chem. Soc.*, **127**, pp. 3090–3099.

Ferraris, A. J., Cowan, D. O., Walatka, V., Perlstein, J. H. (1973). Electron transfer in a new highly conducting donor-acceptor complex, *J. Am. Chem. Soc.*, **95**, pp. 948–949.

Förster, T. (1959). Transfer mechanism of electronic excitation, *Discuss. Faraday Soc.*, **27**, pp. 7–15.

Franken, P. A., Hill, A. E., Peters, C. W., Weinreich, G. (1961). Generation of optical harmonics, *Phys. Rev. Lett.*, **7**, pp. 118–119.

Galán-Mascarós, J. R., Coronado, E., Goddard, P. A., Singleton, J., Coldea, A. I., Wallis, J. D., Coles, S. J., Alberola, A. (2010). A chiral ferromagnetic molecular metal, *J. Am. Chem. Soc.*, **132**, pp. 9271–9273.

Gareth Williams, J. A., Wilkinson, A. J., Whittle, V. L. (2008). Light-emitting iridium complexes with tridentate ligands, *Dalton Trans.*, pp. 2081–2099.

Gatteschi, D., Sessoli, R., Villain, J. (2006). *Molecular Nanomagnets* (Oxford University Press, Oxford).

Gillin, W. P., Curry, R. J. (1999). Erbium (III) tris(8-hydroxyquinoline) (ErQ): a potential material for silicon compatible 1.5 mm emitters, *Appl. Phys. Lett.*, **74**, pp. 798–799.

Glover, P. B., Bassett, A. P., Nockemann, P., Kariuki, B. M., Van Deun, R., Pikramenou, Z. (2007). Fully fluorinated imidodiphosphinate shells for visible- and NIR-emitting lanthanides: hitherto unexpected effects of sensitizer fluorination on lanthanide emission properties, *Chem. Eur. J.*, **13**, pp. 6308–6320.

Goodenough, J. B. (1966). *Magnetism and the Chemical Bond* (Interscience Publisher, New York).

Goodwin, C. A. P., Ortu, F., Reta, D., Chilton, N. F., Mills, D. P. (2017). Molecular magnetic hysteresis at 60 kelvin in dysprosocenium, *Nature*, **548**, pp. 439–442.

Graham, M. J., Zadrozny, J. M., Shiddiq, M., Anderson, J. S., Fataftah, M. S., Hill, S., Freedman, D. E. (2014). Influence of electronic spin and spin-orbit coupling on decoherence in mononuclear transition metal complexes, *J. Am. Chem. Soc.*, **136**, pp. 7623–7626.

Guo, F-S., Day, B. M., Chen, Y-C., Tong, M.-L., Mansikkamäki, A., Layfield, R. A. (2017). A dysprosium metallocene single-molecule magnet functioning at the axial limit, *Angew. Chem. Int. Ed.*, **56**, pp. 11445–11449.

Halcrow, M. A. (ed.) (2013). *Spin-Crossover Materials: Properties and Applications* (Wiley-WCH, Hoboken).

Hiraga, H., Miyasaka, H., Nakata, K., Kajiwara, T., Takaishi, S., Oshima, Y., Nojiri, H., Yamashita, M. (2007). Hybrid molecular material exhibiting single-molecule magnet behavior and molecular conductivity, *Inorg. Chem.*, **46**, pp. 9661–9671.

Horrocks, W. DeW., Sudnick, D. R. (1979). Lanthanide ion probes of structure in biology. laser-induced luminescence decay constants provide a direct measure of the number of metal-coordinated water molecules, *J. Am. Chem. Soc.*, **101**, pp. 334–340.

Hoffmann, R. (1989). *Solid and Surfaces: A Chemist's View of Bonding in Extended Structures* (Wiley-VCH, New York).

Jérome, D., Mazaud, M., Ribault, M., Bechgaard, K. (1980). Superconductivity in a synthetic organic conductor $(TMTSF)_2PF_6$, *J. Phys. Lett.*, **41**, pp. 95–98.

Jiles, D. (1998). *Introduction to Magnetism and Magnetic Materials* (Chapman & Hall, London, New York).

Judd, B. R. (1962). Optical absorption intensities of rare-earth ions, *Phys. Rev.*, **127**, pp. 750–761.

Kaiser, W., Garrett, C. G. B. (1961). Two photon absorption of $CaF_2{:}Eu^{2+}$, *Phys. Rev. Lett.*, **7**, pp. 229–230.

Kido, J., Nagai, K., Ohashi, Y. (1990). Electroluminescence in a terbium complex, *Chem. Lett.*, **19**, pp. 657–660.

Kini, A. M., Geiser, U., Wang, H. H., Carlson, K. D., Williams, J. M., Kwok, W. K., Vandervoort, K. G., Thompson, J. E., Stupka, D. L., Jung, D., Whangbo, M. H. (1990). A new ambient-pressure organic superconductor, κ-$(ET)_2Cu[N(CN)_2]Br$, with the highest transition temperature yet observed (inductive onset T_c = 11.6 K, resistive onset = 12.5 K), *Inorg. Chem.*, **29**, pp. 2555–2557.

Kurmoo, M., Graham, A. W., Day, P., Coles, S. J., Hursthouse, M. B., Caulfield, J. L., Singleton, J., Pratt, F. L., Hayes, W., Ducasse, L., Guionneau, P. (1995). Superconducting and semiconducting magnetic charge transfer salts: $(BEDT\text{-}TTF)_4AFe(C_2O_4)_3{\cdot}C_6H_5CN$ ($A = H_2O$, K, NH_4), *J. Am. Chem. Soc.*, **117**, pp. 12209–12217.

Kuroda-Sowa, T., Yu, Z., Senzaki, Y., Sugimoto, K., Maekawa, M., Munakata, M., Hayami, S., Maeda, Y. (2008). Abrupt spin transitions and LIESST effects observed in FeII spin-crossover complexes with extended π-conjugated schiff-base ligands having N_4O_2 donor sets, *Chem. Lett.*, **37**, pp. 1216–1217.

Kwiatkowski, J. J., Nelson, J., Li, H., Bredas, J. L., Wenzel, W., Lennartz, C. (2008). Simulating charge transport in tris(8-hydroxyquinoline) aluminium (Alq_3), *Phys. Chem. Chem. Phys.*, **10**, pp. 1852–1858.

Launay, J.-P., Verdaguer, M. (2014). *Electrons in Molecules: From Basic Principles to Molecular Electronics* (Oxford University Press, Oxford).

Lehn, J. M. (1990). Perspectives in supramolecular chemistry—from molecular recognition towards molecular information processing and self-organization, *Angew. Chem. Int. Ed. Engl.*, **29**, pp. 1304–1319.

Liddle, S. T., van Slageren, J. (2015). Actinide single-molecule magnets, in *Lanthanides and Actinides in Molecular Magnetism,* Layfield, R. A., Murugesu, M., eds. (Wiley-VCH, Weinheim).

Malta, O. L., Gonçalves e Silva, F. R. (1998). A theoretical approach to intramolecular energy transfer and emission quantum yields in coordination compounds of rare earth ions, *Spectrochim. Acta A,* **54**, pp. 1593–1599.

Mannini, M., Sainctavit, P., Sessoli, R., Cartier dit Moulin, C., Pineider, F., Arrio M.-A., Cornia, A., Gatteschi, D. (2008). XAS and XMCD investigation of Mn_{12} monolayers on gold, *Chem. Eur. J.,* **14**, pp. 7530–7535.

Mannini, M., Pineider, F., Sainctavit, P., Joly, L., Fraile-Rodríguez, A., Arrio, M.-A., Cartier dit Moulin, C., Wernsdorfer, W., Cornia, A., Gatteschi, D., Sessoli, R. (2009). X-ray magnetic circular dichroism picks out single-molecule magnets suitable for nanodevices, *Adv. Mater.,* **21**, pp. 167–171.

Mannini, M., Pineider, F., Sainctavit, P., Danieli, C., Otero, E., Sciancalepore, C., Talarico, A. M., Arrio, M.-A., Cornia, A., Gatteschi, D., Sessoli, R. (2009). Magnetic memory of a single-molecule quantum magnet wired to a gold surface, *Nat. Mater.,* **8**, pp. 194–197.

Mathew, S., Yella, A., Gao, P., Humphry-Baker, R, Curchod, B. F. E., Ashari-Astani, N., Tavernelli, I., Rothlisberger, U., Nazeeruddin, M. K., Grätzel,

M. (2014). Dye-sensitized solar cells with 13% efficiency achieved through the molecular engineering of porphyrin sensitizers, *Nat. Chem.*, **6**, pp. 242–248.

Miller, J. S., Calabrese, J. C., Epstein, A. J., Bigelow, R. W., Zhang, J. H., Reiff, W. M. (1986). Ferromagnetic properties of one-dimensional decamethylferrocenium tetracyanoethylenide (1:1), *J. Chem. Soc., Chem. Commun.*, pp. 1026–1028.

Miller, J. S. (2006). Organic and molecule-based magnets, *Indian Acad. Sci.*, **67**, pp. 1–16.

Miskowski, V. M., Houlding, V. H. (1991). Electronic spectra and photophysics of platinum(II) complexes with α-diimine ligands. solid-state effects. 2. Metal-metal interaction in double salts and linear chains, *Inorg. Chem.*, **30**, pp. 4446–4452.

Montes, V A., Li, G., Pohl, R., Shinar, J., Anzenbacher, P. (2004). Effective color tuning in organic light-emitting diodes based on aluminum tris(5-aryl-8-hydroxyquinoline) complexes, *Adv. Mater.*, **16**, pp. 2001–2003.

Mori, T. (1998). Structural genealogy of BEDT-TTF-based organic conductors I, *Bull. Chem. Soc. Jpn.*, **71**, pp. 2509–2526.

Mori, T., Mori, H., Tanaka, S. (1999). Structural genealogy of BEDT-TTF-based organic conductors II, *Bull. Chem. Soc. Jpn.*, **72**, pp. 179–197.

Mori, T. (1999). Structural genealogy of BEDT-TTF-based organic conductors III, *Bull. Chem. Soc. Jpn.*, **72**, pp. 2011–2027.

Moussa, J., Chamoreau, L. M., Gullo, M. P., Degli Esposti, A., Barbieri, A., Amouri, H. (2016). Induced phosphorescence from Pt → Ag and Ag(I)···Ag(I) metallophilic interactions in benzenedithiolatodiimine-Pt_2/Ag_2 clusters: a combined experimental and theoretical investigation, *Dalton Trans.*, **45**, pp. 2906–2913.

Nazeeruddin, M. K., Pechy, P., Grätzel, M. (1997). Efficient panchromatic sensitization of nanocrystalline TiO_2 films by a black dye based on a trithiocyanato-ruthenium complex, *Chem. Commun.*, **18**, pp. 1705–1706.

Nielsen, M., Chuang, I. (2004). *Quantum Computation and Quantum Information* (Cambridge University Press, Cambridge).

Ofelt, G. S. (1962). Intensities of crystal spectra of rare-earth ions, *J. Chem. Phys.*, **37**, pp. 511–520.

Ohkoshi, S., Takano, S., Imoto, K., Yoshikiyo, M., Namai, A., Tokoro, H. (2014). 90-degree optical switching of output second-harmonic light in chiral photomagnet, *Nat. Photon.*, **8**, pp. 65–71.

Orchard, A. F. (2003). *Magnetochemistry* (Oxford University Press, Oxford).

O'Regan, B., Grätzel, M. (1991). A low-cost, high-efficiency solar cell based on dye-sensitized colloidal TiO_2 films, *Nature,* **353**, pp. 737–740.

Kahn, O. (1993). *Molecular Magnetism* (Wiley-VCH, New York).

Paris, J. P. Brandt, W. W. (1959). Charge transfer luminescence of a ruthenium(II) chelate, *J. Am. Chem. Soc.,* **81**, pp. 5001–5002.

Pedersen, K. S., Ariciu, A.-M., McAdams, S., Weihe, H., Bendix, J., Tuna, F., Piligkos, S. (2016). Toward molecular 4f single-ion magnet qubits, *J. Am. Chem. Soc.,* **138**, pp. 5801–5804.

Peierls, R. E. (1955). *Quantum Theory of Solids* (Oxford University Press, Oxford).

Pérez-Bolívar, C., Takizawa, S., Nishimura, G, Montes, V. A., Anzenbacher, P., Jr. (2011). High-efficiency tris(8-hydroxyquinoline)aluminum (Alq_3) complexes for organic white-light-emitting diodes and solid-state lighting, *Chem. Eur. J.,* **17**, pp. 9076–9082.

Pilkington, M., Decurtins, S. (2002). Oxalate-based 2D and 3D magnets, in *Magnetism: Molecules to Materials II: Molecule-Based Materials,* Miller, J. S., Drillon, M., eds. (Wiley-VCH, Weinheim).

Pop, F., Auban-Senzier, P., Canadell, E., Rikken, G. L. J. A., Avarvari, N. (2014). Electrical magnetochiral anisotropy in a bulk chiral molecular conductor, *Nat. Commun.,* **5**, p. 3757.

Racah, G. (1942). Theory of complex spectra II, *Phys. Rev.,* **62**, pp. 438–462.

Real, J. A., Gaspar, A. B., Niel, V., Munoz, M. C. (2003). Communication between iron(II) building blocks in cooperative spin transition phenomena, *Coord. Chem. Rev.,* **236**, pp. 121–141.

Rechkemmer, Y., Breitgoff, F. D., van der Meer, M., Atanasov, M., Hakl, M., Orlita, M., Neugebauer, P., Neese, F., Sarkar, B., van Slageren, J. (2016). A four-coordinate cobalt(II) single-ion magnet with coercivity and a very high energy barrier, *Nat. Commun.,* **7**, p. 10467.

Réthoré, C., Fourmigé, M., Avarvari, N. (2005). Tetrathiafulvalene-hydroxyamides and -oxazolines: hydrogen bonding, chirality, and a radical cation salt, *Tetrahedron,* **61**, pp. 10935–10942.

Robertson, N., Yee, G. T. (2010). Molecular magnetic materials, in *Molecular Materials,* Bruce, D. W., O'Hare, D., Walton, R. I., eds. (Wiley-WCH, Weinheim).

Robin, M. B., Day, P. (1968). Mixed valence chemistry-a survey and classification, *Adv. Inorg. Chem. Radiochem.,* **10**, pp. 247–422.

Sessoli, R., Gatteschi, D., Caneschi, A., Novak, M. A. (1993). Magnetic bistability in a metal-ion cluster, *Nature,* **365**, pp. 141–143.

Singh, K., Banerjee, S., Patra, A. K. (2015). Photocytotoxic luminescent lanthanide complexes of DTPA–bisamide using quinoline as photosensitizer, *RSC Adv.*, **5**, pp. 107503–107513.

Shultz, D. A. (2002). Valence tautomerism in dioxolene complexes of cobalt, in *Magnetism: Molecules to Materials II: Molecule-Based Materials*, Miller, J. S., Drillon, M., eds. (Wiley-VCH, Weinheim).

Sloof, L. H., van Blaaderen, A., Polman, A., Hebbink, G. A., Klink, S. I., Van Veggel, F. C. J. M., Reinhoudt, D. N., Hofstraat J. W. (2002). Rare-earth doped polymers for planar optical amplifiers, *J. Appl. Phys.*, **91**, pp. 3955–3980.

Strickler, S. J., Berg, R. A. (1962). Relationship between absorption intensity and fluorescence lifetime of molecules, *Chem. Phys.*, **37**, pp. 814–822.

Takahashi, K., Cui, H.-B., Okano, Y., Kobayashi, H., Einaga, Y., Sato, O. (2006). Electrical conductivity modulation coupled to a high-spin-low-spin conversion in the molecular system $[Fe^{III}(qsal)_2][Ni(dmit)_2]_3{\cdot}CH_3CN{\cdot}H_2O$, *Inorg. Chem.*, **45**, pp. 5739–5741.

Takahashi, K., Cui, H.-B., Okano, Y., Kobayashi, H., Mori, H., Tajima, H., Einaga, Y., Sato, O. (2008). Evidence of the chemical uniaxial strain effect on electrical conductivity in the spin-crossover conducting molecular system: $[Fe^{III}(qnal)_2][Pd(dmit)_2]_5{\cdot}$ acetone, *J. Am. Chem. Soc.*, **130**, pp. 6688–6689.

Tamaki, H., Zhong, Z. J., Matsumoto, N., Kida, S., Koikawa, M., Achiwa, N., Hashimoto, Y., Okawa, H. (1992). Design of metal-complex magnets. Syntheses and magnetic properties of mixed-metal assemblies $\{NBu_4[MCr(ox)_3]\}_x$ (NBu_4^+ = tetra(*n*-butyl)ammonium ion; ox^{2-} = oxalate ion; M = Mn^{2+}, Fe^{2+}, Co^{2+}, Ni^{2+}, Cu^{2+}, Zn^{2+}), *J. Am. Chem. Soc.*, **114**, pp. 6974–6979.

Tanabe, Y., Sugano, S. (1954a). On the absorption spectra of complex ions I, *J. Phys. Soc. Jpn.*, **9**, pp. 753–766.

Tanabe, Y., Sugano, S. (1954b). On the absorption spectra of complex ions II, *J. Phys. Soc. Jpn.*, **9**, pp. 766–779.

Tanabe, Y., Sugano, S. (1956). On the absorption spectra of complex ions III, *J. Phys. Soc. Jpn.*, **11**, pp. 864–877.

Tanner, B. (1995). *Introduction to the Physics of Electrons in Solids* (Cambridge University Press, Cambridge).

Tang, C. W., Van Slyke, S. A. (1987). Organic electroluminescent diodes, *Appl. Phys. Lett.*, **51**, pp. 913–915.

Troiani, F., Ghirri, A., Affronte, M., Carretta, S., Santini, P., Amoretti, G., Piligkos, S., Timco, G., Winpenny, R. E. P. (2005). Molecular engineering of antiferromagnetic rings for quantum computation, *Phys. Rev. Lett.*, **94**, p. 207208.

Van Vleck, J. H. (1937). The puzzle of rare-earth spectra in solids, *J. Phys. Chem.*, **41**, pp. 67–80.

Weissman, S. I. (1942). Intramolecular energy transfer the fluorescence of complexes of europium, *J. Chem. Phys.*, **10**, pp. 214–217.

Warner, B., Oberg, J. C., Gill, T. G., El Hallak, F., Hirjibehedin, C. F., Serri, M., Heutz, S., Arrio, M.-A., Sainctavit, P., Mannini, M., Poneti, G., Sessoli, R., Rosa, P. (2013). Temperature- and light-induced spin crossover observed by X-ray spectroscopy on isolated Fe(II) complexes on gold, *J. Phys. Chem. Lett.*, **4**, pp. 1546–1552.

West, A. R. (1999). *Basic Solid State Chemistry*, 2nd ed. (John Wiley & Sons, Chichester,).

White, K. A., Chengelis, D. A., Gogick, K. A., Stehman, J., Rosi, N. L., Petoud, S. (2009). Near-infrared luminescent lanthanide MOF barcodes, *J. Am. Chem. Soc.*, **131**, pp. 18069–18071.

Wieghardt, K., Pohl, K., Jibril, I., Huttner, G. (1984). Hydrolysis products of the monomeric amine complex $(C_6H_{15}N_3)FeCl_3$: the structure of the octameric iron(III) cation of $\{[(C_6H_{15}N_3)_6Fe_8(\mu_3\text{-}O)_2(\mu_2\text{-}OH)_{12}]Br_7(H_2O)\}Br{\cdot}8H_2O$, *Angew. Chem. Int. Ed. Engl.*, **23**, pp. 77–78.

Winpenny, R. E. P. (ed.) (2012). *Molecular Cluster Magnets* (World Scientific Publishing, Singapore).

Winpenny, R. E. P., McInnes, E. J. L. (2010). Molecular nanomagnets, in *Molecular Materials*, Bruce, D. W., O'Hare, D., Walton, R. I., eds. (Wiley-VCH, Weinheim).

Uji, S., Shinagawa, H., Terashima, T., Yakabe, T., Terai, Y., Tokumoto, M., Kobayashi, A., Tanaka, H., Kobayashi, H. (2001). Magnetic-Field-induced superconductivity in a two-dimensional organic conductor, *Nature*, **410**, pp. 908–910.

Zare, D., Suffren, Y., Guénée, L., Eliseeva, S. V., Nozary, H., Aboshyan-Sorgho, L., Petoud, S., Hauser, A., Piguet, C. (2015). Smaller than a nanoparticle with the design of discrete polynuclear molecular complexes displaying near-infrared to visible upconversion, *Dalton Trans.*, **44**, pp. 2529–2540.

Zadrozny, J. M., Niklas, J., Poluektov, O. G., Freedman, D. E. (2015). Millisecond coherence time in a tunable molecular electronic spin qubit, *ACS Cent. Sci.*, **1**, pp. 488–492.

Suggested Reads

Optics and Optical Materials:

Artizzu, F., Mercuri, M. L., Serpe, A., Deplano, P. (2011). NIR-emissive erbium-quinolinolate complexes, *Coord. Chem. Rev.*, **255**, pp. 2514–2529.

Artizzu, F., Quochi, F, Serpe, A., Sessini, E., Deplano, P. (2015). Tailoring functionality through synthetic strategy in hetero-lanthanide assemblies, *Inorg. Chem. Front.*, **2**, pp. 213–222.

Binnemans, K. (2015). Interpretation of europium(III) spectra, *Coord. Chem. Rev.*, **295**, pp. 1–45.

Bünzli, J.-C. G., Chauvin, A.-S., Kimb, H. K., Deiters, E., Eliseeva, S. V. (2010). Lanthanide luminescence efficiency in eight- and nine-coordinate complexes: role of the radiative lifetime, *Coord. Chem. Rev.*, **254**, pp. 2623–2633.

Carter, R. L. (1998). *Molecular Symmetry and Group Theory* (Wiley, NY).

Demas, J., DeGraff, B. (1991). Design and applications of highly luminescent transition metal complexes, *Anal. Chem.*, **63**, pp. 829–837.

Di Bella, S. (2001). Second-order nonlinear optical properties of transition metal complexes, *Chem. Soc. Rev.*, **30**, pp. 355–366.

Dixon, I. M., Collin, J.-P., Sauvage, J.-P., Flamigni, L., Encinas, S., Barigelletti, F. (2000). A family of luminescent coordination compounds: iridium(III) polyimine complexes, *Chem. Soc. Rev.*, **29**, pp. 385–391.

Hung, L. S., Chen, C. H. (2002). Recent progress of molecular organic electroluminescent materials and devices, *Mater. Sci. Eng. R*, **39**, pp. 143–222.

Kalyanasundaram, K., Grätzel, M. (1998). Applications of functionalized transition metal complexes in photonic and optoelectronic devices, *Coord. Chem. Rev.*, **77**, pp. 347–414.

Marder, S. R., Sohn, J. E., Stucky, G. D. (eds.) (1991). *Materials for Nonlinear Optics: Chemical Perspectives* (American Chemical Society, Washington).

Vos, J. G., Kelly, J. M. (2006). Ruthenium polypyridyl chemistry; from basic research to applications and back again, *Dalton Trans.*, pp. 4869–4883.

Wing-Wah Yam, V., Man-Chung Wong, K. (2011). Luminescent metal complexes of d^6, d^8 and d^{10} transition metal centres, *Chem. Commun.*, **47**, pp. 11579–11592.

Magnetism and Magnetic Materials:

Benelli, C., Gatteschi, D. (2015). *Introduction to Molecular Magnetism: From Transition Metals to Lanthanides* (Wiley-VCH, Weinheim).

Blundell, S. (2001). *Magnetism in Condensed Matter* (Oxford University Press, Oxford).

Coey, J. M. D. (2010). *Magnetism and Magnetic Materials* (Cambridge University Press, Cambridge).

Carlin, R. L. (1986). *Magnetochemistry* (Springer-Verlag, Berlin, Heidelberg, New York, Tokyo).

Conductivity and Conducting Materials:

Fourmigué, M., Ouahab, L. (eds.) (2009). *Conducting and Magnetic Organometallic Molecular Materials* (Springer, Berlin).

Index

For Product Safety Concerns and Information please contact our EU
representative GPSR@taylorandfrancis.com
Taylor & Francis Verlag GmbH, Kaufingerstraße 24, 80331 München, Germany

www.ingramcontent.com/pod-product-compliance
Ingram Content Group UK Ltd.
Pitfield, Milton Keynes, MK11 3LW, UK
UKHW022159190926
475702UK00003B/113

* 9 7 8 9 8 1 4 7 7 4 7 6 5 *